T0215359

# Advances in
# 3D Integrated Circuits
# and Systems

# Series on Emerging Technologies in Circuits and Systems

**Series Editor:** Kiat-Seng Yeo *(Singapore University of Technology & Design, Singapore)*

*Published*

Series on Emerging Technologies in Circuits and Systems

# Advances in
# 3D Integrated Circuits
## and Systems

## Hao Yu • Chuan-Seng Tan

*NTU, Singapore*

**World Scientific**

NEW JERSEY • LONDON • SINGAPORE • BEIJING • SHANGHAI • HONG KONG • TAIPEI • CHENNAI

*Published by*

World Scientific Publishing Co. Pte. Ltd.

5 Toh Tuck Link, Singapore 596224

*USA office:* 27 Warren Street, Suite 401-402, Hackensack, NJ 07601

*UK office:* 57 Shelton Street, Covent Garden, London WC2H 9HE

**Library of Congress Cataloging-in-Publication Data**
Yu, Hao (Electrical engineer)
  Advances in 3D integrated circuits and systems / by Hao Yu (NTU, Singapore), Chuan-Seng Tan
(NTU, Singapore).
    pages cm. -- (Series on emerging technologies in circuits and systems)
  Includes bibliographical references and index.
  ISBN 978-9814699006 (hardback : alk. paper) -- ISBN 978-9814699013 (pbk. : alk. paper)
  1. Three-dimensional integrated circuits. I. Tan, Chuan Seng. II. Title.
  TK7874.893.Y83 2015
  621.3815--dc23
                              2015020144

**British Library Cataloguing-in-Publication Data**
A catalogue record for this book is available from the British Library.

In-house Editor: Amanda Yun

Typeset by Stallion Press
Email: enquiries@stallionpress.com

Printed in Singapore

# Preface

In the past few decades, the design of computers has been primarily driven by improving performance with faster clock frequency of single-core processor using transistor scaling. The transistor scaling towards high performance of fast clock frequency is, however, stuck recently due to the constraint of power density. By exploiting parallelism, multi-core processor based design of computers has emerged to scale up performance of throughput under power budget. As such, the scaling paradigm has changed to integrate as many processor cores as possible on one single chip. In the traditional 2D based memory-logic integration, the scalability of many-core integration is limited by the communication between the cores and memory by I/O interconnections, which have posed stringent requirements for high utilization efficiency of both bandwidth and power towards operating thousands of microprocessor cores on chip.

Stacking one layer (with core or memory blocks) above the other with short-distance RC interconnects of through-silicon vias (TSVs), 3D integration has become one promising alternative for many-core memory-logic integration with high bandwidth and low power. The other possible alternative for many-core memory-logic integration is 2.5D integration, to place core and memory blocks both on one common substrate, interconnected with the aid of middle-distance transmission lines (underneath the substrate) of through-silicon interposers (TSIs). Even though 3D and 2.5D integrations are potential candidates for many-core memory-logic integration, there are a few critical issues to be addressed as shown in this book.

From the device fabrication perspective, TSV involves a complex process due to its smaller dimension and demanded accuracy with existence of liner material around metal-fill. On the other hand, fabrication of TSI involves fabrication of TSV, microbumps and other metal layers, thus involving additional complexity. The modeling of TSV/TSI can be quite different from the traditional RC interconnects. Existence of liner material around TSV metal-fill and also thermal-mechanical stress due to mismatch of coefficient of thermal expansion (CTE) between Si and TSV metal-fill can have critical effect on the active device parameters in the vicinity of TSV. In addition, as the fabricated TSVs may result in a poor yield, careful physical design and testing needs to be carried out. However, due to the existence of thousands of TSVs, a low complex physical design is required.

From the system management perspective, thermal management is one of the key challenges to be addressed in 3D integration. In addition to allocating dummy TSV for cooling, recent microfluidic channel based active cooling is also introduced. Microfluidic channels can be etched beneath the substrate of each layer in 3D and the microfluid flowing through the channel can help in cooling the substrate layer. However, an effective management of microfluid has to be performed such as flow-rate control to avoid unnecessary overhead. Similarly, dummy TSVs can be inserted to reduce the thermal and stress gradients across the chip. In addition to thermal management, power management needs to be performed in 3D many-core processors as well to avoid dark silicon dilemma. Many multi-core processor designs make use of off-chip power converters, which may not be scalable for the surge of the supply current demanded by many-core processors. As such, an effective power management of power I/Os needed to be carried with less number of power converters.

What is more, signal I/Os are utilized for the communication between memory and logic blocks. The available limited communication bandwidth is often the bottleneck. Bandwidth utilization of 2.5D TSI I/Os is less compared to that of 3D TSV I/Os. To improve the bandwidth utilization, memory controller needs to be explored with configurability. Previous works on memory controllers are either static or non-scalable for many-core processors. Further, with the increase in communication between memory and logic blocks, communication power also increases. Reduction in I/O voltage swing can help in reducing the communication power at the expense of bit-error-rate (BER). As some types of workloads can tolerate a certain amount of BER, I/O communication can thereby be performed with reduced voltage swing by the use of signal I/O management.

This book is intended to address all the above mentioned challenges. The need for many-core processors with according design challenges faced in the traditional 2D integration is presented in Chapter 1. The rest of the book is divided into five parts as follows.

Part 1 deals with the 3D-IC device modeling. We introduce fabrication methodology utilized for TSVs and TSIs and also evaluate their performance with different kinds of materials in Chapter 2. We further develop TSV and TSI device models in Chapter 3.

Part 2 presents the 3D-IC physical design and testing. We introduce macromodeling to reduce the physical design complexity of thousands of TSVs, which is followed by the corresponding TSV allocation methods for power grid and clock in Chapter 4 and Chapter 5 respectively. Moreover, to effectively evaluate the reliability of TSVs under test, we further propose the TSV testing with compressive sensing as discussed in Chapter 6.

Part 3 illustrates the thermal management of 3D-IC many-core processors. We introduce the system level thermal and power models in Chapter 7. We further discuss microfluid based thermal management in Chapter 8.

Part 4 deals with the power and signal I/O management of the 3D-IC many-core processor. To improve power-converter utilization, we present the power I/O management by space-time multiplexing in Chapter 9. To improve I/O channel utilization, we also present the signal I/O management by the space-time multiplexing and the adaptive voltage-swing in Chapter 10.

Part 5 presents 3D-IC design examples. We first show the 3D integration of MEMS sensor and CMOS readout circuit in Chapter 11. Next, we present the 2.5D I/O design such as pre-emphasis, adaptive driver, and clock-data recovery in Chapter 12. We further discuss an 8-core microprocessor with accelerator and memory blocks integrated by 2.5D I/Os. Lastly, we discuss the 3D integration of the non-volatile memory in Chapter 14.

Lastly, the authors would like to thank their students at Nanyang Technological University: Sai Manoj P. D., Lin Zhang, Hongyu Li, Hantao Huang, Kanwen Wang, Yuhao Wang, Dongjun Xu, Xiwei Huang, Hanhua Qian, Shunli Ma, Yang Shang, Jie Lin and Shikai Zhu, for their contribution to this book. Sai Manoj P. D. in particular has made the significant help for this book. The authors would also like to thank their colleagues: Prof. Paul Franzon, Prof. Sungkyu Lim, Prof. Dennis Sylvester, Prof. Sheldon Tan, Prof. Yuan Xie, Prof. Eby Friedman, Prof. Chiphong Chang, Prof. Zhiyi Yu and Prof. Wei Zhang, for their kind suggestion and collaboration. The relevant research is funded by MOE Tier-2, A*STAR PSF, DIRP and NRF CRP grants from Singapore.

*Hao Yu*
*Chuan Seng Tan*
Singapore

# Contents

## Part 2.   Physical Design          99

## 4.  Macromodel         101

## 5.  TSV Allocation         113

# Chapter 1

# Introduction

## 1.1 Thousand-core On-chip

With the advancement in cloud computing with big-data analytic capability, the amount of data to be handled by the data centers are increasing tremendously and reaching Exa-scale ($10^{18}$ Flops) [1]. To process such a large amount of data, numerous microprocessor cores can be integrated on a single chip for the Exa-scale data processing as accelerators. As an example, as shown in Figure 1.1, let's calculate the resource requirement for a data center to have Exa-scale processing capability. We assume that each core can have four floating point units (FPUs), each having 1.5GHz processing speed. As such, each core can reach performance of 6GFlops. To perform 1Exa-Flop operations, large numbers of such cores and memory blocks need to be integrated. The integration can be arranged in a hierarchical manner, namely, by chips, nodes and racks. One needs to integrate 742 cores on a single chip. Moreover, a set of 16 DRAMs, each DRAM of 1GB capacity, and one chip form a node. Furthermore, 12 such nodes connected with routers form a group. As such, one needs to combine 32 groups of 1.7Peta-Flop processing capability as a rack. Lastly, in order to achieve 1Exa-Flop performance, 583 such racks need to be integrated together. This setup demands 100Gbps bandwidth and 68MW power, with 20,000sft area. The resource to build many Exa-scale data centers will obviously require too much consumption of bandwidth, power and space that may be beyond the capability of the current human society.

A similar situation actually happened in history when humans built their first electronic computer. The first electronic computer, i.e., electronic numeric integrator and calculator (ENIAC), [2] was built in 1946, made of vacuum tubes, as shown in Figure 1.2. It had a processing capability of only

**Interconnect for intra and extra cabinet links**

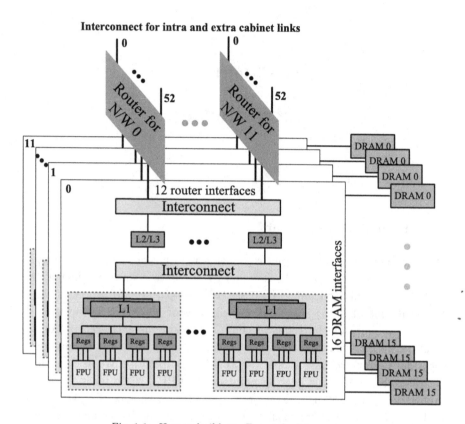

Fig. 1.1   How to build one Exa-scale data center.

5KHz, but consumed 150KW power of within an area of $167\,\text{m}^2$. Due to the advancement in CMOS transistor scaling and VLSI integration, one of the recent (2008) microprocessors from Intel, an i7 quad-core processor as shown in Figure 1.2, only occupies an area of $263\text{mm}^2$ and consumes 130W of power, but has a processing capability of 1GHz. As such, it motivates us with a similar question: can we integrate an Exa-scale data center on chip, shown in Figure 1.3? Figure 1.4 shows the prediction of roadmap of the number of cores that will be integrated on chip. One can observe that it is predicted that thousands of cores will be integrated on a single chip, along with large amounts of memory in next 10 years.

With the integration of thousand-cores on a single-chip, a large amount of power is consumed for communication between cores compared to the computation power. The use of traditional 2D interconnects such as on-chip wire, off-chip PCB trace etc., will result in a large power dissipation and

ENIAC-The first electronic computer (1946)
Power: 150KW
Area:167m²
Speed:~5KHz

Intel microprocessor core i7 (2008)
Power: 130W
Area:263mm²
Speed:~3.9MHz

Fig. 1.2   Scaling of single-core electronic computer by VLSI integration.

Fig. 1.3   Can we integrate an Exa-scale data center on chip?

latency with degraded scaling performance for integrating thousand-core on-chips [3]. The ITRS road map [4] for the power index and RC-delay of 2D on-chip interconnects for $1mm$ length of Copper (Cu) is presented in Figure 1.5, which indicates that in the next few years, the power index and delay of global interconnects both become non-scalable. Moreover, 2D off-chip interconnects such as back-plane PCB traces are obviously with lossy channels that require overdesign of I/O equalization with huge power overhead. On the other hand, optical interconnects can provide high-speed communication but always come with an additional cost of optical-to-electronic conversion with no CMOS based light source, detector and modulator. In this book, to meet the high-speed, low power and high bandwidth demands, we will explore the future thousand-core memory-logic integration by 3D integration. 3D integration by short-distance through-silicon via (TSV) interconnects [5–23] and 2.5D integration by middle-distance

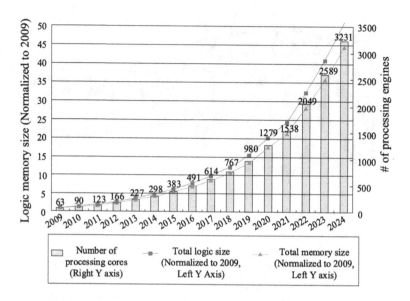

Fig. 1.4    Roadmap of thousand-core on-chip.

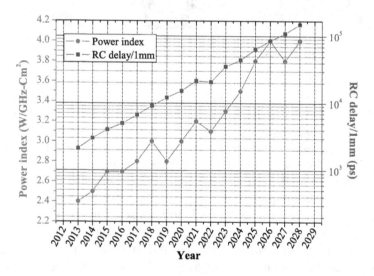

Fig. 1.5    ITRS projection for power index and RC-delay for 2D interconnects [4].

through-silicon interposer (TSI) interconnects [3, 24–27] can be cost efficient in meeting large bandwidth, low power and latency requirements.

In this chapter, we first present state-of-the-art many-core microprocessor designs, followed by the analysis and comparison for memory-logic
integrations of 2D, 3D and 2.5D, respectively.

## 1.2 State-of-the-Art Many-core Microprocessors

Due to simplicity of individual cores in a multi-core microprocessor, the
multi-core microprocessor outperforms a single large core microprocessor
in terms of area, power, speed, number of functional units and energy efficiency [28]. Multi-core processor is power efficient because of integration
of clock frequency demands for proportional increases in voltage levels.
Higher voltage levels may increase the power consumption nearly cubic
to the increase in frequency [29]. In the past few years, a large number of
many-core microprocessors have been developed. In this section, we provide
a glimpse of a few many-core microprocessor examples.

In a many-core microprocessor, the communication between core-
core/memory is often the bottleneck. The interconnection for communication can be through centralized structures like bus interconnects, ring
structures, or by a fixed interconnection. This kind of communication structures may be feasible for a small number of cores, but may not be scalable
for thousand-core chips, which is expected to happen in the next decade or
so. A distributed communication can be performed between cores, with
the help of routers. Tiled architectures are introduced, which are composed of small computational cores instead of one large core. The small
cores are simple, area- and energy-efficient compared to one large core.
A few of the multi-core processors include the 16-core Raw processor [30]
which is commercialized in Tilera, TRIPS [33], Tilera's TILE 64 [34,35] and
Intel's Tera-scale research processor [36], SPARC SoC processor from Sun
Microsystems [37], 48/80-core microprocessors by Intel [31,32], and 64-core
microprocessor by IBM [38]. It is also good to mention that Nvidia Tesla
K80 servers are designed with nearly 2GPUs each consisting of 2496 CUDA
cores [39]. A few of the tiled multi-core processors from industry can be seen
in Figure 1.6. Each of the cores is connected with a router, for the purpose
of effective communication [29]. Tiled cores with distributed computation
and communication structures are efficient and scalable [34–36]. Unlike centralized on-chip communication where a large global interconnect is used,

distributed communication is only routed through the on-chip network and uses only components between the source and destination. In tiled microprocessors, the efficiency can be further improved by deciding the mapping of tasks to cores and how best to utilize them. In a tiled architecture, each core can be of different size. Addition of new cores in tiled design is easy compared to that of a centralized interconnect scheme [29].

A 16-core Raw microprocessor [30], which is a prototype tiled multi-core microprocessor, is shown in Figure 1.6(a), fabricated by MIT in 2003. This Raw microprocessor is commercialized in Tilera microprocessors. The Raw microprocessor in [30], designed at 180nm technology node, outperforms P3 processor [40] with an average throughput advantage of 10.8× (by cycles)

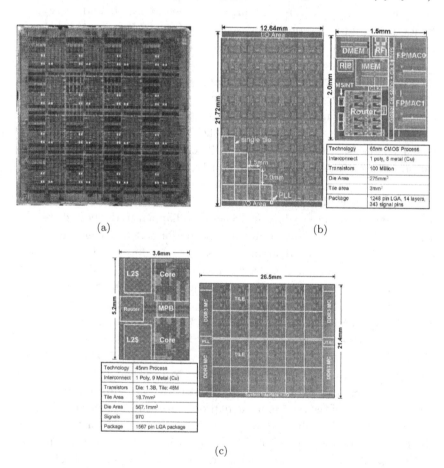

Fig. 1.6 (a) MIT 16-core RAW microprocessor [30]; (b) Intel 80-tile microprocessor [31]; (c) Intel 48-tile microprocessor [32].

and 7.6× (by time). This high performance is achieved due to the high pin bandwidth available to the off-chip memory [30].

An NoC architecture of 80-tile architecture implemented at 65nm CMOS process and arranged in the form of a 2D mesh (10×8) is shown in Figure 1.6(b). The chip is expected to run at a maximum frequency of 3.13GHz with peak performance of 1.0TFLOPS at 1V, and 4GHz maximum frequency with 1.28TFLOPS peak performance at 1.2V, respectively. The whole chip consumes an area of $275mm^2$, with each tile consuming $3mm^2$ of area. A 48-core processor is fabricated by Intel [32] in 2010. The processor is implemented in 45nm CMOS process with a total die area $567mm^2$, with each tile having 2 cores and area of $18.7mm^2$, as shown in Figure 1.6(c). An operating frequency of 1GHz for core and 2GHz for router at voltage of 1.14V is observed.

## 1.3   Memory-logic Integration

For many-core microprocessor design, one needs to integrate the cores with memory through interconnect. The cores load data from the memory, perform computation and then write data back to the memory. Cores need both on-chip and off-chip interconnects to access the on-chip memory (cache) and off-chip memory, resulting in a long-distance communication and larger power consumption in 2D. More I/O pins and more channels are required to access memory for high bandwidth requirement. In the following sections, we show the challenges faced in the traditional memory-logic integration, and further present the solutions by the proposed 2.5D and 3D memory-logic integrations.

### 1.3.1   *2D Integration Challenges*

Undoubtedly the introduction of many-core microprocessors has revolutionized the design of computers. For example, workloads on processors can be distributed and designers can start thinking of parallelizing workloads to increase the processing throughput. Unfortunately, the available 2D integration technique is inefficient for many-core microprocessors at a large scale. A few of the challenges in traditional 2D integration are briefly discussed below.

#### 1.3.1.1   *Scalability*

Reduction in transistor size has fueled the processing speeds of logic blocks. Memory-logic integration in traditional 2D manner will have limited on-chip memory with a large off-chip memory due to the large area occupied by

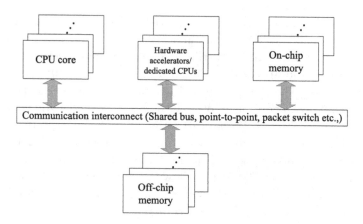

Fig. 1.7   Typical multi-core system architectural view.

cores and other logic blocks (Figure 1.7). With the advancement in technology nodes, interconnect delay started to dominate the delay, because 2D RC interconnects cannot be scaled at the same rate as gates. It is expected that the latency due to interconnects can be cut down by nearly 50% by moving off-chip memory blocks to on-chip [41]. Furthermore, to cope with increasing application demands, a large amount of on-chip memory is projected by ITRS [4]. However, the amount of available space is limited in 2D integration with cores integrated.

### 1.3.1.2   *Channel Loss*

The off-chip interconnection between cores and memory blocks in 2D integration is typically performed by PCB with backplane [42] containing sockets into which other boards and corresponding components can be plugged. As discussed, this methodology is not scalable for large number of cores due to routing and wire-length limitations. More importantly, long trace ($\geq 25cm$) and non-ideal vias on PCB can cause severe loss on the backplane. To compensate the losses and achieve high data rate, current starved circuits are needed with area overhead and also power overhead.

Figure 1.8(a) shows the interconnection of multiple chips using the wireline backplane PCB interconnects. The channel loss for the corresponding integration is depicted in Figure 1.8(b). One can observe nearly $24dB$ channel loss at a frequency of 5GHz [42]. Further, due to the limited area available, it is not possible to accommodate a large number of I/O pins, hence,

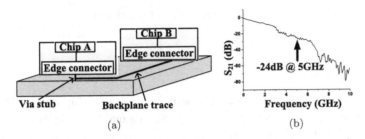

(a)  (b)

Fig. 1.8  (a) PCB based interconnection of two dies with backplane trace; (b) corresponding channel loss.

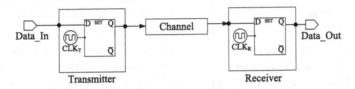

Fig. 1.9  Block diagram of a chip-to-chip communication channel. The main components are: the transmitter, the receiver, and the lossy channel that is composed of the RDL routes, the package traces, and the PCB traces.

the number of I/Os are limited as well and an upper bound is put on the bandwidth available in 2D interconnect.

### 1.3.1.3 *I/O Circuit Design*

A proper design of the interface channel (Figure 1.9) must adhere to a complex set of specifications and constraints. These include voltage levels and noise, bit-error rate (BER), signal jitter and slew rate. Proper design and optimization of these circuits are needed to meet the design specifications at an acceptable cost in terms of area and power [43–47].

To meet the performance and signal integrity constraints for the I/O drivers, good models of the package and board are needed to account for the capacitive loading and coupling as well as the inductive coupling in the system [12, 21, 22]. Early in the design cycle where the package is not yet routed or where no board models exist, good estimations are needed. In high-end designs, it is no longer sufficient to use ad-hoc metrics like rules of thumb to make decisions on the type and size of the I/Os as well as on the capacitive and inductive values of the load that the driver sees. Good virtual models that can capture the package and board effects are needed.

Moreover, complicated compensation of data and clock signals at receiver is required.

#### 1.3.1.4 *Testing*

Testing is one of the critical and prominent phases in chip design. Detection of fault after fabrication will cost more. In a multi-core microprocessor, each core has to be tested before performing the integration, which leads to additional level of testing hierarchy. To avoid interconnection and functional failures, each core and its interconnection with its surrounding cores or memory blocks need to be carried out separately. For a memory-logic integrated many-core microprocessor, each of the individual blocks needs to be tested and further the post-bond testing has to be carried. Use of traditional scan chain methods may not be efficient [48]. In addition to testing the functionality of the cores, the interconnects also have to be tested. Electron migration in interconnects is highly possible due to high usage of interconnects. Due to the large number of I/Os, testing becomes complex and time consuming.

#### 1.3.1.5 *Thermal Management*

The operating temperature of the chips depends on the ambient temperature, power density of cores and the heat-dissipation capacity of the chip [17,25,48]. In a multi-core microprocessor, a large number of hot-spots can be observed due to larger power density. When the on-chip temperature increases beyond the specified operating temperature, the functionality of the chip is questionable. Moving workloads from core with hot-spots to others may not always be an efficient solution and may involve overhead.

Moreover, if one observes the state-of-the-art microprocessors, memory occupies more area than processing units and consumes more power as well due to the static power dominated by leakage current. With the scaling down of transistor size, the transistor cannot be completely turned off. As such, even in the low/off state, leakage current is present. This leakage current can form a positive feedback loop with temperature to further increase temperature, resulting in a thermal runaway failure [25,49,50].

Figure 1.10 shows the occurrence of thermal runaway failure with the increase in temperature and leakage current in a memory-logic integrated system. When the on-chip temperature increases, it forms a positive feedback loop with the leakage current and further increases temperature.

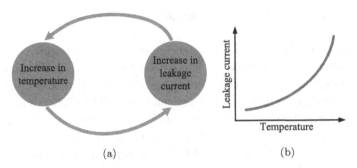

Fig. 1.10 (a) Positive loop between leakage current and temperature; (b) trend showing leakage current versus temperature.

Beyond a certain temperature, the chip will cross its operating temperature range, indicating a thermal runaway failure and malfunction.

### 1.3.1.6 *Power Management*

A thousand-core integrated chip with memory blocks consumes a large amount of power, resulting in an energy-inefficient data center on chip. As the demands from cores are time–varying, providing a uniform voltage-level to all cores results in high power density. As such a dynamic voltage and frequency scaling (DVFS) [51–56] based power management has to be adopted. DVFS is efficient in terms of power management, but it may not be efficient at large scale due to the involvement of off-chip power converters in traditional 2D integration. Additionally, it is complicated to deliver multiple voltage supplies to core and memory blocks in 2D integration.

### 1.3.1.7 *I/O Management*

In a many-core memory-logic integrated system, a large number of I/Os are required. Cores communicate with the external environment such as memory through local and global I/O interconnects. The achieved bandwidth depends on the number of I/Os utilized and the extent to which they are utilized. To improve the effective bandwidth and utilize the available I/Os effectively, proper scheduling techniques with the aid of a memory controller need to be carried out, which becomes complicated due to limited space and routing concerns in 2D integration. Improper scheduling results in stalling of requests, and affects the quality-of-service (QoS) [57]. Furthermore, inter processor/memory communication is often the bottleneck limiting the overall performance in a multi-core microprocessor [58]. Use of point-to-point or shared bus communication may not be scalable and efficient at large scales.

To overcome the above mentioned challenges, designers started exploring new integration techniques for many-core memory-logic integration. The possible integration techniques that are scalable towards thousand-core on-chips can be 3D integration by through-silicon via (TSV) interconnects and 2.5D integration by through-silicon interposer (TSI) interconnects. The 3D integration is carried out by vertical stacking of one layer over an other. Short distance through-silicon vias (TSVs) are utilized for the vertical interconnections. TSVs can help in signal or power delivery, as well as heat-dissipation from the top layers to the heat-sink. Another possible many-core integration technique is 2.5D integration using the middle-distance TSIs. In 2.5D integration, the dies are placed on one common substrate and connected through TSIs. We introduce the 3D and 2.5D integration techniques respectively below.

### 1.3.2  *3D Integration*

Stacking of dies, termed as *three dimensional* (3D) integration, can be seen in Figure 1.11(a), which shows that the basic idea of 3D integration is to vertically connect Chip A and Chip B using the cylindrical TSVs.

Thanks to the recent advancement in 3D integration technology [5–23], which can reduce the physical distance between the memory and the logic blocks, there is a great potential to integrate thousands of cores with scaled

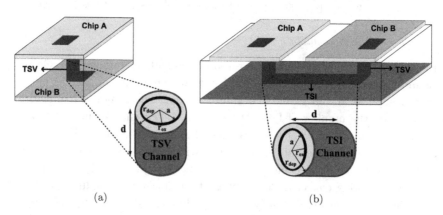

(a)                                    (b)

Fig. 1.11  (a) 3D integration by vertical TSV interconnect; (b) 2.5D integration by TSI interconnect.

performance superior to that of the 2D integration. 3D integration is identified as key and promising path, not only to facilitate the continuation of the conventional scaling, but also to enable the "more-than-Moore" heterogeneous integration of vast different functionalities into a system in a single chip form. Conventional wire-bond interconnections for 3D stacked packages could not satisfy the requirements for low power consumption and signal integrity because of its larger inductance and conductance loss. Hence, it is clear that shorter interconnection technologies are required. Flip-chip technology presents one of the possible solutions, but it is difficult to meet the high chip density requirement due to the large ball size and pitch. In 3D integration, traditional RC-interconnects cannot be utilized, because the scaling of global on-chip wires is not proportionate with the scaling of other features. As discussed previously, at smaller technology nodes, interconnect delays dominate the gate delays. Hence, vias are formed for inter layer communication in vertical direction. Thus, through-silicon via (TSV) emerged as the next potential technology which could provide shorter interconnection; this translates to a lower inductance and conductance loss. In general, the height of TSVs is a few tens of $\mu m$ and has a diameter of a few $\mu m$. TSV can also assist in heat removal, which is a critical challenge faced in 3D integration.

As such, multiple device layers can be vertically connected by TSVs, resulting in not only a shorter length but also a higher integration density. Obviously, the system size can be dramatically reduced with diverse components compactly integrated. Potentially superior performance can be obtained from 3D stacking in terms of smaller footprint size, higher clock speed and heterogeneous integration, which in turn can reduce the chip cost. Also, numerous new design freedoms are given by the 3D stacking. As a result, the heterogeneous components fabricated by analog/RF, MEMS, or digital process can be integrated together into one system with a low fabrication cost and a high yield-rate. More importantly, as there is a significant boost in communication bandwidth, the 3D integration is well suited for I/O-centric systems with concurrent data processing, ideally towards the thousand-core on-chip system integration.

The use of TSV reduces the interconnect lengths, resulting in reduced latency, increase in bandwidth due to small RC values, and also consumption of less power. It has been stated in [41], that for every doubling of stack height, performance per Watt increases by 20–30%. A simplified process of building 3D IC from traditional 2D dies is shown in Figure 1.12(a). Firstly, the dies which need to be vertically stacked are prepared and then thinned

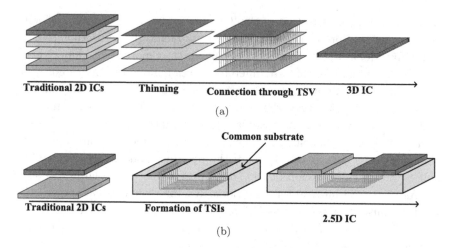

Fig. 1.12    Simplified flow to form (a) 3D IC; (b) 2.5D IC from traditional 2D ICs.

so that vias can be formed easily and the thickness of the integrated die is not so large. Once the thinning is performed, vias are formed and the connections to other layers are made. Finally, the whole integrated die is sent for packaging and other processes. Fabrication of 3D TSV with more details will be given in next chapters.

It is inherent that due to the vertical stacking, heat gets accumulated in the top layers because of the long heat-dissipation path to heat-sink. This accumulated heat can endanger the operational reliability of 3D IC. Another challenge imposed by TSV is its inherent parasitic capacitance which could add additional delay to signal transmission.

### 1.3.3    *2.5D Integration*

Though the benefits of 3D integration are too lucrative to ignore, thermal reliability concerns in 3D integration [10, 22, 25, 59] required designers to look for an alternative for many-core memory-logic integration. The recent introduction of through-silicon interposer (TSI) opened doors for 2.5D integration. In 2.5D integration different dies, i.e., multi-core microprocessor and memory blocks, are integrated along the horizontal direction with the aid of through-silicon interposer (TSI) interconnects on one common substrate [3, 24–27]. In contrast to 3D TSVs, which are usually designed as short RC-interconnects for inter-layer communication, TSIs are designed as a transmission line (T-line) targeted for high-speed, middle-distance communication. Compared to 2D RC-interconnects with repeaters, T-lines demonstrate better latency, speed and bandwidth but with large area over-head [60]. As TSIs are deployed underneath the common substrate, the area overhead is actually mitigated in the 2.5D integration.

Figure 1.11(b) shows a basic idea of 2.5D integration. Chips A and B are placed on one common substrate and TSI realized by T-line is deployed underneath the substrate to connect them. As all the dies are at uniform distance from the heat sink, the thermal and mechanical reliability concerns are dramatically relaxed in 2.5D integration. Compared to 3D integration, the integration density of 2.5D integration is lower with interconnect length of a few *mm*. Figure 1.12(b) shows a brief overview of the 2.5D integration. A substrate on which the 2.5D integration has to be performed is chosen and the vias are drilled to form the T-line based TSI underneath the substrate. The dies are placed on the common substrate and the connection between the vias and the pads of the die is performed by the bonding. Fabrication of 2.5D TSI with more details will be given in the next few chapters.

## 1.4 Organization of the Book

The rest of the book is divided into 5 parts and is organized as follows. Part 1 discusses the device modeling of 3D and 2.5D ICs. Fabrication method-ology utilized for TSVs and TSIs is discussed along with their performance evaluation using different materials in Chapter 2. Device models for TSVs and TSIs are presented in Chapter 3.

Part 2 of this book presents the 3D IC physical design and testing. A macromodeling methodology to reduce the computational complexity, followed by corresponding dummy TSV allocation, is presented in Chapter 4 and Chapter 5 respectively. To assess the reliability of TSVs with less amount of testing data in a large system, a compressive sensing based TSV testing is discussed in Chapter 6.

Part 3 discusses the thermal management of the 3D many-core micro-processor. To perform effective thermal management, power and thermal system models for 3D integrated system are needed, and are presented in Chapter 7. Based on the developed system models, a microfluid–based ther-mal management technique is presented in chapter 8.

In 3D-IC design there is a need for I/O management due to the scale-up of number of cores, which is described in Part 4. Power supply to the cores needs the management of power I/Os. To avoid dark silicon dilemma in a many-core microprocessor, power I/O management by space-time multiplexing is dicussed in Chapter 9. For signal I/Os, bandwidth management by space-time multiplexing based memory controller is presented, followed by communication power reduction using adaptive voltage-swing tuning in Chapter 10.

The last part of this book presents 3D-IC design examples. Firstly, the integration of heterogeneous devices such as MEMS and CMOS readout circuit by Cu-Cu bonding is demonstrated in Chapter 11. Secondly, the design of low power TSI buffers, adaptive TSI buffers and TSI clock-data recovery circuit with measurement results are discussed in Chapter 12. A design of 8-core microprocessor integrated with hardware accelerator and memory blocks by 2.5D TSI I/Os is presented Chapter 13. Lastly, a design of a non-volatile memory for 3D hybrid memory is presented in Chapter 14.

# PART 1
# Device Modeling

# Chapter 2

# Fabrication

## 2.1 Introduction

The idea of three-dimensional (3D) integration emerges in order to overcome the scaling limits posed by traditional two-dimensional (2D) integration and to provide lower signal latency, lower power consumption, smaller form factor, higher bandwidth, higher density and heterogeneous integration [61]. 3D integration employs through-silicon via (TSVs) which replace the conventional interconnections to provide high density and short vertical interconnection between stacked layers. However, high power density can affect the performance of 3D integrated circuit (IC) adversely, and thermal management becomes critically important [62]. TSV is embedded in silicon substrate and commonly fabricated by utilizing the following processes – high aspect ratio silicon deep reactive ion etching (DRIE), lining with dielectric layer and copper super conformal filing [63].

In this chapter, the fabrication process flow of TSV structures for electrical characterizations and thermal property studies are presented first. The fabrication process of TSV mainly consists of a deep reactive ion etch (DRIE) for via formation, followed by a dielectric liner deposition using plasma-enhanced chemical vapor decomposition (PECVD) or thermal oxidation for oxide liner. Thereafter, the barrier and seed layer can be deposited by physical vapor deposition (PVD) or electrografting (eG) process, via filling by Cu ECP, and finally a CMP process for Cu plating overburden removal [64]. We evaluated TSV structures of different sizes, pitches and densities. In order to have a robust TSV structure for the subsequent characterization purposes, each of the TSV process steps must be fully understood and optimized. In addition, the process observations and optimizations done to mitigate the issues encountered are presented.

As TSV is embedded in Si substrate, its electrical coupling effect with the Si substrate through the circular metal-oxide-semiconductor (MOS) structure formed can be achieved, and is discussed in the later part of this chapter. Using the analogy of a typical metal-oxide-semiconductor capacitor (MOSCAP), the metal core (Cu) is the gate, while the surrounding dielectric and the Si substrate which it is embedded in represent the oxide and semiconductor respectively. Hence, the TSV's capacitance-voltage ($CV$) curve will exhibit the same accumulation, depletion and inversion regions. For a power TSV, due to the power distribution noise, such as the simultaneous switching noise, the $CV$ curve follows the high frequency (HF) curve in the inversion region. For a signal TSV, due to the fast digital signal change, the $CV$ curve follows the deep depletion characteristics in the inversion region [65]. TSV parasitic capacitance plays a very important role in the circuit operation, hence, the TSV capacitance behavior must be fully understood and controlled. In addition, a good TSV used for signal transmission should possess the ability to retain the signal. This signifies that a small leakage is desired. Hence, current-voltage ($IV$) measurement which monitors the leakage of the various dielectric materials integrated as the TSV liner should be well understood as well.

In addition to the great benefits achieved by the integration of TSV in Si, it poses a challenge of thermo-mechanical stress to the device which is evident in operation [66]. Cu-TSV exerts thermo-mechanical stress on Si in the immediate vicinity of TSV due to a mismatch in the coefficient of thermal expansion (CTE) of Cu and Si. During the annealing process and subsequent cooling of the Cu-TSV, the Si-surface is under strong compressive stress in the close vicinity of the TSV. This stress level decreases exponentially and becomes constant as a function of radial distance from the TSV edge. Similar observation has also been reported by Trigg *et al.* in their recent report [67]. This is due to the fact that the high temperature annealing step initiates the grain growth inside the Cu pillar [68]. The thermo-mechanical stress leads to detrimental effects such as cracking, delamination and malfunctioning of individual transistors. Furthermore, it can also result in mobility variation and interconnect distortion [69]; the carrier mobility can be varied by more than 7% with every $100MPa$ of stress [70]. The International Technology Roadmap for Semiconductors (ITRS) has invariably proposed the scaling of TSV diameter to as low as $2\mu m$ in the future. Lowering the TSV diameter could minimize the overall thermal stress due to a smaller Cu volume [67, 71].

However, rigorous scaling of the TSV diameter results in an extremely stringent aspect ratio that is technically challenging for void-free Cu filling during the electroplating process. Lastly, we will discuss the use of a dielectric material with a lower elastic modulus, such as low-k [72] which can act as a compliant layer to cushion the thermo-mechanical stress induced by the TSV.

## 2.2 TSV Structure and Fabrication

The design of signal/power and thermal TSVs are discussed first in this section. Fabrication process and results of TSVs with $5\mu m$ of diameter and aspect ratio $1 : 2$ to $1 : 3$ are presented in the later part of this section.

### 2.2.1 *Structure Design*

#### 2.2.1.1 *Wafer Layout and Mask Design*

High aspect ratio TSV is required to meet the demand for high density input/output (I/O). Therefore, in order to fabricate a void free TSV, the process conditions and steps in achieving a super-conformal filling of TSV are desired. In this work, we are targeting a TSV diameter of $5\mu m$ with an aspect ratio of 2-3. We chose this aspect ratio for ease of fabrication, and although higher aspect ratio is always desired, this aspect ratio is sufficient for the study of liner properties. Figure 2.1 shows the process flow of the TSVs' fabrication on an 8" p-type Si wafer which has a resistivity range from 6–9 $\Omega cm$. Also, a deep trench in the Si wafer is etched using the BOSCH process by using DRIE process, followed by a dielectric isolation (SiO$_2$) layer deposition. Thereafter, PVD of Ta barrier and Cu seed layer is done before filling the trench using a Cu ECP process and finally Cu CMP to remove the Cu overburden [65].

Figure 2.2 shows the wafer map which is fully patterned with $12mm \times 12mm$ reticles while Figure 2.3 shows the magnified layout of the reticle. The reticle contains five mask layers and consists of all the major components or structures to be characterized − TSV structures for electrical characterization, TSV structures for thermal property studies, dummy TSV structures for process optimization and failure analysis. As shown in Figure 2.3, different TSV structures are grouped in different areas of the reticle, designed systematically so that each TSV structure can be located easily while performing the test.

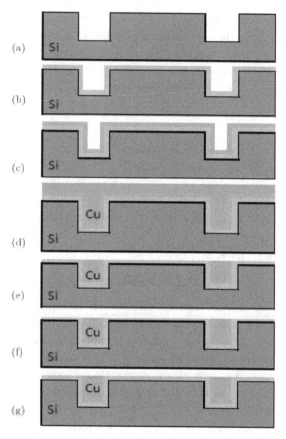

Fig. 2.1  TSV fabrication process flow: (a) DRIE Si etch; (b) dielectric liner deposition; (c) Ta barrier/Cu seed layer sputtering; (d) Cu ECP; (e) Cu CMP; (f) passivation; (g) TSV opening [65].

### 2.2.1.2  *Electrical Structure Design*

The TSV structures studied in this work − single TSV, single TSV row, double TSV rows, TSV arrays with different TSV sizes, TSV pitches and densities are designed to characterize the electrical properties, such as parasitic capacitance and leakage current, of TSV as a function of size, pitch and arrangement. Figure 2.4 shows some of the designed TSV structures which incorporate the variations as described. Typically, the TSVs are surrounded by $p^+$ Boron doping, where one of the Al pads is connected to the TSV arrays while the other Al pad is connected to the $p+$ implant area to provide a stable ground for the Si substrate.

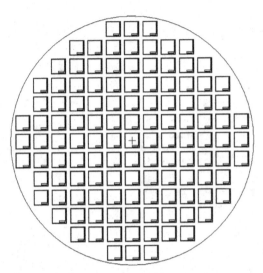

Fig. 2.2   Full wafer map of an 8" *p*-type Si wafer which has a resistivity range from 6–9$\Omega cm$.

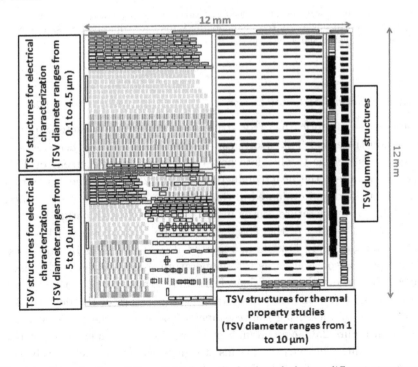

Fig. 2.3   Zoom–in view of the reticle shows die level mask design; different structures are grouped in to different locations of the die.

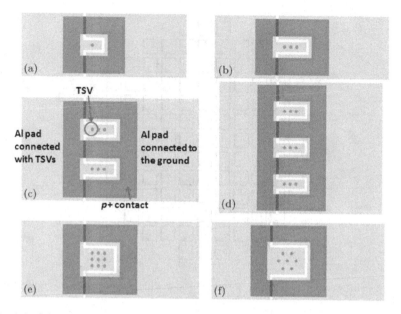

Fig. 2.4   Schematic top view of designed TSV structure for electrical characterizations: (a) single TSV; (b) single TSV row; (c) double TSV rows; (d) triple TSV rows; (e) square TSV array; (f) hexagonal TSV array.

### 2.2.1.3   *Thermal Structure Design*

The purpose of fabricating TSV structures for thermal characterization is to demonstrate that TSV arrays can be used as an effective heat removal element, as well as to elucidate how the density and placement of TSV arrays can affect the heat conduction. Figure 2.5 shows the test structures fabricated and evaluated in this work. To study the heat dissipation capability, a temperature sensor in the form of Kelvin structure is fabricated in the vicinity of the TSV array. Temperature sensors such as $p$-$n$ diode and gate poly have been reported in [73] and [74]. In this work, doped poly-silicon is used as the temperature sensor on top of the oxide layer. Since the resistance value of the doped poly-silicon lines changes with temperature, one can calculate the temperature of the doped poly-silicon lines by measuring the resistance of the lines. The test structures shown in Figure 2.6 have non-uniform poly-Si line to simulate the hotspot in the circuit. The purpose of this design is to see how effectively the TSV can remove the heat when the heating power is varied across the entire chip.

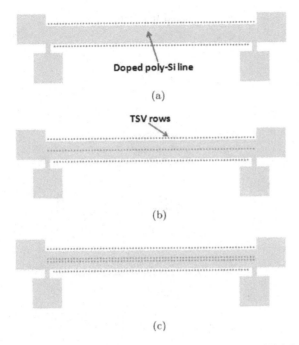

Fig. 2.5 Schematic of the uniform temperature sensor in the form of Kelvin structure for TSV thermal property studies: (a) only TSV rows surrounded the poly-Si line; (b) TSV rows surrounded and one TSV row beneath the poly-Si line; (c) TSV rows surrounded and two TSV rows beneath the poly-Si line.

#### 2.2.1.4 *Dummy TSV Blocks*

In order to have a robust TSV and to optimize the TSV process, failure analysis (FA) tools such as scanning electron microscope (SEM), focus ion beam (FIB) and transmission electron microscope (TEM) has to be performed along the way to monitor the process outcome. Since a single TSV alone is very difficult to locate for FA purposes, the dummy TSV blocks are designed for the ease of finding the locations of TSV. Figure 2.7 shows one example of a TSV block used for FA. A random cut will definitely hit a TSV, hence reducing the time and effort needed to search for a TSV.

### 2.2.2 **Fabrication Process**

#### 2.2.2.1 *Electrical Structure Fabrication Process*

Figure 2.8 shows schematically the process flow of the TSV structure for electrical characterization. The TSV structures are fabricated on a *p*-type

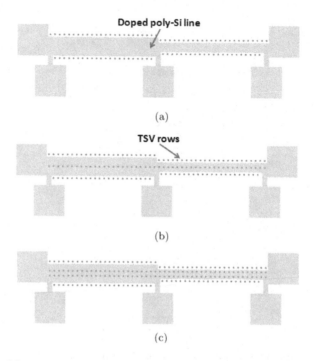

Fig. 2.6   Schematic of the thermal structure with non-uniform temperature sensor.

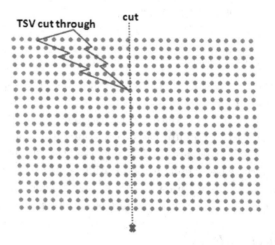

Fig. 2.7   Dummy TSV block for the ease of locating a TSV for failure analysis. A random cut will definitely cut through a TSV.

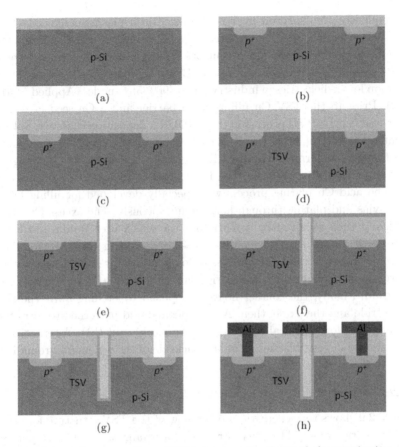

Fig. 2.8 Schematic process flow of TSV structure for electrical characterization: (a) 100Å thermal oxide deposition; (b) boron implantation; (c) 8kÅ CVD oxide; (d) via formation by DRIE Si etch and using oxide as the hard mask; (e) liner deposition; (f) Ta barrier/Cu seed deposition followed by Cu ECP, Cu CMP to remove the Cu overburden and nitride passivation deposition; (g) contact open; (h) Al metallization and patterning.

8" Si wafer with a resistivity in the range of 6–9$\Omega cm$. Firstly, a 100Å thermal oxide layer is on top of the Si wafer to protect the wafer from the subsequent $p^+$ doping implantation, then Boron implantation is done to act as an electrical ground. The implant condition used is $2 \times 10^{16} cm^{-2}$ at $150 keV$ and activation is performed at $1050°C$ for 60 min. Following this, a 8kÅ CVD oxide is deposited, as shown in Figure 2.8(a), (b) and (c). To form the via, the oxide is patterned using photolithography and etched away to be used as the hard mask for the via formation by DRIE Si etch which is a cyclic etching-passivation BOSCH process (Figure 2.8(d)).

Subsequently, the isolation liner is deposited in a plasma-enhanced CVD (PECVD) chamber at temperature $< 400°C$ using tetraethylorthosilicate (TEOS) and Black Diamond precursors for oxide and low-k liner respectively, as shown in Figure 2.8(e). The Black Diamond used is one of the common low-k dielectrics in industry for the $90/65nm$ nodes (Applied Materials). Prior to the TSV Cu filling, a Ta barrier and a Cu seed layer are sputtered using a PVD process in the ENDURA deposition chamber. Subsequently, Cu electroplating (ECP) is carried out till the via hole is fully filled and the TSV is formed. We use RENA system and Enthone or Shanghai XinYang chemical solution for the ECP process. It is worth pointing out that the acid Cu plating process was specially developed for filling blind micro-vias and plating through-holes simultaneously. The excess Cu overburden is then removed by a CMP process which uses Rohm and Haas high polish rate slurry, a colloidal silica-based used in Cu CMP for bulk Cu removal. After the CMP process, a low temperature nitride deposition as well as a Cu annealing are carried out (Figure 2.8(f)) to stabilize the Cu grain. Finally, the contact holes are opened by etching away the silicon nitride and the oxide, then, Al is deposited and patterned to form the probing pad for electrical probing (Figure 2.8(g) and (h)). Alternatively, one can also deposit blanket Al at the backside of the wafers for grounding.

### 2.2.2.2   *Thermal Structure Fabrication Process*

Figure 2.9 shows the schematic process flow of the TSV structure for thermal property study. The TSV structures are fabricated on a $p$-type 8" Si wafer with a resistivity range of $6$–$9\Omega cm$, just like the wafer used for fabrication of TSV structure for electrical characterization. Firstly, $2k\mathring{A}$ thermal oxide is grown on top of the Si wafer and after that, a $5k\mathring{A}$ poly-Si is deposited at the temperature of $580°C$. In the poly-Si deposition process, $PoCl_3$ is used as the doping precursor, and the annealing process after deposition is carried out at the temperature of $900°C$ for 30 min. Subsequently, the doped poly-Si is patterned and etched to form resistor liner as a heat generator (Figure 2.9(a), (b) and (c)) in the form of a Kelvin structure. Then, 1 $\mu m$ CVD oxide is deposited using PECVD at the temperature of $350°C$. Oxide CMP is then carried out to achieve a surface with no topography, following which the standard TSV fabrication process as described earlier is done. Via is firstly formed by etching away the oxide and doped poly-Si layer and then put through DRIE BOSCH process for deep Si etch (Figure 2.9(d) and (e)). Subsequently, TSV liner is deposited

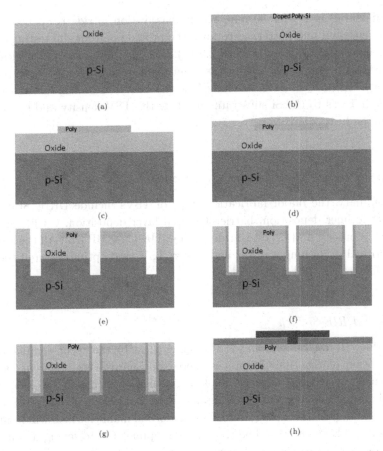

Fig. 2.9 Schematic process flow of TSV structure for thermal property study: (a) $2k\text{Å}$ thermal oxide deposition; (b) $5k\text{Å}$ poly-Si deposition, $PoCl_3$ doping and annealing; (c) doped poly-Si etch; (d) $1\mu m$ CVD oxide deposition; (e) oxide CMP to flatten the surface and via formation by DRIE Si etch; (f) liner deposition and Ta barrier/Cu seed layer sputtering; (g) Cu ECP followed by Cu CMP to remove Cu overburden; (h) nitride passivation, contact open and Al metallization and patterning.

in a PECVD chamber at temperature of $< 400°C$ using TEOS and Black Diamond precursors for oxide and low-k liner respectively, which is the same as that step of fabrication of TSV for electrical characterization. Ta barrier and Cu seed layer sputtering are then carried out by using the ENDURA deposition chamber, and Cu ECP is carried out till all the TSVs are fully filled and CMP is carried out by using the same process as that of the electrical structures (Figure 2.9(f) and (g)). After Cu annealing is carried out at 200°$C$ in N$_2$ for 30 min, finally, passivation nitride is opened and Al is deposited and patterned to form the probing pad for the testing and to complete the process (Figure 2.9(h)). The thermal structure basically consists of a doped poly-Si line as a heat generator surrounded with or beneath TSVs. It is used to study how the heat can be conducted away through TSVs to the Si substrate, and how the TSV density and size can affect the heat conduction.

### 2.2.3  *Process Control and Optimization*

In summary, the major fabrication steps of TSVs include DRIE Si etch, dielectric liner deposition, barrier/Cu seed layer deposition, Cu filling by Cu ECP process and Cu overburden removal by Cu CMP [75]. All the steps are critical in forming a stable and robust TSV, and optimization of these to achieve this will be essential.

#### 2.2.3.1  *DRIE Si Etch*

The sidewall roughness from the scalloping effect by the cyclic etching and passivation of the BOSCH process and oxide hard-mask undercut (also known as over-hang) must be kept as small as possible. If the sidewall roughness or the undercut is too significant, it would be a challenge for the subsequent dielectric liner to form in a conformal manner which will induce a higher leakage current. The same reason applies for achieving a conformal Ta barrier and Cu seed layer as well which will lead to incomplete Cu filling or voids inside the TSV, and thus reliability. Both side effects are detrimental to TSV yield and reliability. From the initial run, the average sidewall scalloping is $\sim 120nm$, and undercut is $\sim 240nm$, as shown in Figure 2.10(a). By adding 20 sccm of C$_4$F$_8$ during the etching cycle, the sidewall scalloping and undercut are both improved. The improved result shows that the sidewall roughness is $\sim 70nm$, and undercut is $\sim 60nm$, as shown in Figure 2.10(b). To eliminate the undercut completely, a double-mask step can also be used to etch the oxide hard-mask and Si separately. Figure 2.11 is a summary of TSV DRIE Si etching results. When a single photo-resist mask is used to etch both oxide hardmask and Si, the addition of a small quantity of C$_4$F$_8$ during the etching cycle improves both roughness and undercut (Process A vs. Process B). To eliminate the undercut completely, a double-mask step can be used to etch the oxide hard-mask

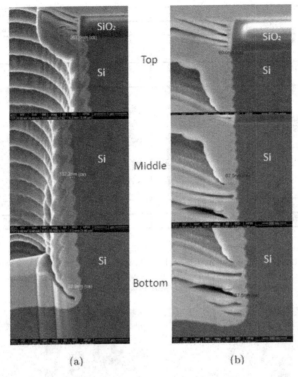

Fig. 2.10  Unfilled TSV after DRIE Si etch: (a) initial results show sidewall scalloping is $\sim 120nm$ and undercut is $\sim 240nm$; (b) improved results show side wall scalloping $\sim 70nm$ and undercut $\sim 60nm$.

and Si separately (Process C). In this process, the opening in the second mask layer is designed to be smaller than the first one.

### 2.2.3.2  *Dielectric Liner Deposition*

In this work, several materials and material combinations are utilized and integrated as the dielectric liner of the TSV structure. Oxide is the commonly used material as the dielectric liner of TSV due to its good dielectric properties and complementary metaloxidesemiconductor (CMOS) process compatibility. Figure 2.12(a) shows the FIB image of the oxide liner deposited at $400°C$ in a PECVD chamber. The step coverage of the oxide liner is continuous and is around 35%, and the average thickness is $\sim 200nm$.

| Process | A | B | C |
|---|---|---|---|
| Mask | Single | Single | Double |
| Etching Gas | $SF_6$, $O_2$ | $C_4F_8$, $SF_6$, $O_2$ | $C_4F_8$, $SF_6$, $O_2$ |
| Passivation Gas | $C_4F_8$ | $C_4F_8$ | $C_4F_8$ |
| Etch Profile<br><br>*(SEM image of unfilled TSV)* | | | |
| Roughness (nm) | 107-129 | 68-75 | 50-75 |
| Undercut (nm) | 239 | 60 | ~0 |

Fig. 2.11  Comparison of TSV etching processes and profiles. When a single mask step is used, the addition of $C_4F_8$ during the etch cycle improves the scallop sidewall roughness as well as oxide hard-mask undercut. The undercut is negligible of a double-mask etching step is used to etch the $SiO_2$ hard-mask and Si separately.

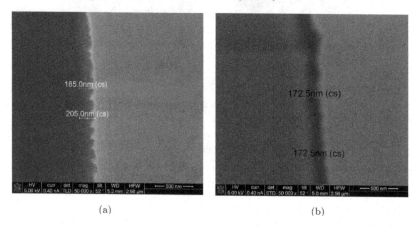

(a)                                                    (b)

Fig. 2.12  FIB image of TSV structure shows: (a) conformal oxide liner deposition; (b) conformal low-k liner deposition. The average thickness of the oxide liner is $\sim 200nm$.

When TSV is used for signal transmission, parasitic capacitance has the most predominant impact on the circuit operation [76] and must be addressed carefully. A large capacitance can result in large signal latency and power consumption. It is therefore imperative to reduce the TSV capacitance either by selecting the optimum dimension or appropriate material. Low-k dielectric can significantly reduce the CMOS backend interconnect capacitance and provide many advantages in circuit performance in deep submicron technology nodes [72]. Since a low-k dielectric and process is compatible with the TSV process, it can be used in principle to reduce the TSV capacitance. In this work, low-k material is successfully integrated in to the TSV structure as the dielectric liner, which has an average thickness of $\sim 200nm$, as shown in Figure 2.12(b). It is carried out at $350°C$ in a PECVD chamber using black diamond as the precursor.

Smaller TSV capacitance is always desired in signal transmission to reduce signal latency and power consumption. On the other hand, capacitance stability is also an important consideration in an integrated circuit (IC) design. In case of a non-uniform hotspot heating across an area, ideally, a temperature-independent TSV capacitance is desired. The negative fixed charge in the oxide liner can be utilized to effectively shift the TSV $CV$ curve so that the stable accumulation capacitance is maintained within the voltage of interests. The density of the fixed charge depends on the oxidizing ambient, temperature, silicon orientation, cooling rate, and subsequent annealing cycle. A negative fixed charge in CVD oxide has previously been observed in other reports [77,78]. By process tuning, such as the ratio of gas flow and power, the desired amount of negative fixed charges is obtained. Other than the process tuning, the desired $CV$ shift can also be achieved by an $Al_2O_3$-induced negative fixed charge. A very thin layer of $Al_2O_3$ is successfully integrated in between the Si substrate and the dielectric liner by utilizing the atomic-layer-deposition (ALD) process. As shown in Figure 2.13, a $\sim 100\mathring{A}$ $Al_2O_3$ layer is inserted in between the Si and oxide liner. An energy dispersive X-ray spectroscopy (EDX) performed on the different material layers confirmed that the atomic ratio of Al to O is 27.2 to 72.8. Similarly, $Al_2O_3$ can also be inserted between Si and low-k liner as shown in Figure 2.14.

### 2.2.3.3 *Ta Barrier/Cu Seed Layer Deposition and Cu ECP*

Ta barrier/Cu seed layer deposition and Cu ECP are very critical for TSV structure fabrication. Non-conformal Ta barrier and Cu seed layer deposition may lead to incomplete Cu filling of the via hole. Cu ECP-related issues are very commonly seen during the process, which includes bottom voids, pinch off and Cu protrusion [79]. In this work, the Ta barrier and Cu seed are sputtered using a PVD process in the ENDURA deposition chamber. Subsequently, Cu electroplating (ECP) is carried out to fully fill the TSV, and we use RENA system and Enthone chemical solution for

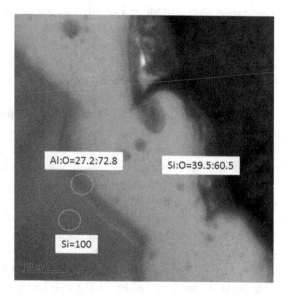

Fig. 2.13   HRTEM image of liner. Conformal deposition of $Al_2O_3$/oxide dielectric liner is achieved with ALD and PECVD. Atomic ratio of Al:O=27.2:72.8 of the $Al_2O_3$ layer is obtained from EDX.

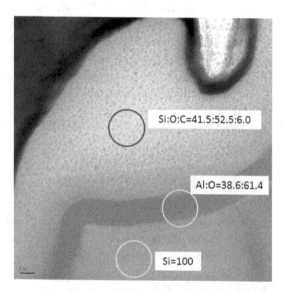

Fig. 2.14   HRTEM image of the liner stack. The $Al_2O_3$ layer is deposited with ALD and low-k dielectric (SiOC) liner is deposited conformally using PECVD at a temperature $< 400°C$.

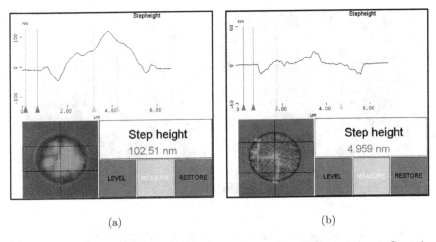

(a)                          (b)

Fig. 2.15    Cu protrusion is controlled with proper annealing to allow optimum Cu grain growth, the Cu protrusion is reduced from (a) $102.5nm$ to (b) $5.0nm$.

the ECP process. The acid copper plating process was specially developed for filling blind micro-vias and plating through-holes simultaneously in this work. In order to control the Cu protrusion, a combination of annealing and CMP touch up is used. In Figure 2.15, AFM surface profiles show that Cu protrusion is reduced from $102.5nm$ (Figure 2.15(a)) to $5.0nm$ (Figure 2.15(b)) with this optimization. Figure 2.16 shows the FIB images of Cu filled TSV with different material selections as the dielectric liner. Excellent step coverage is achieved and all the TSVs are filled with Cu without any voids or delaminations.

### 2.2.3.4    *Cu CMP*

Cu overburden from the previous Cu ECP process is then polished, in the Cu CMP process. In this work, Rohm and Haas high polish rate slurry colloidal silica-based is used in Cu CMP for bulk Cu removal. Figure 2.17(a) shows the plan view of a TSV array with diameter of $5\mu m$ after removing the Cu overburden by CMP from the microscope. It shows that the Cu overburden is completely polished. Figure 2.17(b) shows the final electrical MOS structure formed by TSV for electrical probing testing. From the image of the microscope, it is shown that the Si substrate is grounded by using a $p^+$ contact. The probe pad capacitance, which is a source of secondary parasitic during CV measurement on the TSV, is kept low by

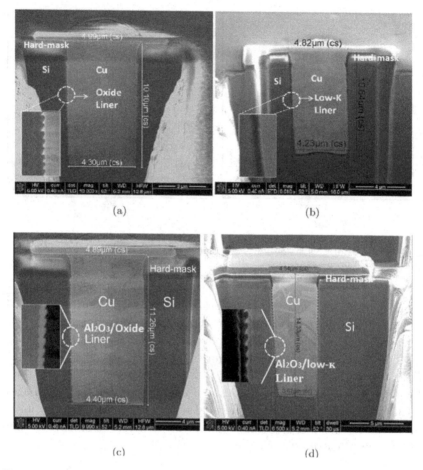

Fig. 2.16 FIB images of TSVs with $\sim 5\,\mu m$ diameter and $\sim$10-15 $\mu m$ depth, filled with Cu and lined with: (a) PETEOS oxide; (b) black diamond low-k material; (c) bi-layer of $Al_2O_3$ and oxide; (d) bi-layer of $Al_2O_3$ and low-k as the isolation liners. Conformal deposition is achieved. No voids or delaminations observed.

using thicker oxide below the Al probe pad and the formation of TSV array to increase the effective TSV area. The oxide below the Al probe pad is much thicker ($\sim 1600nm$) than the TSV liner ($\sim 200nm$). It is also possible to de-embed the probe pad capacitance by forming dummy pads without TSV.

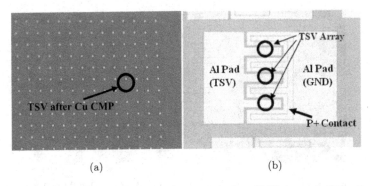

(a)　　　　　　　　　　　　(b)

Fig. 2.17 (a) Plain view of TSV array ($\phi = 5\mu m$) after CMP removal of the Cu over-burden and no sign of Cu debris is found from optical microscope; (b) electrical MOS structure formed by TSV for probing. The Si substrate is grounded using $p^+$ contact. The probe pad capacitance is kept low by using thicker oxide below the Al probe pad and formation of TSV array to increase the effective TSV area.

## 2.3 TSV Electrical Characterization

Till now, wafer level design of the TSV structures for electrical characterization and thermal properties as well as dummy TSV blocks are discussed along with the detailed fabrication process flows. Based on the fabricated TSVs, the electrical characterization and the $CV$ characteristics of TSVs are studied in this section.

### 2.3.1 *Measurement Setup*

The measurement setup image for electrical $CV$ measurement is shown in Figure 2.18, where the measurements are carried out using an Agilent $CV$ meter. For characterization, the input voltage applied is swept from $-25V$ to $25V$ on the Al pad connected to the TSV arrays while the other Al pad is connected to the Si substrate utilizing the $p^+$ contact. The probe pad capacitance, which is a source of secondary parasitic during $CV$ measurement on the TSV, is kept low by using a layer of thicker oxide below the Al probe pad and formation of TSV array to increase the effective TSV area. It is also possible to de-embed the probe pad capacitance by forming dummy pads without TSV. Moreover, the electrical field is stressed up to $3MV/cm$ for the electrical $IV$ measurement.

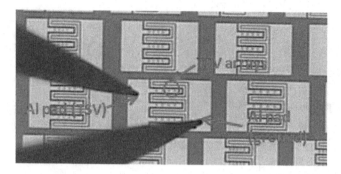

Fig. 2.18  Microscope image of electrical measurement setup. One Al pad is connected to the TSVs, while the other Al pad is connected to $p^+$ ground. Input Voltage is swept from $-25V$ to $25V$.

### 2.3.2  *Conventional PETEOS Oxide Liner*

Plasma Enhanced Tetraethylorthosilicate (PETEOS) oxide is the conventional material to be utilized as the dielectric liner of TSV. This is because of several advantages, such as, good dielectric properties, conformality, and CMOS process compatibility. In our experiments, TSV structures with PETEOS oxide as the dielectric liner will be the baseline structures.

#### 2.3.2.1  *Electrical CV Measurement*

Figure 2.19 shows the measured $CV$ curve at $1MHz$ for two different deposition conditions of the PETEOS oxide liner. The temperature of the two deposition processes are the same ($\sim 180°C$), while other parameters such as ratio of gas flow and power are adjusted to achieve the desired $CV$ results. From Figure 2.19, we can see that the TSV capacitance changes from the accumulation to depletion to inversion region as the voltage applied was swept from $-30V$ to $30V$. This $CV$ characteristic demonstrates a MOS capacitor on a $p$-type Si substrate behaviour. In comparison with process A, the $CV$ curve of process B shifts more to the positive X-axis so that the accumulation capacitance region is of interest within the operating voltage of interests ($\sim$0-5$V$). The shift of the $CV$ curve is caused by the fixed charges induced during the PETEOS oxide deposition process. Observation of a negative fixed charge in PECVD oxide has previously been observed in other reports [77, 78]. In order to calculate the fixed charge density induced during the deposition process, knowledge of the flatband voltage in the $CV$ curve is required. From a $C_{ox}$ of $\sim 1.8 \times 10^{-8} F/cm^2$ from the

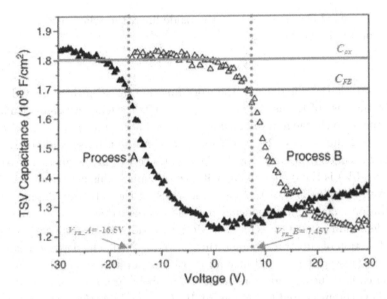

Fig. 2.19 *CV* characteristic of TSV MOS structure showing clear regions of accumulation, depletion, and inversion. With proper process tuning, negative fixed charge induced by oxide liner can cause flatband voltage shift to stabilize the TSV capacitance at accumulation capacitance within the operating voltage region of interest ($\sim$ 0-5$V$). Accumulation capacitance of a single TSV with 5$\mu m$ diameter and 10$\mu m$ depth is $\sim$ 28.3$fF$.

measurement, the effective oxide liner thickness is calculated to be $\sim$192$nm$, given by

$$C_{ox} = \frac{2\pi\varepsilon_{ox}L}{\ln\frac{r_1}{r_0}} \qquad (2.1)$$

where $L$ is the TSV depth; $r_0$ is the radius of the Cu and $r_1$ is the radius (as etched). Following, the flatband capacitance ($C_{FB}$) can be estimated using the equation below [80]

$$\frac{C_{FB}}{C_{ox}} = \frac{1}{1 + \frac{136\sqrt{T/300}}{t_{ox}\sqrt{N}}} \qquad (2.2)$$

where $T$ is the temperature in Kelvin; $N$ is the silicon substrate doping concentration ($\approx 1 \times 10^{16} cm^{-3}$), and $t_{ox}$ is the oxide liner thickness. By using this equation, we can estimate $C_{FB} \approx 0.963 C_{ox} \approx 1.7 \times 10^{-8} F/cm^2$. Hence the locations of the flatband voltage ($V_{FB}$) on the CV plots are $\sim -1.65V$ and $\sim 7.3V$ for processes A and B respectively, as shown in Figure 2.19.

When the fixed charge ($Q_f$) is in the oxide liner, the relationship between $V_{FB}$ and $Q_f$ is shown below

$$V_{FB} = \phi_{ms} - \frac{Q_f}{C_{ox}} \qquad (2.3)$$

where $\phi_{ms}$ is the difference in the work function between the gate and the Si. In the case of Cu as the gate, $\phi_{ms} = -0.22V$. Hence we can calculate that the fixed charge induced during the oxide liner deposition to be $\sim 1.8 \times 10^{12}cm-2$ and $\sim 8.43 \times 10^{11}cm^{-2}$ for process A and process B respectively.

For baseline process A, the TSV capacitance within the operating voltage ($\sim$0-5V) is the depletion capacitance. Although this capacitance value is lower, its value is not stable as it depends on the substrate doping concentration, small signal frequency and the temperature. The accumulation and depletion capacitances of TSV-MOS embedded in the $p$-Si are plotted as a function of substrate temperature in Figure 2.20. The depletion capacitance is susceptible to and increases with temperature while the accumulation capacitance is independent of the temperature. Figures 2.21(a) and (b) show the measured CV curves at $100KHz$ and $2MHz$ respectively on a planar MOS structure (dielectric thickness is $\sim 200nm$). The CV curves show that at a smaller frequency signal ($100KHz$) when the temperature is higher ($> 175°C$), the minor charge carrier electrons with enough energy can respond fast enough to form an inversion layer, so that the capacitance is prevented from decreasing further. But at a higher frequency signal ($2MHz$), the electrons cannot respond fast enough to form the inversion layer, so the depletion continues and capacitance continues decreasing as the voltage is swept. In a 3D circuit with TSV distribution across the entire

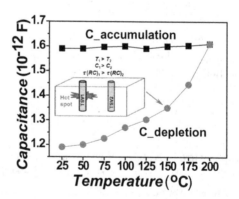

Fig. 2.20 While accumulation capacitance is stable, depletion capacitance is susceptible and increases rapidly with substrate temperature.

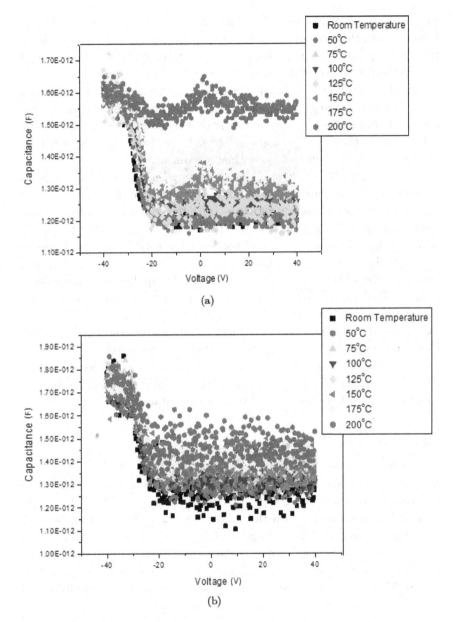

Fig. 2.21 Measured CV characteristics of a planar MOS at a small signal frequency of: (a) $100KHz$; and (b) $2MHz$. It shows that the depletion capacitance varies when the substrate temperature changes and it will complicate the circuit design.

chip, it is essential to ensure that the TSV capacitance is stable. This is because when there is emergence of hot spots due to non-uniform heating, if the TSV capacitance varies with the temperature, it will lead to undesired TSV performance variation across the entire chip and complicate the circuit design. Therefore, process B satisfies this requirement by inducing enough negative fixed charge during the liner deposition process and hence causing a positive shift in the $V_{FB}$ so as to ensure that the accumulation capacitance is kept within the operating voltage. The negative fixed charge successfully results in a $+7.5V$ shift in $V_{FB}$ in this experiment.

Operating the TSV in a stable but higher capacitance will not increase the overall parasitic capacitance significantly in applications where the TSV length is only a small fraction of the on-chip interconnect length. In high-end multi-core processor application, the TSV density is estimated to occupy $\sim 3\%$ of the silicon area [81], and therefore the total TSV capacitance is much smaller compared with the on-chip interconnect capacitance. Hence, operating the TSV in the accumulation region for applications that require a stable TSV capacitance at the expense of a slightly higher value, is an important tradeoff to stabilize the TSV parasitic and to avoid spatial performance variation of the TSV. The TSV capacitance can still be kept small by methods such as TSV design to meet the parasitic capacitance requirement. In actual application, the process tuning can be further optimized to shift the $V_{FB}$ toward higher voltages to minimize the potential process variation and to have a more stable oxide capacitance.

### 2.3.2.2    *Electrical IV Measurement*

The *IV* measurement is used to monitor the leakage of the PETEOS oxide liner using the same samples used for *CV* measurement. From Figure 2.22, the leakage current is $\sim 1.2 \times 10^{-6} A/cm^2$ at an electric field of $2MV/cm$; this value is comparable with the value reported in [82]. In an actual on-chip TSV application, the voltage is of the order of $5V (\sim 0.25MV/cm)$, and therefore, the leakage current is about two orders of magnitude smaller at $\sim 10^{-8} A/cm^2$. The *IV* result also shows that at least up to an electric field of $3MV/cm$, there is no abrupt breakdown observed for the PETEOS oxide liner.

### 2.3.3    **Black Diamond Low-k Liner**

In order to enhance the 3D IC performance, TSV should introduce a small electrical parasitic such as capacitance. In the previous subsection, we have

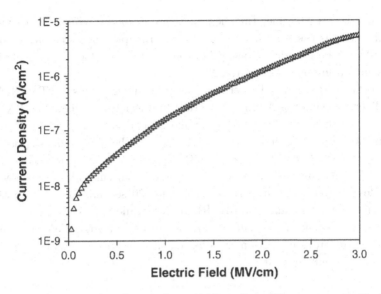

Fig. 2.22 Leakage current through the oxide liner from $IV$ measurement. No hard breakdown is observed. The leakage current density at an electric field of $2MV/cm$ is $\sim 1.2 \times 10^{-6} A/cm^2$ which is comparable with reported value.

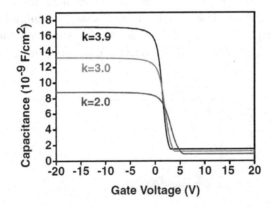

Fig. 2.23 Capacitance reduction in MOS structure using low-k dielectric. Dielectric thickness is $200nm$ and the substrate doping, Na, is $1 \times 10^{16} cm^{-3}$.

discussed the use of the accumulation capacitance within the operating voltage of interests ($\sim$0-5$V$) since it is independent of the substrate temperature fluctuations, but with the expense of a slightly higher capacitance value. The simulated $CV$ characteristics of MOS structure at high frequency as a function of dielectric constant (k) is shown in Figure 2.23. In this simulation, the liner thickness is taken as $200nm$ while the substrate doping (Na) is taken as $1 \times 10^{16} cm^{-3}$. In principle, the capacitance value of TSV can be lowered by using liner with small k value as predicated in the simulation result in Figure 2.23.

In this work, the low-k liner is successfully integrated as the TSV dielectric liner by using the Applied Material Producer Black Diamond PECVD technology to lower the TSV accumulation capacitance. The low-k liner deposition process includes two major steps: a) the deposition step is carried out at $350°C$ for 49.5 seconds with a pressure of 4 torr, a $600W$ radio-frequency (RF) power and 100 sccm $O_2$ and 600 sccm TMS gases flow; b) the Helium treatment step is performed for 20 seconds at $350°C$ with a pressure of 8.7 torr and a $750W$ RF power, while the Helium gas flow is 1300 sccm. Low-k liner is successfully integrated into the TSV structures as shown in previous section, with excellent step coverage and no void or delamination filling with Cu.

### 2.3.3.1   *Electrical CV Measurement*

Figure 2.24 shows the comparison between the measured $CV$ characteristics at $100KHz$ for both PETEOS oxide and low-k liners at an average liner thickness of $\sim 200nm$. The effective low-k value is estimated to be $\sim 2.88$ by using the equation (2.1). The dielectric capacitance value is lowered by $\sim 27.6\%$ by comparing the PETEOS oxide and low-k liner. We can

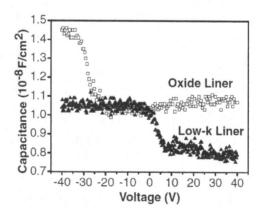

Fig. 2.24   $CV$ measurement (100KHz) on TSV structures formed using PETEOS oxide liner and low-k liner. A $\sim 27.6\%$ reduction in accumulation capacitance is obtained by replacing PETEOS oxide liner with low-k liner.

Table 2.1 Comparison between liner and minimum capacitance value (per unit length) based on analytical calculation and $CV$ measurement (for $r_0 = 2.205\mu m$ and $r_1 = 2.410\mu m$). Dielectric thickness is set as $200nm$ and the substrate doping, $N_a$ is $1e16 cm^{-3}$.

| | PETEOS ($/cm$) | Low-k ($/cm$) |
|---|---|---|
| Liner Capacitance | $22.4\,pF$ | $17.5\,pF$ |
| $C_{ox} = \frac{2\pi\varepsilon_{ox}}{\ln(r_0/r_1)}$ | $22.1\,pF\ (CV)$ | $15.9\,pF\ (CV)$ |
| Si Depletion Radius ($r_2$) | $r_2 - r_1 = 265\,nm$ | $r_2 - r_1 = 265\,nm$ |
| $\frac{qNa}{2\varepsilon_s}\left(r_2^2 \ln\frac{r_2}{r_1} - \frac{r_2^2 - r_1^2}{2}\right) = 2\phi_F$ | | |
| Si depletion capacitance | $62.4\,pF$ | $62.4\,pF$ |
| $C_{dep} = \frac{2\pi\varepsilon_s}{\ln(r_2/r_1)}$ | | |
| $C_{min}$ | $17.5\,pF$ | $13.7\,pF$ |
| $C_{min} = (1/C_{dep} + 1/C_{ox})^{-1}$ | $15.6\,pF\ (CV)$ | $12.1\,pF\ (CV)$ |

also observe that the $CV$ curve of the PETEOS oxide liner TSV shifted negatively in the $V_{FB}$ due to the positive fixed charge induced during the liner deposition step, which is reported to be $\sim 1.8 \times 10^{12} cm^{-2}$ in the previous section. The low-k liner deposition process is observed to have a much less induced fixed charge. In Table 2.1, the measured capacitance values are compared with the analytical data based on the MOS model [83]. A small discrepancy is seen and this is attributed to the non-uniformity in the liner thickness in actual TSV structure arising from process-induced sidewall roughness.

Low-k material can significantly reduce the TSV accumulation capacitance. However, due to its inherent porosity, several reliability issues such as mechanical reliability as well as dielectric reliability are becoming more important as devices scale down [72]. Leakage current is one of the important dielectric reliability issues that we must consider. In the next subsection, improving on the leakage current by annealing will be presented. Therefore, it is important to monitor the possible change in the permittivity value of the low-k liner when annealing is used in order to retain its role in capacitance reduction. Figure 2.25 shows a small reduction in the effective k value under various annealing conditions which are applied to control the leakage current. In addition, annealing is also effective in reducing the density of fixed charge as seen from the negative shift in the flatband voltage to the ideal value of $-0.27V$.

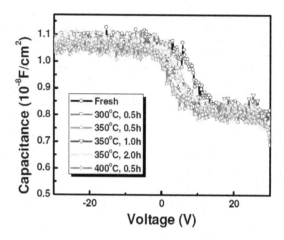

Fig. 2.25  $CV$ characteristic under various annealing conditions. The flatband voltage is shifted to the left signifying a reduction in negative fixed charge density. The k value changes slightly from 2.88 to 2.85 ($300°C$ and $350°C$ annealing) and then to 2.80 ($400°C$ annealing).

### 2.3.3.2  *Electrical IV Measurement*

While it is encouraging that the TSV capacitance can be reduced by the incorporation of low-k material as the liner, its implication on the leakage current to the Si substrate needs to be investigated for a complete assessment. Electrical $IV$ measurements were performed to monitor the leakage of the TSV dielectric liner. The isolation property of the dielectric liner must be preserved during the TSV operation. A good dielectric liner should keep leakage as small as possible. Figure 2.26 shows the leakage current density versus the electric field. The results show that for an electric field up to $3MV/cm$ (corresponding to $60V$), there's no abrupt breakdown observed for both PETEOS oxide and low-k liners. This shows that the liner thickness is sufficient to prevent dielectric breakdown for the usual operating voltage below $60V$. However, Figure 2.26 also reveals that the low-k liner obviously suffers from higher leakage current as compared to PETEOS oxide liner due to its structural porosity.

A careful study on the leakage current was then carried out and the result is presented in the cumulative plot in Figure 2.27. From Figure 2.27, it can be observed that low-k suffers from a higher leakage current (at an electric field of $2MV/cm$) as compared with the PETEOS oxide liner by as much as 2.5 times at mid-distribution. It was found that annealing of the TSV in forming gas ($N_2/H_2$) at $350°C$ for 30 min can improve the

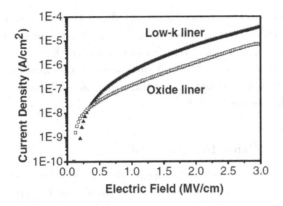

Fig. 2.26 *IV* measurements on TSV structures. Low-k liner is found to suffer from higher leakage current as compared with PETEOS oxide liner with PETEOS oxide and Black Diamond low-k liners.

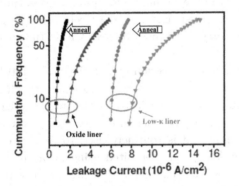

Fig. 2.27 TSV leakage current measurements in an electric field of $2MV/cm$. The low-k liner experiences much higher leakage as compared to the PETEOS oxide liner. The leakage current is reduced after annealing at $350°C$ for 30 min in forming gas for both liners.

leakage current for both the PETEOS oxide and low-k liners. For post annealing, the leakage current at mid-distribution is $\sim 1.2 \times 10^{-6}$ and $\sim 6.8 \times 10^{-6} A/cm^2$ for the PETEOS oxide and the low-k liner respectively in an electric field of $2MV/cm$. The leakage current value for the PETEOS oxide liner is comparable with the value reported by Archard et al [82]. A higher leakage current is measured for the low-k liner in comparison with the PETEOS oxide liner due to a higher porosity in the low-k film. The higher leakage current value observed for the low-k liner can also be caused by imperfections such as the scalloping roughness (as a result of BOSCH etching), which causes difficulty in the deposition of a conformal low-k liner on the TSV sidewall. Therefore, for the low-k liner to be effectively used in TSV application, further optimization in the leakage current is required. Possible solutions include optimization in annealing (with higher temperature or longer duration), optimization in scalloping roughness, and to use slightly thicker low-k liner.

### 2.3.4 $Al_2O_3/Oxide$ Bi-layer Liner

In the previous section, we have discussed the operation of the TSV in the stable accumulation capacitance region to overcome the spatial TSV performance variations caused by non-uniform hotspot heating. The depletion capacitance is susceptible to, and increases with, temperature. TSV with tolerance to spatial variation is desirable and was a successfully achieved previously by PETEOS oxide liner deposition process tuning. This caused the induced negative fixed charge to effectively shift the flatband voltage, and the accumulation capacitance obtained within the operating voltage of interests ($\sim$0-5$V$).

Here, a novel material selection of thin $Al_2O_3$ ($\sim 100 \overset{\circ}{A}$) is successfully integrated and inserted between the Si substrate and the dielectric liner. The flatband voltage shift is achieved by the negative fixed charge in the thin $Al_2O_3$ layer at the $p$-Si/liner interface. $Al_2O_3$ is known to induce a negative fixed charge and is routinely used to passivate $p$-Si in Si solar cells [84, 85]. The $Al_2O_3$ was deposited using an atomic layer deposition (ALD) process at $300°C$. The deposition process involves the sequential surface reactions of $Al(CH_3)_3$ (tri-methylaluminum (TMA)) and $H_2O$. An advantage of using ALD in the deposition of $Al_2O_3$ is the ability to attain perfect step-coverage on high aspect ratio vias, and it has been discussed that ALD has the capability of filling high aspect ratio vias till the extent of 40:1 [86]. Due to the higher k value of the $Al_2O_3$, if we assume a constant total liner thickness, a larger thickness of $Al_2O_3$ will result in an increase of the final TSV capacitance, which is undesirable. Therefore, a thin layer of $Al_2O_3$ is preferred. In view of our PETEOS liner thickness of 200$nm$, we then decided to adopt an $Al_2O_3$ thickness of 100$\overset{\circ}{A}$, which is 5% of total dielectric thickness, as a start.

#### 2.3.4.1 Electrical CV Measurement

The simulated $CV$ characteristic of MOS structure at high frequency on $p$-Si is shown in Figure 2.28. In principle, a negative fixed charge ($Q_f$)

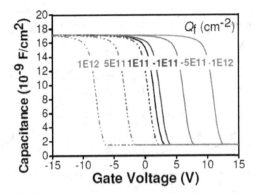

Fig. 2.28 The role of fixed charges in shifting the flatband voltage of MOS structure from simulation. (Dielectric thickness is $200nm$ and the substrate doping, Na, is $1e16cm^{-3}$).

results in a positive flatband voltage $V_{FB}$ shift, and one can use it to operate the TSV-MOS in the accumulation region within the operating voltage of interests ($\sim$0-5$V$). This is achieved experimentally by inserting a thin layer of $Al_2O_3$ between the Si substrate and the PETEOS oxide liner. A positive shift in $V_{FB}$ is obtained as shown in Figure 2.29(a)–2.29(c) under various annealing conditions which were applied to control and improve the leakage current. A smaller accumulation capacitance ($C_{ox}$) is measured for the $Al_2O_3$/PETEOS oxide bi-layer liner due to a larger effective liner thickness, such as 16.1 vs. 17.0 $nF/cm^2$ with $N_2/H_2$ annealing in Figure 2.29(b). Without the presence of the oxide fixed charge, $V_{FB}$ is simply the difference in the work functions between the gate and Si, and is approximately $-0.27V$. A fixed charge ($Q_f$) in the liner can effectively shift the $V_{FB}$ according to the equation (2.3) discussed in previous section. The effect of annealing in forming gas $N_2/H_2$ and inert gas $N_2$ on PETEOS oxide liner and $Al_2O_3$/PETEOS oxide bi-layer liner is presented in Figure 2.30(a) and 2.30(b). The PETEOS oxide liner contains a large number of positive fixed charge. The charge density is on the order of $\sim 9.70 \times 10^{11} cm^{-2}$ based on the calculation equations in previous section. This results in a negative shift in $V_{FB}$ ($V_{FB} = -9.60V$ for the sample annealed in $N_2/H_2$). Therefore, TSV is operated in the depletion and unstable within the operating voltage of interests ($\sim$0-5$V$). With the insertion of $Al_2O_3$ layer, a large amount of negative fixed charge is induced. The charge density calculated is on the order of $\sim -7.44 \times 10^{11} cm^{-2}$ which causes a positive shift in $V_{FB}$ ($V_{FB} = 7.10V$ for the sample annealed in $N_2/H_2$). As a result, the TSV is in the stable accumulation region between $\sim$0-5$V$ and does not fluctuate with temperature. Figure 2.31 summarizes the $CV$ characteristics of

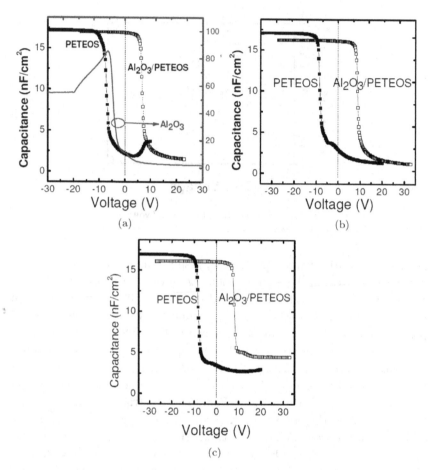

Fig. 2.29   (a) *CV* characteristic of TSV-MOS before annealing; (b) *CV* characteristic of TSV-MOS with annealing in forming gas ($N_2/H_2$) at $300°C$ for 30 min; (c) *CV* characteristic of TSV-MOS with annealing in inert $N_2$ at $300°C$ for 30 min.

TSV structures with the PETEOS oxide liner and $Al_2O_3$/PETEOS oxide bi-layer liner after annealing in forming gas ($N_2/H_2$) and inert $N_2$ gas at $300°C$ for 30 min at $100kHz$.

As demonstrated in Figure 2.32, for the $Al_2O_3$/PETEOS oxide bi-layer liner after gas annealing, the $C_{ox}$ value is stable when the measurement temperature is changed between 25 to $100°C$. However, the depletion capacitance increases with temperature. The accumulation capacitance is higher than the depletion capacitance, and the value is estimated to be $\sim 25.3fF$. However, this value is still below the requirement of $< 100fF$ proposed

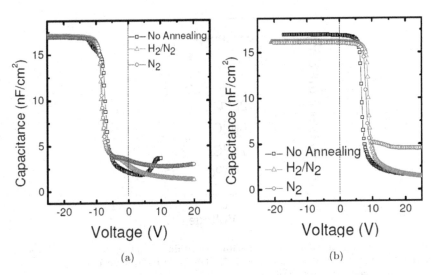

Fig. 2.30 (a) Summary of $CV$ characteristic of PETEOS oxide liner with various anneal-ing ambient. The fixed charge density $Q_f$ is $\sim 9.7 \times 10^{11} cm^{-2}$ and flatband voltage $V_{FB}$ is $-9.6V$; (b) Summary of $CV$ characteristic of $Al_2O_3$/PETEOS oxide bi-layer liner with various annealing ambient. The fixed charge density $Q_f$ is $\sim -7.44 \times 10^{11} cm^{-2}$ and flatband voltage $V_{FB}$ is $7.1V$ (annealing in $N_2/H_2$).

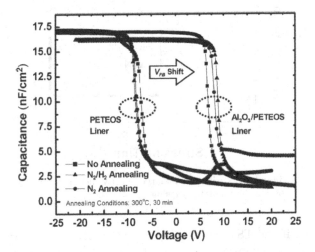

Fig. 2.31 Summary of $CV$ characteristics of TSV structures with $Al_2O_3$/PETEOS oxide bi-layer liner and PETEOS oxide liner after annealing in inert $N_2$ gas and forming gas $(N_2/H_2)$ at $300°C$ for 30 min.

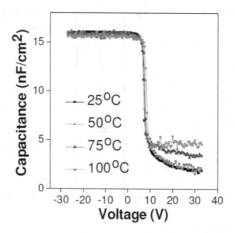

Fig. 2.32 Stable accumulation capacitance within operating voltage of interests (~0-5 $V$) is achieved with the $Al_2O_3$/PETEOS oxide bi-layer liner. Depletion capacitance increases with temperature.

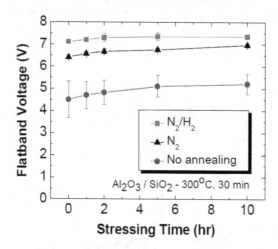

Fig. 2.33 $Al_2O_3$-induced negative fixed charge density increases initially and becomes stable after ~ 5 hours of accumulated biasing at $2MV/cm$.

in [87]. Even if the TSV depth is increased to $25\mu m$ as reported in [88] which has an aspect ratio of ~ 1 : 5, the estimated accumulation capacitance of ~ $63.3fF$ is still within the requirement.

Since the positive shift in $V_{FB}$ is achieved by the $Al_2O_3-$ induced negative fixed charge, the stability of the negative fixed charge density is important. Figure 2.33 shows the stability of the negative fixed charge as seen

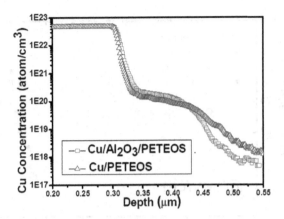

Fig. 2.34 SIMS analysis shows that $Al_2O_3$ (100$\mathring{A}$) is effective in suppressing Cu diffusion.

from the $V_{FB}$ change. After an accumulated constant biasing $(2MV/cm)$ for 10 hours, the negative fixed charge density is found to increase initially and plateau after $\sim 5$ hours. It confirms that within the operating voltage of interests $(\sim 0\text{-}5V)$, the stable accumulation capacitance is maintained. In addition, $Al_2O_3$ can effectively act as extra barrier against Cu diffusion in TSV as confirmed with secondary ions mass spectroscopy (SIMS) analysis on blanket wafers in Figure 2.34.

### 2.3.4.2  *Electrical IV Measurement*

Another important electrical property of the dielectric liner is the leakage current from the Cu core through the liner to the Si substrate. *IV* measurement is performed to evaluate the leakage performance of the liner. Figure 2.35 shows the leakage current density of the $Al_2O_3$/PETEOS oxide bi-layer liner and PETEOS oxide liner after annealing in inert $N_2$ gas and forming gas $(N_2/H_2)$ at $300°C$ for 30 min under an electric field of $2MV/cm$. The result shows that annealing at $300°C$ for 30 min can significantly improve the leakage current, and it can be noted that annealing in forming gas $(N_2/H_2)$ brings about the largest improvement. The largest improvement of the leakage current observed after annealing in forming gas is due to the reduction of defect states of the Si-dielectric interface. Owing to the small atomic size of hydrogen, it could effectively diffuse into the Si-dielectric interface and complete the Si dangling bond. In contrast, $N_2$ gas with its much larger atomic size, is not able to diffuse as effectively

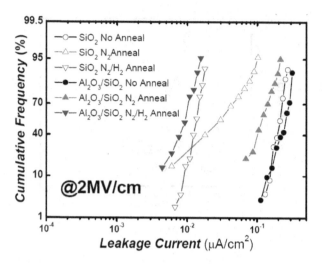

Fig. 2.35  *IV* measurement results of $Al_2O_3$/PETEOS oxide bi-layer liner and only PETEOS oxide liner. Leakage current density is improved by $\sim 10\times$ after annealing in forming gas ($N_2/H_2$) at $300°C$ for 30 min.

as $H_2$ used in forming gas. Hence, the use of forming gas has the best impact with regard to the improvement of the leakage current. By post-annealing in forming gas, at mid-distribution, the leakage current density of the $Al_2O_3$/PETEOS oxide bi-layer liner is on the order of $10^{-8} A/cm^2$; this has been reduced by $\sim 10\times$ as compared to that of the unannealing samples. It is worth pointing out that this leakage current density level is suitable for actual TSV applications [82].

## 2.4    TSV Thermal Characterization Results

By taking advantage of the excellent thermal conduction properties of silicon, we demonstrate experimentally that the use of TSV arrays in circuits can provide an additional heat conduction path through the Cu core to the silicon substrate in this section; this has the benefit of enhanced heat dissipation which helps in the thermal management of 3D integration.

### 2.4.1    *Measurement Setup*

The purpose of this experiment is to demonstrate that TSV arrays can be used as an effective heat removal element, and to show how the density

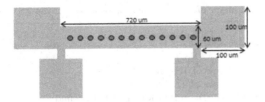

Fig. 2.36 Schematic of the Kelvin structure fabricated for electrical measurement. The length is 720$\mu m$ and width is 60$\mu m$. Four probe pins are needed for the testing: current is injected to pad "$I_{in}$", pad "GND" is connected to ground pin, pads "$V_1$" and "$V_2$" are connected to other two pins to measure the voltage difference across the poly-silicon line.

and placement of TSV arrays could affect the heat conduction. The Kelvin structures fabricated for electrical measurement is shown in Figure 2.36.

### 2.4.2 *Cu-TSV Thermal Modeling*

In our design, the TSV arrays are placed such that they cover the surroundings of and beneath the doped poly-silicon line. The resistance of the doped poly-silicon line is measured in a temperature range of 25-200°$C$ by placing the wafers on the heated probe chuck. By applying a test current (1$mA$), the resistance can be estimated from:

$$R = \frac{(V_1 - V_2)}{I} \qquad (2.4)$$

where $R$ is the resistance of line; $I$ is the current injected and $V$ is the voltage.

Measurements are performed at a steady state to eliminate the inaccuracies, and the test current is chosen to be small enough to avoid unwanted heating that causes the temperature of the doped poly-line to increase. The relationship between the line resistance and the line temperature for different TSV placements and densities (in terms of TSV pitches) is summarized in Table 2.2 and plotted in Figures 2.37(a)–2.37(g). It is observed that approximate linear relationships can be established.

During the second step of the characterization, the doped poly-silicon line temperature is maintained at 25°$C$, while stressing current ranging from 1$mA$ to 112.5$mA$ (corresponding power density ranges from ~0 to $8 \times 10^7 W/m^2$) were injected. By calibrating the resistance value to the temperature value based on the linear relationship established in Figures 2.37(a)–2.37(g), it is easy to estimate the doped poly-silicon line temperature under various stressing currents. The result in Figure 2.38

Table 2.2   Summary of the placement and density of the TSV structures.

|   | TSV placement | TSV density (pitch) |
|---|---|---|
| 1 | No TSV | N. A. |
| 2 | 1 row of TSV beneath | 15 $\mu m$ |
| 3 | 2 rows of TSV beneath | 15 $\mu m$ |
| 4 | 1 row of TSV beneath and 2 rows of TSV surround | 15 $\mu m$ |
| 5 | 2 rows of TSV beneath and 2 rows of TSV surround | 15 $\mu m$ |
| 6 | 1 row of TSV beneath | 10 $\mu m$ |
| 7 | 1 row of TSV beneath | 20 $\mu m$ |

shows that the temperature of the doped poly-silicon line is significantly reduced when there are TSV arrays placed beneath the line. When there were one and two TSV rows beneath the doped poly-silicon line (TSV pitch is 3 times the diameter), we measured a temperature reduction of $\sim 32°C$ and $\sim 44°C$ respectively at a power density of $8 \times 10^7 W/m^2$. It is worth pointing out, however, that the TSV arrays surrounding the line do not help significantly in the temperature reduction. Figure 2.39 shows the effect of TSV pitch on the temperature reduction. It shows that when the pitch is smaller, the temperature reduction is more significant. This observation is expected given that when the TSV pitch is smaller, the TSV densities will be higher and hence more Cu cores can be used to conduct the heat away through the Cu core to the silicon substrate.

### 2.4.3   *Cu-TSV Induced Stress Modeling*

A 3D finite element analysis (FEA) model was constructed to determine the thermal stress distribution in the TSV samples, with the material physical properties used summarized in Table 2.3. The model consists of TSVs with diameter ranging from 4 to 10 $\mu m$, and a depth of 10 $\mu m$ on the Si (100) surface. The conformal $200 nm$ dielectric liner, Ta diffusion barrier and Cu filling in the TSVs were also taken into account. A thermal load from the reference temperature ($200°C$, assumed to be stress-free) to room temperature ($25°C$) of $\Delta T = 175°C$ was considered. For simplicity, a liner elastic and isotropic model was assumed to obtain a first-order intuition. As a result of the thermal load during cooling, the Cu core tended to contract at a higher rate than that of Si due to the larger CTE of Cu. Since Cu was confined in the TSV, thermo-mechanical stress was induced on the

surrounding Si. In reality, such near-surface stresses could possibly be estimated via $\mu$-Raman analysis using a $442nm$ laser line with an approximate
penetration depth of $0.16\mu m$ into the Si surface. The stress state in the Si
surface is considered to be nearly biaxial at this depth. Thus, the biaxial
stress $(\sigma_r + \sigma_\theta)$ profile was extracted at a depth of $0.16\mu m$ from the above
simulation. The stress profile is presented in the next section.

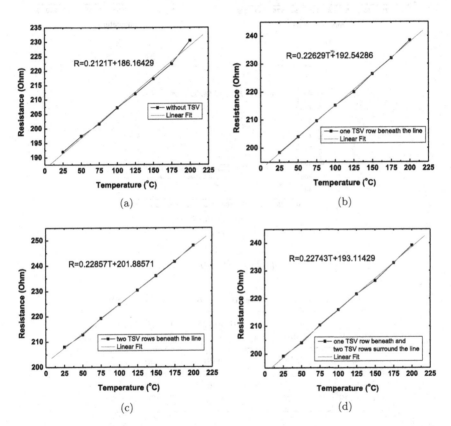

Fig. 2.37   Poly-silicon line resistance vs. line temperature: (a) structure without TSV;
(b) structure with one TSV row beneath the line (TSV pitch=3 times TSV diameter);
(c) structure with two TSV rows beneath the line (TSV pitch=3 times TSV diameter); (d) structure with one TSV row beneath and two TSV rows surround the line
(TSV pitch=3 times TSV diameter); (e) structure with two TSV rows beneath and two
TSV rows surround the line (TSV pitch=3 times TSV diameter); (f) structure with
one TSV row beneath the line (TSV pitch=2 times TSV diameter); (g) structure with
one TSV row beneath the line (TSV pitch=4 times TSV diameter).

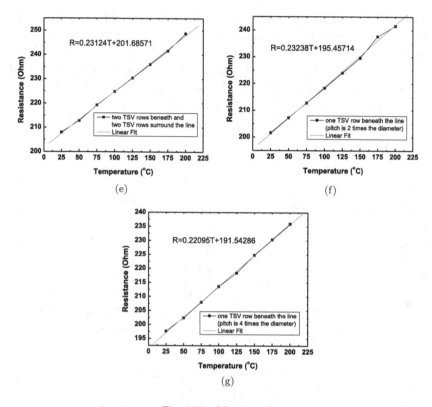

Fig. 2.37   (*Continued*)

Table 2.3   Physical properties of materials in TSV.

| Material | Young's modulus (GPa) | Poisson ratio | Density $(Kg/m^3)$ | CTE $(ppm/K)$ |
|---|---|---|---|---|
| Silicon | 170 | 0.28 | 2329 | 2.6 |
| Copper | 120 | 0.34 | 8960 | 16.5 |
| Low-k | 4.2 | 0.30 | 1500 | 12.0 |
| PETEOS | 70 | 0.17 | 2200 | 0.5 |

At the very beginning of this section, we have discussed that scaling down the TSV size could improve the thermo-mechanical stress exerted on the Si substrate due to the smaller volume of the Cu core. Our simulation study FEA on a single Cu-TSV which is shown in Figure 2.40(a) has clearly revealed that the larger the TSV diameter, the larger the compressive stress in the Si substrate in the close vicinity of the TSV row.

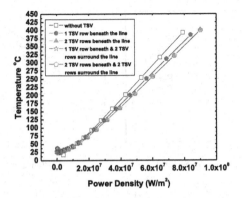

Fig. 2.38 Line temperature at different power density. It shows that TSV arrays beneath the line can significantly reduce down the temperature. The temperature reduction as high as $\sim 44°C$ at a power density of $8 \times 10^7 W/m^2$ can be achieved when there are two TSV rows beneath the doped poly-silicon line (TSV pitch is 3 times the diameter).

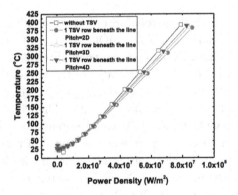

Fig. 2.39 Line temperature at different power density. It shows that TSV arrays with smaller pitch have more significant effect on the temperature reduction since the TSV density is higher.

Two-dimensional (2D) finite element analyses on Cu and dielectric liner interconnect structures have shown that the triaxial tensile stress in Cu arises significantly due to the stiff barrier layers surrounding the Cu line [89, 90]. This stress is transmitted to the peripheral Si crossing through the liner barrier. Thus, the use of a softer liner with a higher compliance in principle imparts a lower stress on the Si. Simulation on a poly (p-xylylene)-type organo polymeric compound [91] (commonly known as parylene with Young's modulus of $\sim 4GPa$) as a liner material in a TSV interconnect shows that it could relieve the thermal stress distribution in

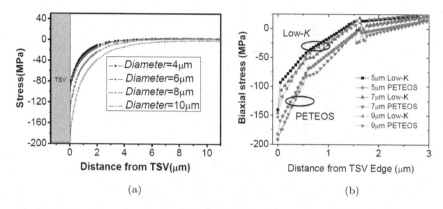

Fig. 2.40    FEA simulation curves of the thermo-mechanical stress exerted by a single Cu-TSV on Si substrate as a function of distance from the edge of TSV. Larger TSV has exerted higher compressive stress on the Si substrate.

Si by $75 MPa$. Theoretical postulation reveals that the incorporation of a low-k dielectric (carbon-doped oxide, SiOC) in place of a traditional oxide-based dielectric significantly reduces the triaxial tensile stresses in Cu but enhances plastic deformation, particularly in the via and its vicinity [92]. SiOC is commonly known as carbon-doped porous silica where the average porosity is in the range of $13 nm$ [93, 94]. Therefore, the application of SiOC as a liner material for TSV could potentially alleviate the thermal stress. There is limited experimental evidence of the buffering of the thermo-mechanical stress effect on Si by using low-k SiOC as the liner material in TSV. Owing to a low Young's modulus of $\sim 4.2 GPa$ for the low-k dielectric as compared to $70 GPa$ for conventional plasma-enhanced tetraethyl orthosilicate (PETEOS) dielectric, a significantly lower transmission of thermal stress to the Si bulk in the close vicinity of Cu-TSV is expected.

FEA simulation results presented in Figure 2.40(b) show that the low-k liner with a smaller elastic modulus acts as a compliant layer to cushion the Cu-TSV induced stress on the Si as compared with the PETEOS oxide liner. The stress-free temperature is set at $200°C$ and the system is cooled down to $25°C$ for this simulation.

It was observed from Figure 2.40(b) that for a TSV structure with a similar diameter, the near-surface stress in Si surrounding the Cu-TSV decreases substantially when a low-k liner is used rather than the PETEOS oxide liner.

### 2.4.4 Cu-TSV Induced Stress Measurement by Micro-Raman Analysis

Various techniques have been adopted to investigate the thermo-mechanical stress in the Cu interconnect structures. They include

(i) synchrotron radiation X-ray diffraction [95],

(ii) the bending beam technique [96] (i.e., the measurement of curvature change under thermal annealing), cross-sectional transmission electron microscopy (TEM) with convergent beam electron diffraction [97] and

(iii) $\mu$-Raman analysis which is the most effective method [71, 98].

The high-resolution TEM accompanied by electron diffraction could accurately extract the localized stress information. However, the TEM sample preparation is a destructive process that invariably causes partial stress relaxation that leads to a less accurate measurement. In contrast, the $\mu$-Raman analysis is non-destructive and the measurement accuracy level can reach a spectral resolution of $0.05cm^{-1}$, which corresponds to $21.7MPa$ in the stress variation [98]. In addition, changing the Raman laser spectral line could possibly extract the stress depth profile [71]. In this work, $\mu$-Raman spectroscopy with high-groove-density ($2400gr/mm$) diffraction grating was used, and resulted in a high spectral resolution of $0.3cm^{-1}/pixel$ for the $442nm$ laser source. Lorentzian peak fitting was then used to further refine the resolution to $0.03cm^{-1}$, which corresponds to the stress sensitivity level of $15MPa$.

$\mu$-Raman spectroscopy with a $442nm$ laser, a $100\times$ objective lens (NA = 0.90), and a focused spot size of $0.8\mu m$ was used to estimate the stress level in Si surrounding the Cu-TSV. Raman measurement was obtained with a line scan along the [011] direction on the Si (100) wafer with a step of $250nm$ as shown schematically in Figure 2.41. To extract this biaxial stress from the Raman shift, the effective proportionality relation was applied as

$$\sigma_r + \sigma_\theta = -470\Delta\omega \ (cm^{-1}) \tag{2.5}$$

where $\Delta\omega$ (i.e., $\omega_i - \omega_0$) is the deviation of the Si-Si Raman vibration peak due to stress, $\omega_i$ is the Si–Si peak position under stress condition, and $\omega_0$ is the reference wave-number position for stress-free bulk Si. The measured Raman curves were fitted by Lorentzian fitting for better resolution similar to the approach reported in [99]. A resolution of $0.03cm^{-1}$ was obtained by fitting. Based on this approach, the stress level in the Si area along the scan line can be calculated.

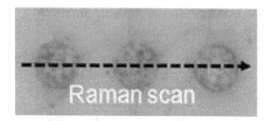

Fig. 2.41   Schematic of high resolution Raman scanning along a TSV row prior to metallization.

First of all, the $\mu$-Raman analysis is carried out on single TSVs with various diameters to examine and confirm the scaling effect on thermomechanical stress. The measurement result matches well with the simulation result presented in the previous section (Figure 2.40(a)). It shows that the larger the TSV diameter, the higher will be the compressive stress in the Si in the close vicinity of the TSV row. For PETEOS oxide liner TSV with a diameter of $10\mu m$, the maximum stress experienced is $220MPa$, whereas those of $8\mu m$, $6\mu m$ and $4\mu m$ are 123, 105 and 85 $MPa$, respectively (shown in Figure 2.42(a)). For the low-k liner TSV, the same observation is found. When the diameter is $10\mu m$ for the low-k liner TSV, the maximum stress is only $123MPa$, whereas the stresses of $8\mu m$, $6\mu m$ and $4\mu m$ are 94, 59 and 42 $MPa$ respectively (shown in Figure 2.42(b)). This is due to the low Cu volume in the smaller diameter TSV, which results in lower stress in the Si surrounding the lower diameter TSV as the stress originates from the large difference in the CTE values for Cu and Si. Similar observation has recently been reported in [96]. By taking into account the laser intensity attenuation, the measured value of stress is a weighted volume average of the near-surface stress up to the laser penetration depth. In our simulation we considered the overall average resultant stress to be at a depth of $160nm$ from the Si surface, which is the penetration depth of the $442nm$ laser. Thus, with the increasing emphasis on miniaturization of modern packaging technology with 3D TSV array interconnect, it is desirable to scale down the TSV diameter to below $5\mu m$ in order to minimize the thermo-mechanical stress issue. Table 2.4 summarizes the $\mu$-Raman analysis results of the single TSV with various diameters and different liner materials.

Furthermore, the effect of a dielectric liner (PETEOS oxide vs. low-k) on the Si stress distribution in the close vicinity of TSV interconnects was then carefully studied with $\mu$-Raman spectroscopy and is shown in Figure 2.43.

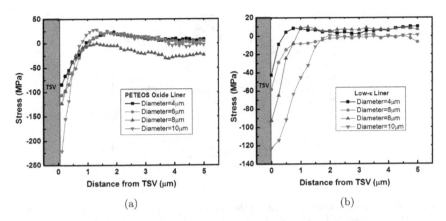

Fig. 2.42  μ-Raman analysis results of the thermo-mechanical stress exerted by a single Cu-TSV with: (a) PETEOS oxide liner on Si substrate; (b) low-k liner on Si substrate as a function of distance from the edge of TSV.

Table 2.4  Summary of μ-Raman analysis results of TSV with various diameters and liner materials.

| TSV diameter ($\mu m$) | TSV with PETEOS oxide liner Stress ($MPa$) | TSV with low-k liner Stress ($MPa$) |
|---|---|---|
| 10 | 220 | 123 |
| 8 | 123 | 94 |
| 6 | 105 | 59 |
| 4 | 85 | 42 |

The measurement results again match well with the simulation results (in Figure 2.40(b)). It was observed that for a TSV structure with the same diameter, the near-surface stress in Si surrounding the Cu-TSV decreases substantially when a low-k liner is used rather than the PETEOS oxide liner. In the immediate vicinity of the Si/TSV interface, the compressive stress decreased from $\sim 221 MPa$ (PETEOS oxide liner) to $\sim 122 MPa$ (low-k liner) for the TSV with a diameter of $10\mu m$. A similar trend was also observed for the $8\mu m$ TSV (stress decreased from $\sim 123$ to $\sim 92$ $MPa$) and $6\mu m$ TSV (stress decreased from $\sim 106$ to $\sim 59$ $MPa$) samples. Hence, it can be concluded that the use of a low-k liner can reduce the compressive stress near the TSV surroundings by 29 to 45%, as compared with the PETEOS oxide counterpart for the TSV diameter presented above. Under our experimental conditions, the observed thermo-mechanical stress relief with a low-k liner could be attributed to the lower Young's modulus of

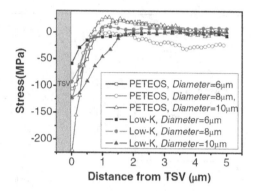

Fig. 2.43   $\mu$-Raman analysis results of the thermo-mechanical stress exerted by a single Cu-TSV on Si substrate as a function of distance from the edge of TSV with PETEOS oxide and low-k dielectric liners. The TSV diameter ranges from 6 to 10 $\mu m$. Measurement results match well with the simulation results and show that low-k liner can act as a compliant layer to reduce the stress on the Si substrate.

$4.2 GPa$ (shown in Table 2.3) and higher porosity than those of the PETEOS oxide liner. The lower Young's modulus and higher porosity of the low-k liner make it more compliant as a buffer layer to cushion the stress exerted by Cu-TSV on the surrounding Si.

In order to verify the biaxial stress exerted on Si substrate by Cu-TSV, high resolution $\mu$-Raman spectroscopy is also performed on a row of 3 TSVs prior to metallization. In Figure 2.44, it is verified that with low-k liner, lower compressive stress is exerted by the Cu-TSV on the Si between the TSVs. This has positive implication on the variability, reliability and keep-out zone (KoZ). It is also worthy to point out that the stress on Si outside the TSV row is due to the liner thin film residual stress on Si which is thickness dependent and can be eliminated with chemical mechanical polishing (CMP).

## 2.5   TSI Structure and Fabrication

In addition to TSV fabrication process, a simplified TSI fabrication process [100] is discussed in this section. As the TSI is expected to consist of a TSV for vertical connection along with horizontal metal interconnects,

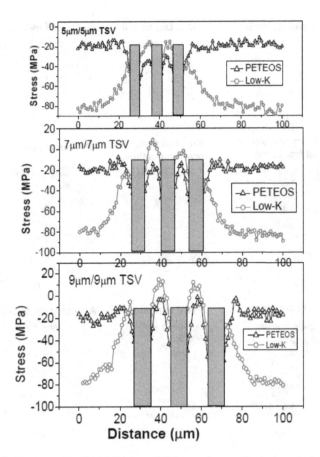

Fig. 2.44 Si Stress profile (biaxial) along TSV rows from $\mu$-Raman analysis. Each row contains 3 TSVs with similar TSV pitch and diameter (5, 7, 9 $\mu m$). The stress profiles show that low-k liner is more compliant and can cushion the stress exerted by Cu-TSV on the Si between TSVs more effectively than PETEOS oxide liner.

fabrication of TSI starts with the formation of TSV followed by extra horizontal interconnects. The height and diameter of TSV can be smaller compared to 3D TSV interconnects, because horizontal interconnects are more prominent in 2.5D TSI I/Os.

### 2.5.1 *Structure Design*

The TSI Fabrication process is shown in Figure 2.45. Initially, a lithography is performed on the substrate, after which a deep reactive ion etching

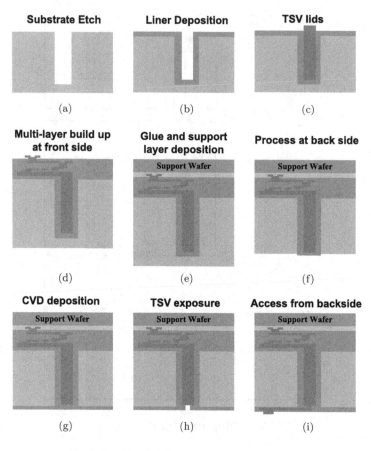

Fig. 2.45   Simplified process of TSI Fabrication.

(DRIE) is done using a plasma etcher to form the via holes (Figure 2.45(a)). The formed via is then coated with the liner material (Figure 2.45(b)), followed by TSV-fill metal material deposition, copper (Cu) in our case. Even though the copper can form electrical interconnect between layers, it is removed because wiring structures have to be structured and Cu may not be compatible with many of the demands. Moreover, most applications need a polymer spacer layer between the silicon and wiring layers [100]. Hence a layer of silicon oxide or liner material is deposited. A lid of thin electro-plated copper (Cu) is created on the top of each filled TSV (Figure 2.45(c)), because TSVs may be too small to be directly accessed by polymer vias. If multi-metal layers are needed, they are created using these Cu pads

(Figure 2.45(d)). After the completion of processing on the front side, high density wiring on the back side needs to be formed, which may be tough in the case of thin substrate layers. To overcome this difficulty, a glue material is applied on the top and a support wafer is attached using wafer to wafer bonding. Similar to the front-end, the substrate is etched on the back side (Figures 2.45(e), (f)) and liner material is deposited using chemical vapor deposition (CVD) technique for having a polymer space (Figure 2.45(g)). A short cleaning of the revealed TSV on the backside and silicon dry etching is performed to create a stand-off between the copper plug and surrounding silicon (Figure 2.45(h)). Finally, after having a small opening to TSV, metal layers are deposited to have electrical connection path (Figure 2.45(i)).

Figure 2.46 shows the components needed to integrate multiple ICs using the TSI interconnect. One can see the microbumps attached to the IC to connect to the TSI substrate. Further, with the aid of microbumps, the connection is provided to the metal layers. The TSV is connected to the

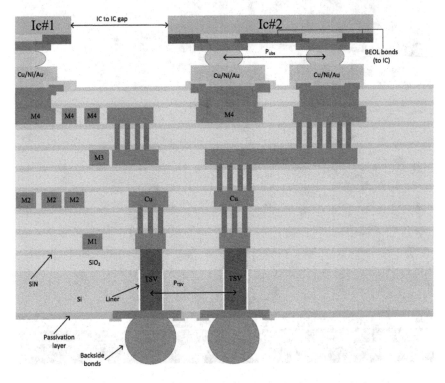

Fig. 2.46   Composition of components in a typical TSI [24].

bottom metal layer and further connected to the package with the help of ball grid arrays. In Figure 2.46, four metal layers are utilized. The number of metal layers vary, based on the fabrication process and the requirements.

### 2.5.2 *Fabrication Process*

A 2 metal layer fabricated TSI with $12\mu m$ TSV, with $100\mu m$ depth on a $300mm$ Si-(100) substrate, is presented in Figure 2.47 [101]. Copper void-free TSV was achieved by electroplating, as discussed in previous sections. Electrical connections were present between metal layers M1-via-M2 in the figure, with M1 in contact with TSV.

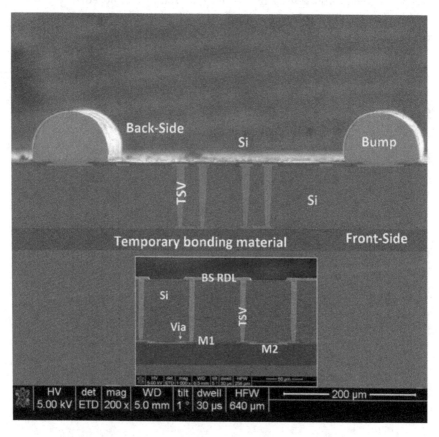

Fig. 2.47   Cross-sectional view of TSI with TSV [101].

## 2.6 Summary

In this chapter, detailed fabrication process flows of power/signal and dummy TSV structures using CMOS process are presented. It was found that the TSV scalloping effect and undercut can be significantly improved for better leakage control and better Cu filling by optimizing DRIE BOSCH process conditions. Electrical characterizations such as *CV*, *IV* measurement results of various TSV liners are presented. The TSV parasitic capacitance is further optimized in terms of capacitance stability and capacitance reduction of 26% by the low-k material. Further, the thermal characterization of TSV as a heat removal element is examined and the results show that TSV arrays can effectively conduct the heat away through the Cu core to the silicon substrate. Significant temperature reduction of up to $\sim 44°C$ is experimentally achieved by placing two rows of TSV with a diameter of $5\mu m$, a depth of $10\mu m$, and a pitch of $15\mu m$ beneath the doped poly-silicon line. Moreover, the simulation and $\mu$-Raman analysis results demonstrate that the miniaturization of the TSV dimension is one way to reduce the thermo-mechanical stress in Si surrounding the TSV structure due to the lower Cu volume. Finally, the fabrication process for a 2 metal layer TSI is presented.

# Chapter 3

# Device Model

## 3.1 Introduction

To overcome the channel loss, latency, energy efficiency and scalability limitations in many-core memory-logic integration of 2D printed circuit board (PCB) [42, 102, 103], 3D integrated circuits (3D ICs) emerged as one of the promising solutions utilizing through-silicon via (TSV) interconnects. Utilization of TSV interconnects in 3D IC significantly reduces the wire length, latency and power dissipation [8, 10, 12–16, 76, 83, 104, 105]. However, a robust 3D IC design requires a careful examination of TSV I/O interconnects in which electrical states such as delays are coupled from multiple physical domains. In 3D integration, accumulation of heat in top layers is a primary concern, due to the long distance between the heat-sink and the top layer [12, 66]. The metal-fill of TSV is surrounded by a liner material to avoid ion diffusion as well as to provide isolation. As such, there exists a large temperature gradient between substrate and TSV, resulting in an electrical-thermal coupling. A nonlinear MOS-capacitance (MOSCAP) is formed between TSV metal-fill and substrate due to the presence of liner material. Owing to the differences in the coefficient of thermal expansion (CTE) of TSV material and substrate material, at high operating temperatures, a large mechanical stress is induced onto the substrate by TSV, leading to the variation in driver delay by the electrical-mechanical coupling. Hence, a physics-based electrical-thermal-mechanical coupled model for TSV needs to be developed. Dissipation of accumulated heat in top layers is one of the biggest concerns of 3D integration, and can even degrade the reliability of IC.

An alternate approach for many-core memory-logic integration is with the recently introduced through-silicon interposer (TSI) [24] I/O interconnect–based 2.5D integration. In contrast to TSV, which is a short RC interconnect for inter-layer communication, TSI is usually designed as a transmission line (T-line), targeted at high-speed medium- or long-distance communication between the main memory and cores on one common substrate. Compared to the RC-interconnect with repeaters, 2D single-ended T-line (STL) or differential T-line (DTL) [106, 107] with current-mode-logic (CML) buffers [3] have demonstrated better latency, power and bandwidth performance, but with large area overhead. As the TSI based T-line can be designed through and under the common substrate in 2.5D integration, area overhead can be mitigated. Ideally, a TSI can be deployed for memory-logic integration with high performance yet low area overhead. Similar to TSVs, the liner material is in contact with the metal-fill in TSIs, hence the TSI capacitance is modeled as nonlinear MOSCAP as well. As memory and logic components are spread on the common substrate close to heat-sink, the thermal reliability concern of the integrated high-performance server is relaxed in 2.5D integration by TSIs.

In this chapter, based on the recent measurement results in [71, 83, 105, 108, 109], nonlinear electrical-thermal-mechanical coupled 3D TSV and 2.5D TSI device physical models are developed. Here, we will present a nonlinear MOSCAP model, followed by delay and power models of 3D TSV and 2.5D TSI, with consideration of electrical, thermal and mechanical coupling.

## 3.2 Nonlinear MOSCAP Model

Traditional 2D interconnects were modeled as RC models with resistance linearly dependent on the temperature $T$ and capacitance as a constant, as given in (3.1).

$$
\begin{aligned}
R_T &= R_0(1 + \alpha \cdot \delta T); \quad \delta T = T - T_0; \\
D_{RC} &= R_0 C_0(1 + \alpha \cdot \delta T)
\end{aligned}
\tag{3.1}
$$

where $R_T$ is the temperature-dependent resistance; $R_0$ represents the resistance at room temperature; $C_0$ represents the capacitance at room temperature; $\delta T$ is the difference between the operating temperature $T$ and room temperature $T_0$; and $\alpha$ is the temperature dependent coefficient for the resistor. This is how the traditional RC interconnect is modeled as linearly–dependent on the temperature.

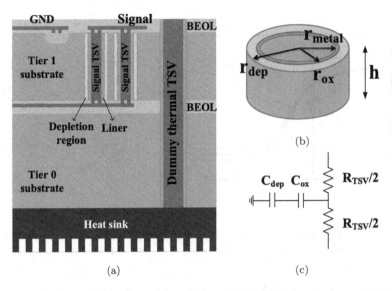

Fig. 3.1 (a) Signal TSV and dummy TSV in 3D IC; (b) 3D view of circular TSV; (c) Equivalent simplified RC-circuit of TSV.

In contrast to RC interconnects, TSVs are different from a physical perspective. TSVs that are deployed for inter-layer signal communication are called signal TSVs. A signal TSV and its cross sectional view is shown in Figures 3.1(a) and (b) respectively. The equivalent and simplified circuit representation of a signal TSV is depicted in Figure 3.1(c). When voltage is applied across a TSV, the existence of liner material around it forms a MOSCAP, and is nonlinearly–dependent on the temperature due to the different CTEs of TSV and substrate [76, 83]. This nonlinear capacitance depends on the biasing voltage ($V_{BIAS}$) as well as temperature [21,22,105]. The difference in work-function between metal-material of TSV and the substrate results in the existence of a depletion region. The radius of depletion region varies with the biasing voltage and temperature, resulting in a nonlinear capacitance. Based on the $CV$ curves presented in previous chapter, a typical $CV$ curve can be drawn as shown in Figure 3.2(a), and one can observe that it is divided into accumulation, depletion and inversion regions separated by flat-band voltage ($V_{FB}$) and threshold-voltage ($V_T$). At higher frequencies ($> 1MHz$), the inversion region can be further divided into deep-depletion and inversion regions. One needs to note that TSIs also consists of TSVs.

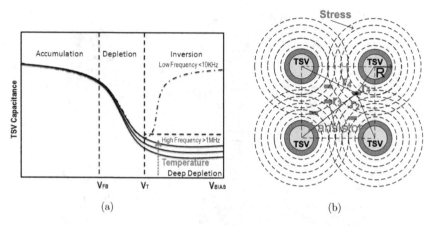

(a)                                                (b)

Fig. 3.2    (a) Typical $CV$ curve of nonlinear MOSCAP model for signal TSV; (b) Exertion of stress on from TSVs to transistors/drivers in a unit square-grid.

Based on this MOSCAP model and the measurement results presented in the previous chapter (Figures 2.19, 2.21, 2.37), the RC-delay parameters of temperature–dependent TSV can be given as

$$R_T = \frac{\rho h}{\pi r_{metal}^2}; \quad \frac{1}{C_T} = \frac{1}{C_{ox}} + \frac{1}{C_{dep}};$$

$$C_{ox} = \frac{2\pi\varepsilon_{ox}h}{\ln(\frac{r_{ox}}{r_{metal}})}; \quad C_{dep} = \frac{2\pi\varepsilon_{si}h}{\ln(\frac{r_{dep}}{r_{ox}})} \tag{3.2}$$

where $R_T$, $C_T$ represents the temperature–dependent resistance and capacitance of TSV respectively; $C_{ox}$ and $C_{dep}$ are the liner and depletion-region capacitances of TSV respectively; TSV height is represented by $h$; $\rho$ is the resistivity of metal-material of TSV; $\varepsilon_{ox}$ and $\varepsilon_{si}$ are dielectric constants of silicon dioxide and silicon respectively; and $r_{metal}$, $r_{ox}$ and $r_{dep}$ are the outer radii of TSV metal, silicon and depletion region respectively, as shown in Figure 3.1(b).

As the thermal conductivity of liner material (SiO$_2$) is nearly one hundred times lower than that of the silicon substrate, liner prevents the dissipation of heat from the substrate. This accumulated heat results in a hotspot at and around signal TSVs. As shown by measurement results in [105] and in the previous chapter, the TSV capacitance approaches liner capacitance at high temperature due to the existence of a hotspot and nonlinear temperature dependence.

## 3.3  TSV Device Model

As discussed previously, the temperature impacts the electrical delay of the TSV. In addition, the temperature also affects the mechanical reliability of the system. The exerted mechanical stress from a TSV has an impact on the mobility of charge carriers at the deep sub-micron level and further the delay of the drivers get affected. As TSVs are connected to the drivers, the combined impact is preferred to that of only TSV characteristics. In this section, we discuss electrical, thermal and mechanical models, followed by coupled delay and power models.

### 3.3.1  *Electrical Model*

We derive the electrical model for the TSV by considering the MOSCAP formed between Cu fill and the silicon substrate of TSV. Even though at higher frequencies, the $CV$ curve of a signal TSV tends to be flat, with the change in $V_{BIAS}$, the TSV capacitance $C_T$ in the deep-depletion region tends to vary nonlinearly with temperature due to $r_{dep}$. The resistance of TSV $R_T$ can be modeled as linearly–dependent with the temperature, experimentally shown in Figure 2.37 and mathematically given in (3.3). The nonlinear temperature–dependent TSV capacitance $C_T$ based on the measurement results from fabricated TSVs in [105] can be given as

$$R_T = R_0(1 + \alpha \delta T); \quad C_T = C_0 + \beta_1 T + \beta_2 T^2 \qquad (3.3)$$

where $C_T$, $C_0$ are TSV capacitance at temperature $T$ and zero-temperature respectively; $R_0$ is the TSV resistance at room temperature $T_0$; $\alpha$ is the temperature dependent coefficient for resistance; $\beta_1$ and $\beta_2$ are the first and second order temperature dependent coefficients of $C_T$ respectively; and $\delta T$ represents difference between $T$ and $T_0$. The values of $\beta_1$ and $\beta_2$ are determined experimentally and reported in [105].

The nonlinear variation of capacitance is different from the traditional via characterization, modeled as linearly variant with the temperature. As such, the nonlinear electrical-thermal coupling of signal TSVs is strong and brings significant impact on the timing with temperature in 3D integration.

Let us compare the nonlinear TSV model and the linearly modeled RC-interconnect delay models. The signal TSV is modeled by considering the nonlinear temperature–dependent capacitance as described in (3.3) and the RC interconnect modeled as in (3.1). A signal TSV with the following

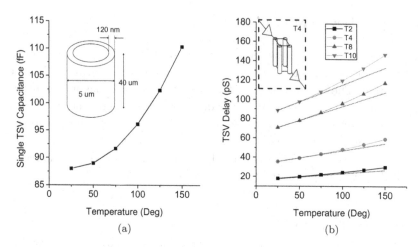

Fig. 3.3    (a) Variation of TSV capacitance with temperature; (b) variation of TSV delay with temperature for different TSV bundles.

parameters is chosen for the purpose of comparison: height $40\,\mu m$, diameter of $15\,\mu m$ with a resistance of $44\,m\Omega$ at room temperature. Based on the results reported in [105], the values for coefficients of temperature–dependent parameters $\alpha$, $C_0$, $\beta_1$ and $\beta_2$ in (3.1) and (3.3) are stated as: $0.00125/K$, $88.8fF$, $0.0667fF/K$ and $0.0014fF/K^2$ respectively. In addition, to confirm reliability, a bundle of TSVs was used for signal distribution instead of one single TSV. A TSV bundle is formed by grouping a few number of TSVs, named as T2, T4, T8 and T10 to represent 2, 4, 8 and 10 TSVs in each bundle, respectively.

The nonlinear variation of a signal TSV MOSCAP based on (3.3) is presented in Figure 3.3(a). It needs to be noted that from Figure 3.3(a), at high temperatures the TSV capacitance varies nonlinearly due to the existence of liner material and difference in CTE. The same is explained mathematically in (3.3). For example, one signal TSV capacitance at room temperature $25^oC$ is $87fF$, at $75^oC$ is nearly $93fF$, and at $150^oC$ is $113fF$, which shows a nonlinear growth. Experiments with process variations in the TSV are carried out by varying the capacitance value by a maximum of 10%. The corresponding variation in delay of signal TSVs is however, less than 3%, and hence is negligible when compared to thermal or mechanical impact.

To study the delay characteristics, an inverter buffer is connected at both the ends of TSV RC interconnect. The length of the input and output

2D wires to signal TSV is assumed to as small as possible. An inverter in 22 $nm$ CMOS process is used as a buffer with following settings: sizing parameters and driver settings are $\frac{R_D}{S_D} = 100\,\Omega$ and $S_D C_P = S_L C_L = 2fF$. The nonlinear effects of temperature on RC-delay at different temperatures for different TSV bundles T2, T4, T8 and T10 are shown in Figure 3.3(b). Though the nonlinear temperature-dependent MOSCAP contributes to a significant amount of delay, the use of signal TSV bundles can help in the reduction of temperature, thereby reducing the overall skew.

It needs to be observed from Figure 3.3(b), that the delay for the T8-bundle at 120°$C$ reaches nearly 100$ps$, which is 67% of the half-clock cycle for a 3.3 $GHz$ multi-processor. At a normal temperature of 75°$C$, the delay for the TSV T8-bundle is nearly 60 $ps$; and at the maximum temperature of 200°$C$, it reaches nearly 140 $ps$, which is nearly 46% of a clock-cycle of 3.3 $GHz$ multi-processor. This delay is of serious concern if no proper cooling technique is applied.

Additionally, it needs to be noted that if the TSV is modeled as the traditional linear coupled model, then the estimated delay will be inaccurate. For example, for the T10-bundle, at a temperature of nearly 125°$C$, the delay with the nonlinear coupled model is nearly 130 $ps$ whereas with the traditional linear coupled model, the delay is 120 $psm$, which is inaccurate. This difference can affect the design considerations as well as timing constraints during the design of the 3D IC.

### 3.3.2 Thermal Model

In a 3D circuit, thousands of TSVs may co-exist to deliver the power and signal. These TSVs are normally filled with metal which has a higher thermal conductivity than silicon. Previous works on thermal/dummy TSVs [11,110] suggest that they can have a significant influence on the steady state temperature of 3D ICs. Such influence is due to the change of thermal conductivity in the heat transfer path when a TSV is inserted, and is highly dependent on the TSV's size and position. For thermal modeling, all TSVs are assumed to be cross-sectionally squarish to ease the modeling process, with the understanding that round TSVs can always be projected to squarish ones of proper sizes.

For two grid cells without TSVs, the thermal conductance between them is formulated as

$$g_{ij} = \frac{kA}{t} \tag{3.4}$$

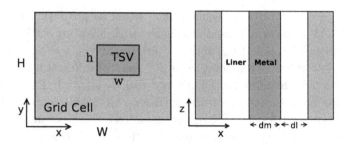

Fig. 3.4   Geometry of grid cell with TSV.

where $k$ denotes thermal conductivity, $A$ is the cross-sectional area through which the heat flux passes and $t$ is the length of the heat transfer path.

When a TSV is inserted as shown in Figure 3.4(a), the equivalent $x$, $y$, and $z$ direction thermal conductances of the grid cell can be computed as

$$g_x = \left(1 + \frac{\omega h(\beta - 1)}{WH(1 + (1 - \omega/W)(\beta - 1))}\right) g_{x0} \tag{3.5}$$

$$g_y = \left(1 + \frac{\omega h(\beta - 1)}{WH(1 + (1 - h/H)(\beta - 1))}\right) g_{y0} \tag{3.6}$$

$$g_z = \left(1 + \frac{\omega h(\beta - 1)}{WH}\right) g_{z0} \tag{3.7}$$

where $g_{x0} = K_{cell}HL/W$, $g_{y0} = K_{cell}WL/H$ and $g_{z0} = K_{cell}WH/L$ are the thermal conductances of the grid cell without TSV and $L$ is the depth of the grid cell. $\beta = K_{TSV}/K_{cell}$ is the ratio of the thermal conductivities between TSV and grid cell. A rectangular TSV is considered in general since a TSV may fall on the boundaries of grid cells so that only a part of it is inside the grid cell. This set of equations reveals that thermal conduction improvement due to TSV insertion is anisotropic, which is more significant vertically and mitigated laterally. Note that the superposition principle is applicable if multiple TSVs are inserted into the same grid cell.

In real implementation, TSVs are normally surrounded by a thin layer of less thermally conductive dielectric liner for insulation purposes, as shown in Figure 3.4(b). To account for the thermal isolation of dielectrics, the equivalent thermal conductivity of TSV with liner is defined as follows.

$$K_{TSV,xy} = \left(1 + \frac{d_m^2(\gamma - 1)}{(d_m + 2d_l)(d_m + 2d_l\gamma)}\right) K_{liner} \qquad (3.8)$$

$$K_{TSV,z} = \left(1 + \frac{d_m^2(\gamma - 1)}{(d_m + 2d_l)^2}\right) K_{liner} \qquad (3.9)$$

where $d_m$ is the side length of the metal core; $d_l$ is the liner thickness and $\gamma = K_{metal}/K_{liner}$ is the ratio of thermal conductivity between metal and liner.

To calculate the thermal conductances between all the solid grid cells with TSV effects, thermal conductances without TSV are first obtained according to (3.4). The sizes and positions of TSVs are then imported to determine their overlaps with all grid cells. Based on the equivalent thermal conductivity of every TSV, their contributions to thermal conductance are then added up to the overlapped solid grid cells according to (3.5)–(3.7). The thermal conductance between two adjacent solid grid cells can then be obtained as half the sum of their thermal conductances.

To validate the proposed thermal model of TSV, we compare our results with commercial tools like COMSOL and Hotspot 5.0. More details about the system settings with TSV are given in Table 3.1. A set of uniform power density (PD) at typical magnitude is assigned to the device layers of 3D IC. The maximum and minimum steady state temperatures from the three simulators are reported in Table 3.2, where power density is measured in $W/cm^2$.

In this experiment, a $16 \times 16$ grid division on each layer is used for Hotspot and our simulator. The mesh in COMSOL is set at the finest granularity to produce the most accurate result. Both our model and Hotspot

Table 3.1   Experimental settings.

| Heat-sink cooling setting | |
| --- | --- |
| TIM thickness | $20\mu m$ |
| TIM thermal conductivity | $4W/mK$ |
| Spreader size $(mm)$ | $30 \times 30 \times 1$ |
| Spreader thermal conductivity | $400W/mK$ |
| Heat-sink size $(mm)$ | $60 \times 60 \times 6.9$ |
| Heat-sink thermal conductivity | $400W/mK$ |
| Heat-sink convective conductivity | $10W/K$ |

Table 3.2   Steady state temperature (K) comparison among COMSOL, HOTSPOT and proposed simulator.

|     | PD  | COMSOL | Hotspot | Ours   | Error |
|-----|-----|--------|---------|--------|-------|
| 50  | Max | 313.72 | 312.83  | 312.83 | 0.3%  |
|     | Min | 308.16 | 307.41  | 307.40 | 0.2%  |
| 100 | Max | 327.45 | 325.66  | 325.66 | 0.5%  |
|     | Min | 316.33 | 314.81  | 314.81 | 0.5%  |
| 150 | Max | 341.17 | 338.49  | 338.49 | 0.8%  |
|     | Min | 324.49 | 322.22  | 322.22 | 0.7%  |
| 200 | Max | 354.90 | 351.32  | 351.32 | 1.0%  |
|     | Min | 332.66 | 329.62  | 329.62 | 90.0% |
| 250 | Max | 368.62 | 364.15  | 364.15 | 1.2%  |
|     | Min | 340.82 | 337.03  | 337.02 | 1.1%  |

Table 3.3   Runtime comparison among COMSOL, HOTSPOT and proposed simulator.

| Model PD | 50    | 100   | 150   | 200   | 250   |
|----------|-------|-------|-------|-------|-------|
| COMSOL   | 42    | 42    | 42    | 42    | 42    |
| Hotspot  | 0.763 | 1.019 | 1.187 | 1.31  | 1.41  |
| Ours     | 0.068 | 0.068 | 0.059 | 0.061 | 0.066 |
| Speedup  | 11×   | 15×   | 20×   | 21×   | 21×   |

adopted the same equation (3.4) to calculate the thermal conductances between solid cells, which results in the same thermal circuit for heat-sink cooled 3D ICs. Thus, it is not surprising that our thermal simulator and Hotspot produce almost identical results although different solvers are adopted. Compared with COMSOL, the deviation of our thermal simulator is in the range of 0.2–1.2% due to coarser vertical granularity, which was set as the total number of sub-layers. Finer vertical granularity is possible by further dividing the sub-layers. However, the tradeoff between accuracy and runtime shall be considered to determine a suitable grid resolution. The deviation in temperature results increases monotonically with the increase of heat flux, because increasing the heat flux amplifies the differences in thermal circuits.

Table 3.3 lists the runtime of the three simulators in *seconds* for this experiment set. Only the speedup against Hotspot is reported in this table because it is unfair to compare the speedup against COMSOL, which has a larger problem size due to much finer granularity. Up to 21× speedup is achieved for this test case. As the recursive multi-grid approach by Hotspot

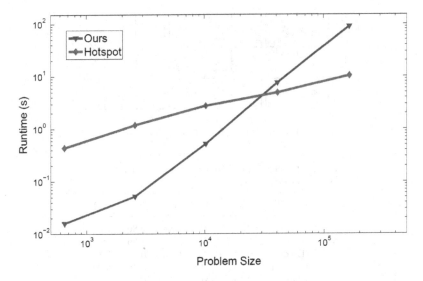

Fig. 3.5   Runtime scaling of our thermal simulator and Hotspot.

scales logarithmically with the problem size, it will have a shorter run-time when the grid dimension is large. Figure 3.5 depicts the runtime scaling of both simulators at a moderate power density of $150W/cm^2$. The five problem sizes correspond to $8 \times 8$, $16 \times 16$, $32 \times 32$, $64 \times 64$ and $128 \times 128$ lateral grid dimensions respectively, with vertical grid resolution equal to the total number of sub-layers, which is 16. It can be seen that hotspot performs better after the fourth point. However, the grid dimension in Hotspot is restricted to powers of two to compromise on the convenience of its recursive solving method, while our model has no restriction on the grid dimension, which is important for modeling microfluid-cooled 3D ICs.

We then examine the accuracy of TSV modeling by inserting three different patterns (Figure 3.6) of TSVs into the circuit. For convenience, all TSVs in COMSOL are assumed to have identical size, and punch through the substrate of the top tier to the device layer of the bottom tier. $Si_3N_4$ is assumed to be the liner material and copper is used as the filling metal. Table 3.4 lists the reduction of maximum temperatures compared with the baseline without TSVs. It shows the estimation error varies with different TSV patterns, where the largest deviation appears on Pattern A. Considering the base temperature of 300K, our TSV thermal model is accurate with the largest deviation of only 0.57K.

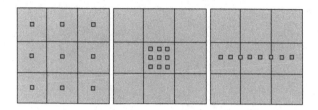

Fig. 3.6   The three TSV patterns.

Table 3.4   Reduction of maximum temperature due to TSV insertion.

| Pattern | PD | 50 | 100 | 150 | 200 | 250 |
|---|---|---|---|---|---|---|
| | COMSOL | 0.20 | 0.40 | 0.60 | 0.80 | 1.00 |
| A | Ours | 0.31 | 0.63 | 0.94 | 1.26 | 1.57 |
| | Error | 0.11 | 0.23 | 0.34 | 0.46 | 0.57 |
| | COMSOL | 0.14 | 0.29 | 0.44 | 0.59 | 0.73 |
| B | Ours | 0.18 | 0.36 | 0.53 | 0.71 | 0.89 |
| | Error | 0.04 | 0.07 | 0.09 | 0.12 | 0.16 |
| | COMSOL | 0.12 | 0.24 | 0.36 | 0.48 | 0.60 |
| C | Ours | 0.16 | 0.31 | 0.47 | 0.63 | 0.78 |
| | Error | 0.04 | 0.07 | 0.11 | 0.18 | 0.18 |

### 3.3.3   *Mechanical Model*

In addition to the thermal and electrical models, we developed a mechanical model of the TSV. Difference in the operating temperatures and mismatch in CTEs between TSVs and the substrate resulted in the exertion of mechanical stress from TSV. The mechanical stress exerted by multiple TSVs on the substrate found by the principle of superposition [108] can be modeled as

$$\sigma_i = -\frac{B\Delta T\Delta\alpha}{2}\left(\frac{R_i}{r_i}\right)^2$$

$$\sigma = -\sum_{i=1}^{n}\sigma_i = -\frac{Bn\Delta\alpha\Delta T}{2}\left(\frac{R}{r}\right)^2; \quad n = \eta A \tag{3.10}$$

where $\sigma_i$ is the stress from $i^{th}$ TSV; $B$ is the biaxial modulus; $\Delta\alpha$ is the CTE difference between TSV material and substrate, dependent on TSV and substrate materials; $\Delta T$ is the temperature difference; $R_i$ is the radius of $i^{th}$ TSV; $r_i$ represents the distance of a transistor or a driver from center

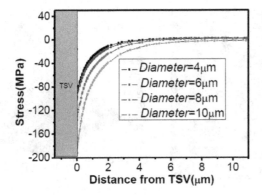

Fig. 3.7 FEA simulation curves of the thermo-mechanical stress exerted by a single Cu-TSV on Si substrate as a function of distance from the edge of TSV.

of $i^{th}$ TSV; and $n$ represents the number of TSVs with a TSV density of $\eta$ in area $A$.

For the simplicity of illustration, all the TSVs are considered to be of same radius $R$, and the drivers are approximated with the same distance $r$ from the center of TSVs. Based on these assumptions, we observe an thermal-mechanical dependence to characterize the mechanical stress between TSV and the substrate. In thermal-mechanical coupling, the main focus is on exertion of mechanical stress with respect to temperature and TSV density.

A quick revisit of the finite element analysis (FEA) for the stress exerted by Cu TSV on the substrate is presented in Figure 3.7. It can be observed that stress decreases with an increase in distance. The results fits the stress model presented in (3.10).

Further, to validate the principle of superposition of thermal-mechanical impact, we follow a similar setup as in Figure 3.2(b). The mechanical stress from TSV is caused by difference of CTEs between TSV and substrate. Different layers of 3D IC have different temperatures, and hence TSVs and substrate will be at different temperatures, resulting in a stress gradient. For the purpose of simulation, all TSVs are considered to have a diameter of $15\mu m$ and a density of $400/mm^2$. The exerted mechanical stress from TSV on device is given by (3.10).

By considering the stress from all TSVs, the exerted mechanical stress on a driver that is placed inside a square surrounded by TSVs is shown in Figure 3.8(a). It can be observed that beyond stress and mobility remains

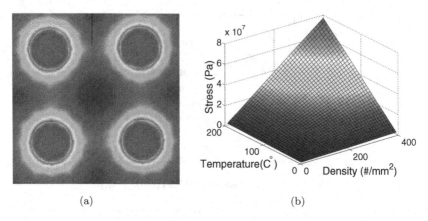

(a)                                    (b)

Fig. 3.8   (a) Variation of TSV stress with distance; (b) variation of TSV stress with temperature and TSV density.

nearly uniform beyond a particular distance, which defines the keep-out-zone (KoZ); but for the area inside the KoZ, there is a significant variation in stress and mobility observed. In our experiment, a keep-out-zone of $3\mu m$ is considered.

The coupled impact of TSV density and temperature gradient on stress is shown in Figure 3.8(b). When the temperature gradient increases, and at high TSV density, the amount of exerted stress on the substrate will be high. Hence, in order to reduce TSV stress on the substrate, the temperature gradient has to be reduced as well. What is more, the amount of stress can vary when the TSV density is different. Let us consider a temperature gradient of $150°C$, when the stress at a TSV density of $200/mm^2$ is $28.93MPa$, and the stress at a TSV density of $400/mm^2$ is $57.87MPa$, indicating stress has been affected by the temperature gradient as well as TSV density. The mechanical stress from TSVs also affects the carrier mobilities of drivers on the substrate. Based on the alignment of the transistor and amount of exerted stress, the carrier mobility varies [71, 108, 109]. This variation in charge carrier mobility adversely affects the delay of drivers. Variation in carrier mobility due to exerted stress [71, 108, 109] is given by

$$\frac{\delta\mu}{\mu} = -\Pi \times \sigma; \quad m = -\Pi_x \tag{3.11}$$

where $\delta\mu/\mu$ is the ratio of mobility variation; $\Pi$ is the tensor of piezo-resistive coefficients; $\sigma$ represents mechanical stress from the TSV; and $m$ represents the maximum value of $\Pi$ among all the directions (x, y, z), in order to capture its most significant impact on the transistor.

For example, $\Pi_x$ is used to represent the maximum value in tensor $\Pi$. $m$ indicates the enhancement factor along the direction that results in maximum stress. The direction of maximum stress is different for NMOS and PMOS devices, resulting in different amount of mobility variations [71, 108, 109] for the same amount of stress, due to the inherent property of materials. The ratio of mobilities with and without stress can be given as

$$\frac{\mu_s}{\mu} = 1 + \frac{\delta\mu}{\mu} = 1 + m\sigma \tag{3.12}$$

where $\mu_s$ and $\mu$ represent the mobility of charge carriers with and without impact of stress respectively; and $\delta\mu/\mu$ is the mobility variation ratio.

With the increase in the amount of exerted stress, the ratio of mobilities with and without stress $\mu_s/\mu$ also increases. This variation in mobility results in change of source resistance of driver, given by

$$R_D^M = \frac{R_D}{1 + \frac{\delta\mu}{\mu}} = \frac{R_D}{1 + m\sigma} \tag{3.13}$$

where $R_D^M$ is the driver resistance with impact of stress; and $R_D$ is the driver resistance without impact of stress.

The simulation results for electrical-mechanical impacts on driver is presented. As the amount of exerted mechanical stress varies, the deformation in lattice structure also varies, resulting in variation of carrier mobility.

For the purpose of illustration, a single TSV having a diameter of $15\mu m$ is considered as the source of stress, the variation in mobility and delay due to the exerted stress with different distance is discussed here. Considering a keep-out-zone of $3\mu m$, inside which variation of mobility and delay with distance for a $22nm$ metal gate PMOS and a NMOS placed on substrate is shown in Figure 3.9(a). The delay of one according driver is shown in Figure 3.9(b). Note that the simulations are carried out in a SPICE simulator [111].

Moreover, one can have the following observations from Figure 3.9(a). The amount of stress exerted varies with the distance, which is in agreement with (3.10). In addition, with the same amount of stress, the variation in hole mobility is higher than electron mobility, due to the material properties. For example, for a distance of $2\mu m$, there is a variation of nearly 4% and 1.8% in mobilities of holes and electrons, respectively.

What is more, the variation of delay of the driver in the presence of mechanical stress is shown in Figure 3.9(b). It can be observed that as the distance increases, the variation in the delay decreases as the exerted stress decreases with distance. There is a decrease of just 1.6% delay at a distance

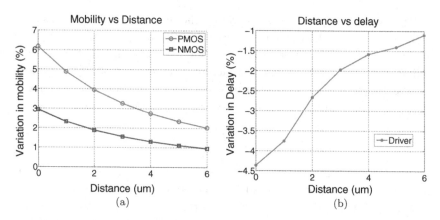

Fig. 3.9　(a) Variation of NMOS/PMOS carrier mobility with distance under TSV stress; (b) Variation of driver delay with distance under TSV stress.

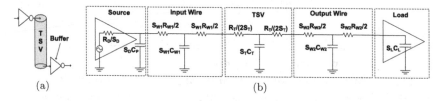

Fig. 3.10　(a) TSV with buffer at its ends; (b) Delay model of 3D clock-tree with nonlinear electrical-thermal-mechanical coupled model.

of $4\mu m$ whereas nearly 3.7% at distance of $1\mu m$. So, if the keep-out-zone is increased, there will be less stress experience with small variation of mobility and delay.

### 3.3.4　Delay Model

Based on the presented thermal and mechanical impacts on the device in the previous section, a coupled model for electrical delay of a TSV I/O with inverter buffers at its both ends, as shown in Figure 3.10(a) is developed.

#### 3.3.4.1　Electrical-Thermal Coupled Delay Model

As aforementioned, temperature has a significant impact on the signal TSV forming a nonlinear MOSCAP as given in (3.2). By considering nonlinear electrical-thermal coupling of signal TSV in (3.3), the signal delay $D_{TSV1}$

for clock-tree shown in Figure 3.10(b) can be calculated as

$$D_{TSV1} = R_{in}\alpha\beta_2 T^3 + R_{in}[(1 - \alpha T_0)\beta_2 + \alpha\beta_1]T^2 + [\alpha(D_0 + R_{in}C_0)$$
$$+ (1 - \alpha T_0)R_{in}\beta_1]T + (1 - \alpha T_0)(R_{in}C_0 + D_0); \tag{3.14}$$

with

$$R_{in} = \frac{R_D}{S_D} + S_{w1}R_{w1} + \frac{R_T}{2S_T};$$

$$D_0 = \frac{1}{2}\left(S_{w1}^2 R_{w1}C_{w1} + S_{w2}R_{w2}S_L C_L\right) + \left(\frac{R_T}{S_T} + S_{w1}R_{w1} + \frac{S_{w2}R_{w2}}{2}\right)$$

$$\left(S_{w2}C_{w2} + S_L C_L\right) + \frac{R_D}{S_D}\left(S_{w1}C_{w1} + S_{w2}C_{w2}S_L C_L + S_D C_P\right) \tag{3.15}$$

where $R_{in}$ is the total resistance counted from TSV capacitor; $C_L$ is the load capacitance; $\alpha$ is the temperature dependent coefficient of TSV resistance $R_T$; $\beta_1, \beta_2$ are the first order and second order temperature dependent coefficients of TSV capacitance $C_T$ respectively; $T$ represents the temperature; $D_0$ is the delay of the circuit shown in Figure 3.10(b) without TSV; and $R_{w1}, R_{w2}$ are the unit wire resistances and their according sizing parameters are $S_{w1}$ and $S_{w2}$.

For TSV I/O delay calculation with electrical-thermal coupling in (3.14), the delay contribution from horizontal metal wires and buffers are also considered. The nonlinearity in delay is mainly introduced due to the nonlinear MOSCAP of TSV, and the horizontal metal wire is modeled as a linear temperature dependent resistor.

### 3.3.4.2 *Electrical-Mechanical Coupled Delay Model*

The impact of mechanical stress on the transistor or driver is studied here. The mechanical stress from TSVs has non-negligible impact on the driver resistance as given in (3.13). Hence the impact of mechanical stress on the delay of the driver connected to TSV has to be considered during modeling for better accuracy. Delay $D_{TSV2}$ with electrical-mechanical coupling is calculated as

$$D_{TSV2} = \frac{D_\sigma}{1 + m\sigma} + D_w; \tag{3.16}$$

with

$$D_\sigma = \frac{R_D}{S_D} \left[ C_P S_D + S_{w1} C_{w1} + S_T C_0 + S_{w2} C_{w2} + S_L C_L \right];$$

$$D_w = S_{w1} R_{w1} \left[ \frac{S_{w1} C_{w1}}{2} + S_T C_0 + S_{w2} C_{w2} + S_L C_L \right] \qquad (3.17)$$

$$+ S_{w2} C_{w2} \left[ \frac{S_{w2} C_{w2}}{2} + \frac{R_0}{S_T} \right] + \frac{R_0 C_0}{2}$$

where $D_\sigma$ and $D_w$ are stress–dependent and independent TSV delays respectively; $R_D$ is the driver resistance without impact of stress $\sigma$; $R_0$, $C_0$ are the temperature independent TSV resistance and capacitance respectively; $m$ is the mobility enhancement factor; $C_L$ is the load capacitance; $\alpha$ is the temperature dependent coefficient of TSV resistance $R_T$; $\beta_1, \beta_2$ are the first order and second order temperature dependent coefficients of TSV capacitance $C_T$; $T$ represents the temperature; and $S_{w1}$, $S_{w2}$ are the scaling parameters of unit wire resistances.

From (3.16), one can observe that $D_\sigma$ is composed of stress–dependent and independent components. Mechanical stress mainly affects the driver. A large amount of stress can deform the substrate, resulting in broken links as well.

### 3.3.4.3  Electrical-Thermal-Mechanical Coupled Delay Model

Till now the delay model of TSV with individual impacts was discussed. Considering the previously mentioned electrical-thermal and electrical-mechanical coupled models, the electrical delay $D_{TSV}$ considering electrical-thermal-mechanical coupling is given by

$$D_{TSV} = D_c + D(T) + D(\sigma) + D(T, \sigma); \qquad (3.18)$$

with

$$D_c = D_0 - D_\sigma + R_0 C_0 (1 + \alpha T_0) - \frac{R_0 (S_{w2} C_{w2} + S_L C_L) \alpha T_0}{S_T};$$

$$D(\sigma) = \frac{R_D}{S_D (1 + m\sigma)} \left[ S_D C_P + S_{w1} C_{w1} + S_T C_0 + S_L C_L + S_{w2} C_{w2} \right];$$

$$D(T) = R_0 \alpha \beta_2 T^3 + R_0 \left[ (1 - \alpha T_0) \beta_2 + \alpha \beta_1 \right] T^2 + \left[ R_0 \beta_1 (1 - \alpha T_0) \right.$$

$$\left. + \alpha R_0 C_0 + S_{w1} R_{w1} S_T \beta_1 + \frac{R_0 \alpha (S_{w2} C_{w2} + S_L C_L)}{S_T} \right] T;$$

$$D(T, \sigma) = \frac{R_D}{S_D(1 + m\sigma)} [S_T \beta_1 T + S_T \beta_2 T^2] \tag{3.19}$$

where $D_c$ is a constant delay composed of $D_0$ and $D_\sigma$ given in (3.14) and (3.16) respectively; $D(T)$ is the delay as a function of temperature only; $D(\sigma)$ represents the delay as a function of stress only; $D(T, \sigma)$ is the delay as a function of both temperature and stress; $C_L$ is the load capacitance; $\alpha$ is the temperature–dependent coefficient of TSV resistance $R_T$; $\beta_1, \beta_2$ are the first order and second order temperature dependent coefficients of TSV capacitance $C_T$; $T$ represents the temperature; $D_0$ is the delay of the circuit shown in Figure 3.10(b) without TSV; $D_\sigma$ and $D_w$ are stress–dependent and independent delays respectively; $R_D$ is the driver resistance without impact of stress $\sigma$; $R_0$, $C_0$ are the temperature independent TSV resistance and capacitance respectively; $m$ is the mobility enhancement factor; and $S_{w1}$, $S_{w2}$ are the scaling parameters of unit wire resistances.

As a summary, delay in 3D case is quite in contrast with the 2D case that has only linear dependence on temperature. The nonlinear dependence on temperature in the 3D case arises from TSV MOSCAP.

TSV delay by considering nonlinear electrical-thermal-mechanical coupling is presented here. Considering a 3D TSV clock-tree similar to the one in Figure 3.10(b), the variation of delay on a single signal TSV with drivers in different technologies are presented in Figure 3.11.

In Figure 3.11, delay values are shown for a single TSV of diameter $15\mu m$ having buffers at both ends placed at $3\mu m$ distance from the TSV (Figure 3.10(a)). One can observe that one signal TSV adds up the delay by nearly 4 times, which indicates that the TSV delay is comparable to the minimum sized driver delay. As the signal TSV is modeled as a nonlinear MOSCAP with nonlinear temperature-dependency, the delay increases with an increase in temperature significantly when electrical-thermal coupling is considered. It can be observed that the delay can increase by nearly 15% for all technologies when the temperature is increased at $200°C$. Then, with the electrical-mechanical coupling considered, the delay with stress gradient

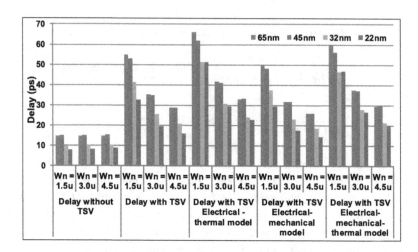

Fig. 3.11   Signal delay of one single TSV under different technologies.

at a reference temperature of $75°C$ is calculated, which is found to be 9% lower than the delay without insertion of TSV, as stress enhances mobility. By considering all these effects, there is approximately 10% delay variation introduced at $200°C$.

### 3.3.5   Power Model

Here we explore the dependence of TSV power on other physical parameters. The dynamic power of a $N$-channel 3D TSV with supply voltage of $V_{DD}$ can be given as

$$
\begin{aligned}
P_{TSV} &= N \cdot C_T \cdot V_{DD}^2 \cdot f \\
&= N \cdot (C_0 + \beta_1 T + \beta_2 T^2) \cdot V_{DD}^2 \cdot f
\end{aligned}
\tag{3.20}
$$

where $P_{TSV}$ is the 3D TSV interconnect power; $f$ is the clock frequency; and $C_T$ is the temperature–dependent 3D TSV capacitance, given in (3.3).

From (3.20), it needs to be noted that the dynamic power of the TSV I/O increases with temperature due to nonlinear MOSCAP. Additionally, the drivers attached to one TSV I/O channel also dissipate power due to its load capacitance $C_L$. Thus, the total power dissipation $P_{TSV\_total}$

considering TSV channel power $P_{TSV}$ can be given as

$$P_{TSV\_total} = P_{TSV} + N \cdot C_L \cdot V_{DD}^2 \cdot f; \qquad (3.21)$$

where $P_{TSV}$ is $N$-channel TSV power, as given in (3.20).

Simulations to obtain the dynamic power of 3D TSV interconnects are carried out under different temperature and length constraints and will be compared with its counterpart TSI in the next section.

## 3.4 TSI Device Model

2.5D integration by TSI realized as a T-line is another alternative for many-core memory-logic integration, in which the thermal reliability concern is alleviated. The heat spreading in 2.5D integration is much stronger than in 3D integration, because the heat-sink is closer to the substrate layer, and there would be less hotspots and a smaller temperature gradient in 2.5D IC, as one can observe in Figure 3.12. Moreover, as the dimension of TSVs in TSIs are smaller, due to lower operating temperatures, mechanical impact can be neglected. However, as liner material is still in presence as a part of TSI, a temperature–dependent delay model for TSI has to be developed similar to that of the TSV. Due to the existence of liner material, TSI capacitance needs to be modeled as a MOSCAP varying nonlinearly with temperature, as given in (3.2) and (3.3).

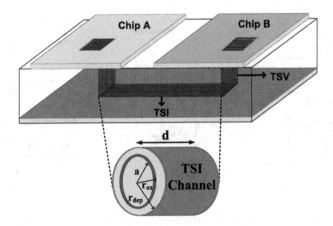

Fig. 3.12   2.5D integration of cores on one common substrate using TSI.

### 3.4.1  *Delay Model*

When compared to the 2D RC-interconnects, TSI realized as transmission-line (T-line) has smaller delay, low power and higher bandwidth but larger area cost. However, TSIs are deployed underneath the substrate with no area overhead (Figure 3.12). As such, it is ideal to realize TSI by T-line for a high speed and low power 2.5D I/Os, as shown in Figure 3.13(a). A T-line can be designed as a single-ended T-line (STL) or differential T-line (DTL) depending on the requirements. Note that the delay of the T-line depends on its characteristic impedance and operating mode. Hence, the characteristic impedance is first determined to model the temperature–dependent delay.

#### 3.4.1.1  *T-line Model*

A single-ended T-line (STL) is shown in Figure 3.13(b)(i). The characteristic impedance $Z_s$ of STL is

$$Z_s = \sqrt{\frac{(R + j\omega L)}{(G + j\omega C_{TI})}}; \quad \gamma = \sqrt{(R + j\omega L)(G + j\omega C_{TI})} \qquad (3.22)$$

where $\gamma$ is the propagation constant; $\omega$ is the angular frequency; $C_{TI}$ is the temperature dependent TSI capacitance ($C_{TI} = C + \beta_1 T + \beta_2 T^2$); and based on [60], $R$, $L$, $C$, $G$, resistance, inductance, capacitance and shunt

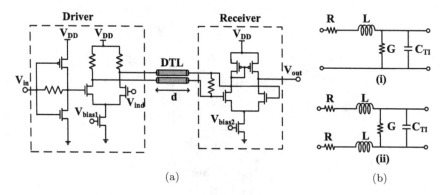

Fig. 3.13   Temperature dependent delay model of one TSI STL or DTL I/O channel.

conductance per unit length can determined as

$$R = \frac{R_s}{\pi a} \left( \frac{(\frac{d}{2a})}{\sqrt{(\frac{d}{2a})^2 - 1}} \right); \quad L = \frac{\mu_0}{\pi} \ln \left( \left[ \frac{d}{2a} \right] + \sqrt{\left( \frac{d}{2a} \right)^2 - 1} \right);$$

$$G = \frac{\pi}{\ln \left( \left[ \frac{d}{2a} \right] + \sqrt{(\frac{d}{2a})^2 - 1} \right)}; \quad C = \frac{\pi \varepsilon}{\ln \left( \left[ \frac{d}{2a} \right] + \sqrt{(\frac{d}{2a})^2 - 1} \right)}. \tag{3.23}$$

Here, $R_s$ $(= \sqrt{\pi \mu_0 f / \sigma})$ is the surface resistance of T-line for skin effect; $a$ and $d$ are the radius and pitch of the T-line respectively; $\mu_0$ and $\varepsilon$ are the permittivity of vacuum and permeability of material respectively; $\sigma$ is the conductivity of the conductor; $\omega$ is the angular frequency; and $f$ is the operating frequency.

For a differential T-line (DTL), there exists a mutual inductive and capacitive coupling, because of which characteristic impedance $Z_{diff}$ becomes

$$Z_{diff} = 2\sqrt{\frac{(L - L_m)}{(C_{TI} + C_m)}} \tag{3.24}$$

where $C_m$, $L_m$ are the mutual inductance and capacitance of DTL respectively; and $C_{TI}$ is temperature–dependent TSI capacitance.

### 3.4.1.2 *Delay of T-line*

For a T-line operating above takeover frequency (lossless LC region), the temperature dependent delay of a T-line $D_{TSI}$ can be given as

$$D_{TSI} = \frac{1}{\omega} \sqrt{Im(\gamma Z_0) \times Im\left( \frac{\gamma}{Z_0} \right)} = \sqrt{LC_{TI}}$$

$$\gamma = \begin{cases} j\omega\sqrt{LC_{TI}} & STL; \\ 2\omega\sqrt{LC_{TI}} & DTL; \end{cases} \quad Z_0 = \begin{cases} \sqrt{\dfrac{L}{C_{TI}}} & STL; \\ 2\sqrt{\dfrac{L}{C_{TI}}} & DTL; \end{cases} \tag{3.25}$$

where $Im(x)$ indicates the imaginary part of $x$; $Z_0$ is the characteristic impedance of the T-line which can be either DTL or STL as in (3.22) and (3.24) respectively; and $C_{TI}$ is the temperature–dependent TSI capacitance.

Thus, the delay of TSI can be written as a function of inductance $L$ and capacitance $C_{TI}$, nonlinearly dependent on temperature $T$. The simplified TSI delay based on (3.25) can be given as

$$D_{TSI} = \sqrt{L(C + \beta_1 T + \beta_2 T^2)}. \tag{3.26}$$

The delay of TSI connected with a driver with load capacitance $C_L$ and a resistance of $R_D$ can be given as

$$D_{25d-io} = R_D C_L + D_{TSI}. \tag{3.27}$$

When compared to the temperature–dependent delay model of 3D TSV I/O in (3.14), one can observe that the temperature dependence of the 2.5D TSI given in (3.27) is much weaker with square-root dependence. This makes the 2.5D TSI delay less prone to temperature variation compared to the 3D TSV. What is more, the temperature gradient $\delta T$ is also much smaller for the 2.5D TSI than the 3D TSV interconnects. The comparison of 3D TSV and 2.5D TSI I/Os delay with temperature and length will be presented later.

### 3.4.2   Power Model

Based on the driver resistance $R_D$ and characteristic impedance $Z_0$ in (3.25), dynamic power $P_{TSI}$ of a $N$-channel TSI I/Os can be calculated as

$$P_{TSI} = \frac{N * V_{DD}^2 * s}{(R_D + Z_0)} f; \quad R_D = \frac{\rho d}{A_c} \tag{3.28}$$

where $s$ is the duration of signal pulse; $R_D$ is the driver resistance of TSI; $d$ and $A_c$ are the length and cross–sectional area of the T-line respectively; and $\rho$ is the resistivity of the material.

The rising trend of the $Z_0$ is linear to the temperature variation which causes the power of TSI interconnect to be less sensitive to the temperature than TSV. By observing (3.28) and (3.21), it can be concluded that the power of the TSI I/O channel is less dependent on temperature (with square-root dependence) than the TSV I/O channel (quadratic dependence). Additionally, the temperature of a 2.5D IC will be lower compared to a 3D IC.

### 3.4.2.1 *TSV and TSI Comparison*

Delay and power trends under different TSV and TSI interconnect lengths are discussed here. TSVs and TSIs are modeled as discussed previously. Figure 3.14(a) shows the result for TSV I/Os. The delay of one channel of TSV is around 22.50ps of 50$\mu m$ length and 40.50ps of 90$\mu m$ length under 25°$C$ respectively, with linear relation. The power of TSV shows a similar relation with 0.11$mW$ of 50$\mu m$ length and 0.20$mW$ of 90$\mu m$ length. Figure 3.14(b) shows the result for TSI I/Os. The delay of one channel of TSI is around 34.52ps of 1.5$mm$ length and 69.04ps of 3.0$mm$ length under 25°$C$ respectively, also with linear relation. The power of TSI is 4.11$mW$ for both lengths at 25°$C$.

Let us discuss the delay and power comparison under different temperatures for TSV and TSI I/Os. As shown in Figure 3.15, one can observe that the nonlinear temperature-dependent capacitance has much stronger coupling for TSV I/Os than TSI I/Os. For example, the delay of TSV and TSI are 22.50ps and 34.52ps at 25°$C$ respectively. When the temperature rises to 130°$C$, the delay of TSV and TSI becomes 31.95ps and 35.53ps, indicating an increase of 42% and 3%, respectively. This shows that the TSV I/O delay is more sensitive to the temperature. Though the power curves look linear or constant, for variation of temperature from 25°$C$ to 130°$C$, the power of TSV and TSI shows incremental change of 20% and 2% respectively.

### 3.4.2.2 *Energy-efficiency Analysis*

The power and energy-efficiency relation with bandwidth is presented in Figure 3.16. The bandwidth can be adjusted by varying the number of I/O

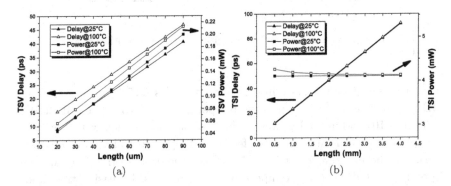

Fig. 3.14 Delay and power comparison under different lengths for (a) TSV I/Os; (b) TSI I/Os.

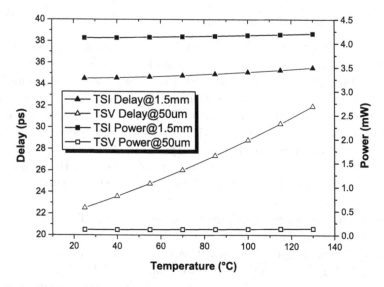

Fig. 3.15   Delay and power comparison under different temperatures for TSV I/Os and TSI I/Os.

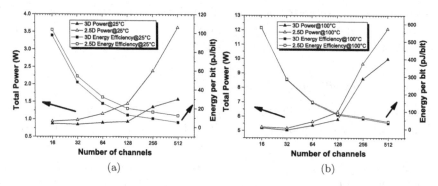

Fig. 3.16   Total power and energy efficiency with different bandwidths: (a) at $25°C$; (b) at $100°C$.

channels. On average, 2.5D integration consumes 62% more power than 3D integration at $25°C$. When the temperature rises to $100°C$, 3D TSVs consume 10% additional power. If heat dissipation is not well designed, temperature will form a positive feedback loop with leakage current and result in thermal runaway failure. Note that the delay of the TSVs will be greatly affected when the temperature is high, thus decreasing the bandwidth. Here the energy-efficiency is defined as energy consumption per bit.

It can be seen that energy efficiency will decrease as the number of channels increase. On average, 2.5D integration consumes slightly more energy per bit than 3D integration at $25°C$. But when the temperature rises to $100°C$, 2.5D integration achieves almost the same energy efficiency as 3D integration. However, with the increase in number of channels, 2.5D TSIs are energy efficient than 3D TSVs.

## 3.5  Summary

3D integration by TSVs and 2.5D integration by TSIs are the possible efficient many-core memory-logic integration techniques on a large-scale. In 3D ICs, signal TSVs utilized for inter-layer signal connections are modeled as MOSCAP, varying nonlinearly with temperature due to the existence of liner material. Delay of TSV varies nonlinearly with temperature due to MOSCAP. Further, the differences in CTE and operating temperatures between TSV and substrate result in the exertion of stress from TSV onto the substrate and the drivers in the vicinity, resulting in modified mobility and delay characteristics. Hence, an electrical-thermal-mechanical coupled delay model is developed for TSV delay calculation. Compared to 3D integration, the heat dissipation is strong in 2.5D integration due to the uniform distance to the heat-sink. Delay and power models for TSIs are presented in this chapter. Delay and power characteristics of TSV is observed to be more prone to temperature compared to that of TSI.

# PART 2

# Physical Design

# Chapter 4

# Macromodel

## 4.1 Introduction

3D integrated systems [13–16] have the facility to have flash, DRAM and SRAM memories integrated, on top of logic devices and even microprocessors. 3D integration helps to achieve larger integration density with good form factor, which is essential for tomorrow's SoCs. However, with the increase in the number of device layers, it becomes challenging to remove the heat and deliver the power supply in 3D ICs.

Figure 4.1 illustrates a typical 3D stacking of multiple device layers within one package. The supply voltage is delivered from the bottom power/ground planes in the package, passed through the vias and C4 bumps, and connected to the on-chip power/ground grid on active device layers. We call the through vias that deliver the supply voltage *power/ground vias*. 3D integration, by definition, has integrated more than one layer of active devices. They draw a much larger current from the package power/ground planes than 2D ICs. This can obliviously result in the IR drop for horizontal on-chip power/ground grid. The surge from the injecting current further leads to a large simultaneous switching noise (SSN) for those I/O drivers at the chip package interface. Figure 4.2 shows a detailed view on how to place signal and power/ground vias through package planes. Clearly, the regions to place power vias or ground vias decide the path of the returned current and the loop area for those signal nets connecting I/Os. They form a number of different sized loop-inductances that have significant couplings with each other. We call the voltage bounce at I/Os *power integrity* in this work. Compared to the allocation of off-chip decoupling capacitors ($mm^2$), placing vias ($\mu m^2$) have a smaller cost of area. This is extremely important for the high density integration in 3D ICs as there

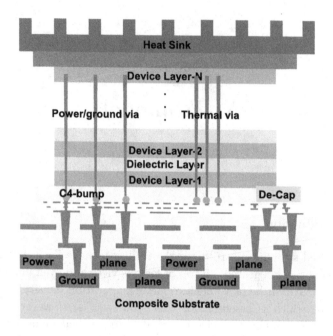

Fig. 4.1   A typical 3D stacking with non-signal through silicon vias.

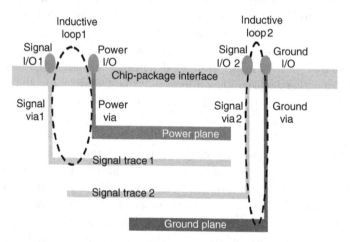

Fig. 4.2   The power delivery by vertical power/ground vias and its impact to inductive current loops.

may be a large number of signal nets to be delivered from the package to the chip.

In a highly dense 3D IC, an accurate electrical or thermal model considering all device layers and package planes can result in millions of unknowns with thousands of time-varying input sources. Moreover, during the physical design process, design parameters such as the device size (or density) and the location further increase the complexity. This makes the design for power and thermal integrity difficult to accomplish in a reasonable time and hence further blocks the physical level optimization. To reduce the simulation cost, macromodel–based approaches [12, 112–114] are used to find a compact representation of the electrical and thermal model. The works in [9,10] further introduces a structured and parameterized macromodel to provide both the nominal response and the sensitivity for the purpose of optimization.

The macromodeling–based TSV allocations in [9,10], however, are not effective when the number of input ports or output ports is large. Generally, there can be thousands of thermal-power sources injected at each of the active layers or hundreds of switching-current source injected I/Os. The size of the macromodel increases with the number of ports, and hence the computational cost to solve the macromodel is still high. Moreover, the effectiveness to apply the macromodel also heavily depends on coping with the design parameters. Here the parameters are the via densities at tracks. Assuming that the locations of tracks are pre-designated, [8] adjust the via density in each track. Yu *et al.*, in [10] further decomposes those tracks into multiple groups described by levels. The vias are then allocated uniformly for those tracks in the same level. However, it is still expensive to probe the integrity and adjust the via density at each track for a large number of pre-designated tracks.

In this chapter, we first discuss the power and thermal integrity concerns in a 3D integration followed by macromodeling to reduce the complexity of physical design.

## 4.2  Power and Thermal Integrity

Due to increased power density and slow heat-convection at inter-layer dielectrics, heat dissipation is another primary concern in 3D ICs. The excessively high temperature can significantly degrade the reliability and performance of interconnects and devices [115,116]. We call the temperature gradient at active device layers *thermal integrity*. As shown in Figure 4.1, a heat-sink is placed at the top of device layers and it is the primary heat-removal path to the ambient air. One observation is that there are through vias delivering supply voltages or signals from the bottom package through each active device layer. Since the metal vias are good thermal conductors, the through vias can provide additional heat-removal paths passing the inter-layer dielectrics to the top heat-sink. This leads to the concept of adding *dummy thermal vias* or *thermal vias* or *dummy vias* directly inside chips [116] to reduce effective thermal resistances. Their physical arrangements are further studied in [7, 8].

In modern VLSI designs, dynamic power management such as clock-gating and uncertainty from the workload can lead to time-varying power inputs. This results in a spatially and temporally variant thermal model. The inputs are time-varying thermal power (see Figure 4.3) [117, 118] defined by the running-average of the cycle-accurate (often in the range of *ns*) power over several thermal time constants (often in the range of *ms*). They are injected at the input ports of each layer. As such, a temporally and spatially variant temperature at output ports can be considered by defining an *integrity integral* with respect to time and space [9]. As a result, the temperature gradient can have either a sharp-transition with a

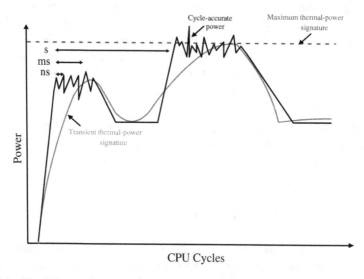

Fig. 4.3 The definitions of the cycle-accurate power, transient thermal-power, and maximum thermal-power at the different scale of time-constant.

large peak value, or a time-accumulated impact on the device's reliability. In addition, different regions can reach their worst-case temperatures at different times.

A dynamic thermal integrity constraint is thereby needed to accurately guide the physical level resource allocation. Since the active device layer at the bottom (see Figure 4.1) has the longest path to the heat-sink at the top, here, *dynamic thermal integrity* is defined as the integrated temperature fluctuation at $p_o$ output ports on the bottom device layer. As shown in [9–12, 115, 116, 119], a dynamic thermal integrity can accurately capture not only the sharp-transition of temperature change due to dynamic power management, but also the impact of time-accumulated temperature that can affect device reliability. Similar to static thermal-integrity analysis, dynamic thermal integrity assumes the worst-case input from a limited number of thermal-power inputs. However, since dynamic integrity has a more accurate transient temperature profile, it leads to a smaller allocation when compared to the static thermal-integrity based design [7,8]. Note that the dynamic power-integrity has already been employed in many on-chip or off-chip power integrity verifications and designs [112, 120, 121]. A similar *dynamic power integrity* is defined as the time-integrated voltage bounce at power/ground I/Os, which are located on the interface between the bottom device layer and the package.

To calculate the dynamic thermal integrity, we represent the 3D ICs with two distributed models: a thermal-$RC$ model for heat-removal and an electrical-$RLC$ model for the power-delivery. They (without power/ground vias) can be described in the state-space by

$$\mathcal{G}x(t) + \mathcal{C}\frac{dx(t)}{dt} = \mathcal{B}\mathbf{I}(t), \ y(t) = \mathcal{L}^T x(t) \tag{4.1}$$

or in frequency $(s)$ domain

$$(\mathcal{G} + s\mathcal{C})x(s) = \mathcal{B}\mathbf{I}(s), \ y(s) = \mathcal{L}^T x(s). \tag{4.2}$$

Note that $\mathcal{B}$ is the topology matrix to describe $p_i$ input ports with injected input sources, and $\mathcal{L}$ is the one to describe $p_o$ output ports for probing thermal or power integrity and adjusting via density.

## 4.3 Macromodeling

To reduce the discretized state-matrices and to tackle computational complexity, macromodeling is performed. To consider dynamic integrity during the design optimization of TSV allocation, the main difficulties in applying the above state-space equation are three-fold. Firstly, there are so many inputs to try and so many outputs to probe. Secondly, the dimensions of the distributed thermal-$RC$ and electrical-$RLC$ models are too large to analyze. Lastly, for the sake of design optimization, we are more interested in the sensitivity than the nominal response. In the following, we will show how to compress the states, then I/Os, and further, how to generate sensitivity for the design automation.

### 4.3.1 *Complexity Compression*

Here, we first discuss the complexity compression for the states, followed by complexity compression for the corresponding I/Os.

#### 4.3.1.1 *Complexity Compression of States*

As the layouts of active device layers and power/ground planes are discretized sufficiently, the size of the resulting state-matrices can be huge. As such, the distributed thermal-$RC$ and electrical-$RLC$ models are difficult to be employed for either integrity verification or optimization.

Model order reduction [122,123] finds the dominant state variables and obtains compact macromodels. As shown in [122,123], the dominant state variables are related to the block Krylov subspace

$$\mathcal{K}(\mathcal{A}, \mathcal{R}) = \{\mathcal{A}, \mathcal{A}\mathcal{R}, \cdots \mathcal{A}^{q-1}\mathcal{R}, \ldots\}, \tag{4.3}$$

constructed from moment matrices

$$\mathcal{A} = (\mathcal{G} + s_0\mathcal{C})^{-1}\mathcal{C}, \quad \mathcal{R} = (\mathcal{G} + s_0\mathcal{C})^{-1}\mathcal{B} \tag{4.4}$$

by expanding the system transfer function

$$H(s) = \mathcal{L}^T(\mathcal{G} + s\mathcal{C})^{-1}\mathcal{B} \tag{4.5}$$

at one frequency $s_0$.

By applying the block Arnoldi iteration [123], a small–dimensioned projection matrix $Q$ ($N \times q \times p_i$) can be found to contain $q$th-order block Krylov subspace

$$\mathcal{K}(\mathcal{A}, \mathcal{R}, q) \subseteq Q. \tag{4.6}$$

Using $Q$ the original system can be reduced by projection

$$\hat{\mathcal{G}} = Q^T\mathcal{G}Q \quad \hat{\mathcal{C}} = Q^T\mathcal{C}Q \quad \hat{\mathcal{B}} = Q^T\mathcal{B} \quad \hat{\mathcal{L}} = Q^T\mathcal{L}. \tag{4.7}$$

Accordingly, the reduced system transfer function becomes

$$\hat{H}(s) = \hat{\mathcal{L}}^T(\hat{\mathcal{G}} + s\hat{\mathcal{C}})^{-1}\hat{\mathcal{B}}. \tag{4.8}$$

As proved in [122, 123], the reduced $\hat{H}$ approximates the original system transfer function $H(s)$ by matching the first $q$ block moments expanded at the frequency-point $s_0$. This procedure can be applied to generate compact macromodels for both thermal-$RC$ and electrical-$RLC$ circuits.

Our thermal-$RC$ and electrical $RLC$ models have multiple inputs and multiple outputs (MIMO). This brings challenges for the projection–based model order reduction. The dimension of the reduced MIMO system $\hat{H}$ ($\in R^{p_o \times p_i}$) depends on the input-port number $p_i$ and $p_o$. In general, there are large numbers of injecting inputs. An accurate monitoring of integrity also needs large numbers of pre-designated regions to probe outputs. Therefore, an effective macromodel needs to further compress the number of ports where $p_i$ and $p_o$ are both large. In the following, we identify a much smaller number of principal input and output ports by studying the correlation.

### 4.3.1.2 *Complexity Compression of I/Os*

Generally, there can be thousands of thermal-power sources injected at each active layer or hundreds of switching-current sources injected at I/Os. The size of the macromodel increases with the number of ports, and hence the computational cost to solve the macromodel is still high. Since the electrical signals may share the same clock and operate within a similar logic function, their waveforms in the time-domain at certain input ports can show a correlation. Similarly, the thermal power may differ significantly between those regions with and without the clock gating, but can be quite similar inside the region with the same mode since inputs have similar duty-cycles over time. Based on the correlation, we can reduce the redundancy in I/Os by identifying those principal ports .

We call this phenomenon *input similarity*. As the input vector

$$\mathbf{I}(t) = \begin{bmatrix} \mathbf{I}_1 & \mathbf{I}_2 & \cdots & \mathbf{I}_{p_i} \end{bmatrix} \in R^{p_i \times 1}. \tag{4.9}$$

is usually known during the physical design, they can be represented by taking a set of 'snapshots' sampled at $\mathcal{N}$ time-points

$$\begin{bmatrix} \mathbf{I}_1(t_0) & \cdots & \mathbf{I}_1(t_\mathcal{N}) \\ \vdots & \ddots & \vdots \\ \mathbf{I}_{p_i}(t_0) & \cdots & \mathbf{I}_{p_i}(t_\mathcal{N}) \end{bmatrix} \tag{4.10}$$

in a sufficiently long period $[0,\ T_p]$. The sampling cycle is in a different time-scale for the thermal-power ($ms$) and switching current ($ns$). According to the POD analysis [124], the similarity can be mathematically described by a correlation matrix (or Grammian), estimated by a co-variance matrix:

$$\mathcal{R} = \frac{1}{\mathcal{N}} \sum_{\alpha=1}^{\mathcal{N}} (\mathbf{I}(t_\alpha) - \bar{\mathbf{I}})(\mathbf{I}(t_\alpha) - \bar{\mathbf{I}})^T \in R^{p_i \times p_i}. \tag{4.11}$$

$\bar{\mathbf{I}}$ is a vector of mean values defined by:

$$\bar{\mathbf{I}} = \frac{1}{\mathcal{N}} \sum_{\alpha=1}^{\mathcal{N}} \mathbf{I}(t_\alpha) \tag{4.12}$$

Usually, the input vector $\mathbf{I}(t)$ is periodic and the waveform in each period can be approximated by the piecewise-linear model.

An *output similarity* is defined for responses at the output ports and measured by an *output correlation matrix*. To extract the output correlation matrix that is independent of the inputs, we assume that the $p_i$ inputs in the input vector $\mathbf{I}(s)$ are all the unit-impulse source $h(s)$ and define an input-port vector $\mathcal{J}(s)$ by

$$\mathcal{J} = \mathcal{B}\mathbf{I}(s), \ \in R^{1 \times N}, \tag{4.13}$$

which has $p_i$ non-zero entries with the unit-value '1'. Accordingly, the $p_o$ output responses $y(s)$ are calculated by

$$\begin{aligned} y(s) &= \mathcal{L}^T (\mathcal{G} + s\mathcal{C})^{-1} \mathcal{J} \\ &= \begin{bmatrix} y_1(s)\ y_2(s) \cdots y_{p_o}(s) \end{bmatrix} \in R^{p_o \times 1}. \end{aligned} \tag{4.14}$$

The corresponding output correlation matrix is extracted in the frequency-domain. Similarly, the output signals can be represented by taking a set of 'snapshots' sampled at $\mathcal{N}$ frequency points

$$\begin{bmatrix} y_1(s_0) & \cdots & y_1(s_{\mathcal{N}}) \\ \vdots & \ddots & \vdots \\ y_{p_o}(s_0) & \cdots & y_{p_o}(s_{\mathcal{N}}) \end{bmatrix} \tag{4.15}$$

in a sufficiently wide band $[0,\ s_{max}]$. The $s_{max}$ locates in a low-frequency range for the temperature and in a high-frequency range for the voltage.

A co-variance matrix is defined in the frequency-domain as follows

$$R = \sum_{\alpha=1}^{\mathcal{N}} (y(s_\alpha) - \bar{y})(y(s_\alpha) - \bar{y})^T \in R^{p_o \times p_o} \tag{4.16}$$

to estimate the correlation matrix among $p_o$ outputs.

$\bar{y}$ is a vector of mean values defined by:

$$\bar{y} = \frac{1}{\mathcal{N}} \sum_{\alpha=1}^{\mathcal{N}} y(s_\alpha). \tag{4.17}$$

Let $\mathcal{V} = [v_1, v_2, \ldots, v_K]$ ($\in R^{N \times K}$) as the first $K$ singular-value vectors of the input correlation matrix $\mathcal{R}$, and $\mathcal{W} = [w_1, w_2, \ldots, w_K]$ ($\in R^{N \times K}$) as the first $K$ singular-value vectors of the output correlation matrix $R$. All singular-value vectors are obtained from the singular-value decomposition (SVD) of $(\mathcal{V}, \mathcal{W})$. A rank-K matrix $P_i$ can be constructed by $P_i = \mathcal{V}\mathcal{V}^T$, and a rank-K matrix $P_o$ can be constructed by $P_o = \mathcal{W}\mathcal{W}^T$. As shown in [124], the correlation matrix $(\mathcal{R}, R)$ is essentially the solution that minimizes the least-square between the original states $(\mathbf{I}(t), y(s))$ and their rank-K approximations $(P_i \cdot \mathbf{I}(t), P_o \cdot y(s))$. As a result, both the input signals $\mathbf{I}(t)$ and the output signals $y(s)$ can be approximated by an invariant (or dominant) subspace spanned by the orthonormalized columns of $V$ and $W$, respectively:

$$\mathbf{I} = \mathcal{V}\mathbf{I}_K, \ y = \mathcal{W}y_K. \tag{4.18}$$

Based on (4.18), it leads to the following equivalent system equation

$$(\mathcal{G} + s\mathcal{C})x_K(s) = \mathcal{B}_K\mathbf{I}_K(s), \ y_K(s) = \mathcal{L}_K^T x_K(s) \tag{4.19}$$

where

$$\mathcal{L}_K^T = \mathcal{W}^T\mathcal{L}^T, \ \mathcal{B}_K = \mathcal{B}\mathcal{V}. \tag{4.20}$$

Therefore, both the dimensions of $\mathcal{L}$ ($\in R^{N \times p_o}$) and $\mathcal{B}$ ($\in R^{N \times p_i}$) are greatly reduced when $K << p_i$ and $p_o$. We call $\mathbf{I}_K$ and $y_K$ *principal inputs and outputs* identified by *principal input-port and output-port matrices* $\mathcal{B}_K$ and $\mathcal{L}_K$, respectively.

### 4.3.2 *Parameterization*

Recall that the design parameter in power and ground design is the via density at one track. Blindly allocating the via for power and thermal integrity by searching all kinds of combinations would be computationally expensive if not impossible. Therefore, we decide the via density based on the changes in outputs, i.e., sensitivities, caused by the change in via density.

To calculate sensitivity, let's first parameterize the nominal system (4.2). The added via at one track is described by two parameters: $n_j$ the via density and $X_j$ the topological matrix that connects the via into the nominal system. As such, a parameterized state-space description can be obtained by

$$(\mathcal{G} + s\mathcal{C} + \sum_{j=1}^{p_o} n_j g_j + s \sum_{j=1}^{p_o} n_j c_j) x_K(\mathbf{n}, s) = \mathcal{B}_K \mathbf{I}_K(s),$$

$$y_K(\mathbf{n}, s) = \mathcal{L}_K^T x_K(\mathbf{n}, s). \tag{4.21}$$

Similar to [9–12,120,121], we expand $x(\mathbf{n}, s)$ in the Taylor series with respect to $n_j$, and introduce a new state variable $x_{ap}$

$$x_{ap} = [x^{(0)}, x_1^{(1)}, \dots, x_{p_o}^{(1)}]^T. \tag{4.22}$$

It contains both the nominal response $x^{(0)}$ and its first-order sensitivities $[x_1^{(1)}, \dots, x_{p_o}^{(1)}]$ with respect to $p_o$ parameters $[n_1, \dots, n_{p_o}]$. The overall response is obtained by

$$x = x^{(0)} + \sum_{j=1}^{p_o} x_j^{(1)}.$$

Substituting (4.22) into (4.21), (4.21) can be reformulated into a parameterized system with an augmented dimension by

$$(\mathcal{G}_{ap} + s\mathcal{C}_{ap})x_{ap} = \mathcal{B}_{ap}\mathbf{I}_K(s), \quad y_{ap} = \mathcal{L}_{ap}^T x_{ap}, \tag{4.23}$$

where $\mathcal{G}_{ap}$ and $\mathcal{C}_{ap}$ show a lower-triangular-block structure and hence $x_{ap}$ can be solved from block-backward-substitution [9–12, 120, 121].

To further compress the dimension of the state-matrices $\mathcal{G}_{ap}$ and $\mathcal{C}_{ap}$, we first construct a lower-dimensioned subspace $Q_{ap}$ from the moment expansion of (4.23), and then transform $Q$ into the block-diagonal form $\mathcal{Q}_{ap}$. After the block-orthonormalization of $\mathcal{Q}_{ap}$, we apply a two-side projection to (4.23) by $\mathcal{Q}_{ap}$ and obtain a dimension-reduced system with a preserved lower-triangular-block structure [9–12,120,121]. We call the resulted macromodel a structured and parameterized macromodel . The accuracy of the macromodel is preserved to match the dominant moments of the original model. More importantly, due to the structure-preservation, both the nominal response and sensitivity with regard to the change in via-density can be calculated simultaneously.

## 4.4 Summary

In 3D integration, the accumulated heat in the top layers result in a thermal gradient. Signal TSVs, also called power/ground vias, are utilized to deliver power; and dummy TSVs can effectively transfer the heat. The wise assignment of power/ground vias is to reduce the size of the loop current and the voltage bounce induced by inductive coupling. By further extending power/ground vias to the top heat-sink, they can be reused to remove the heat. This chapter explains the need for dynamic thermal integrity for high-performance 3D integration. To efficiently solve the problem with dynamic-integrity constraints, compression of ports (parameters) and a parameterized macromodel are developed to effectively and efficiently generate both integrity and sensitivity.

# Chapter 5

# TSV Allocation

## 5.1 Introduction

In 3D integration, as a large number of devices are densely packed in a number of device layers, it brings a significant burden to heat removal and power delivery (supply voltage) in 3D ICs. In this chapter, we discuss an allocation algorithm of through-silicon via (TSV) for thermal integrity in 3D ICs, followed by heat removal with the insertion of dummy TSVs. Firstly, we illustrate the need for a high-performance 3D design driven by dynamic thermal integrity, and then discuss how to allocate TSVs to remove heat. More importantly, to cope with large-scale design complexity, the modern macromodeling technique discussed previously is applied to handle not only large numbers of dynamic inputs/working-loads but also large sizes of interconnection networks that distribute the heat.

Secondly, the optimization of clock-skew in a 3D clock-tree through insertion of dummy TSVs by balancing temperature and stress gradients is presented. One needs to note that dummy/thermal TSVs are utilized to transfer the heat from top layers to heat-sink. Insertion of TSVs can help to reduce temperature, but also increases area overhead. Further, sensitivity analysis is performed to study the variation of temperature and stress gradients with respect to dummy TSV density. A reliable zero-skew 3D clock-tree design is considered and a nonlinear optimization–based dummy TSV insertion is performed to reduce the clock-skew and balance temperature and stress gradients without incurring large area overhead. A nonlinear optimization algorithm is performed to balance thermal and stress gradients considering area overhead constraint. Experiments applying the design automation for 3D integration presented in this chapter show promising results to reduce both runtime and resource.

## 5.2   Power Ground Design

In this section a power/ground design by TSV allocation is discussed. Firstly a power/ground problem is formulated, followed by sensitivity based via allocation for power/ground design along and thermal/power integrity.

### 5.2.1   *Problem Formulation*

We notice that previous thermal via allocations [7,8] assume adding dummy vias to conduct heat. They ignore the fact that power/ground vias can help remove heat as well. Therefore, the reusing of the power/ground via as the thermal via can save the routing resource for signal nets. More importantly, the allocation of power/ground vias can minimize not only dynamic power integrity, i.e., voltage bounce for those I/Os at package and chip interface, but also thermal integrity, i.e., the temperature gradient at those active device layers.

   Similar to the work in [8], this work assumes that the via allocation is after the placement and global routing but before the detailed routing of the signal nets. The power ground vias are placed at centers of tiles between two layers, and follow an aligned path from the bottom package I/Os to the top heat-sink. We call those aligned paths *vertical tracks* or *tracks*. As vias are aligned, the $p_o$ tracks pass both $p_o$ output ports of the electrical-*RLC* model and $p_o$ output ports of the thermal-*RC* model. The density of power/ground vias at each track is the primary design parameter considered in this work. The density is adjusted to satisfy two requirements at output ports. The first is the integrity constraint of the temperature gradient and voltage bounce. The second is the resource constraint with provided signal net congestion.

   Moreover, to capture the sharp-transition of temperature change as well as the time-accumulated temperature impact, we employ the thermal-integrity integral used in [9] as the measure of dynamic thermal integrity at $jth$ $(j = 1, \ldots, p_o)$ output port:

$$f_j^T = \int_{t_0}^{t_p} max[y_j(t), Tc]dt = \int_{t_s}^{t_e} [y_j(t) - T_r]dt, \qquad (5.1)$$

with a pulse-width $(t_s, t_e)$ in a sufficient long time-period $t_p$ all in the scale of thermal-constant $(ms)$. $y_j(t)$ is the transient temperature waveform at the $jth$ output port, and $T_r$ is the reference temperature.

To further consider the spatial difference of the $p_o$ output ports at the bottom device layer, the overall thermal integrity is defined by a normalized summation:

$$f^T = \frac{\sum_{j=1}^{p_o} f_j^T}{t_p^T \cdot p_o}. \tag{5.2}$$

Such a measure of thermal integrity takes into account both the temporal and spatial variation of temperature. Similarly, the power-integrity integral $f_j^V$ is defined at the *jth* power/ground I/O with reference voltages Vdd and ground, and integrated at the period $t_p^V$ in the scale of electrical-constant (ns). The overall power integrity $f^V$ is defined similarly to $f^T$ for $p_o$ power/ground I/Os with the reference voltage $V_r$ (0 for ground vias and Vdd for power vias).

Accordingly, we have the following problem formulation:

*Given the targeted voltage bounce $V_t$ for $p_o$ output ports at power/ground I/Os, and the targeted temperature gradient $T_t$ for $p_o$ output ports at bottom device layer, the via-allocation problem is to minimize the total via number, such that the temperature gradient $f^T$ is smaller than $T_t$ and the voltage bounce $f^V$ is smaller than $V_t$.*

Such a via-allocation problem simultaneously driven by power and thermal integrity can be represented by

$$min \sum_{j=1}^{p_o} n_j$$

$$s.t. \ f^V \leq V_t, \ f^T \leq T_t$$

$$and \ n_{min} \leq n_j \leq n_{max} \tag{5.3}$$

Note that $n_j$ is the via density at the *jth* track and $V_t$ and $T_t$ are the targeted voltage bounce and temperature gradient. $f^V$ and $f^T$ are the metrics of power integrity and thermal integrity, respectively.

As discussed later, $n_j$ is decided according to the power and thermal sensitivities obtained from the macromodel. As our power/ground vias are allocated after the placement and global routing of signal nets at each active device layer, the densities of those inter-layer signal nets are available to calculate a maximum density $n_{max}$ for the power/ground vias. In addition, for the sake of the reliability concern of the large current, the via density $n_j$ on the other hand can not be smaller than the minimum density $n_{min}$. These parameters ($n_{max}, n_{min}, V_t, T_t$) can be estimated and provided by users.

### 5.2.2 Sensitivity based TSV Allocation

To reduce the complexity of resulting state matrices from the modified nodal analysis a macromodeling is performed as discussed in the previous chapter.

To reduce (4.23), a $q$-th order flat projection matrix $\mathbf{Q}_{ap}$ can be obtained from block Arnoldi iteration. To reserve the separated nominal value and its sensitivity during projection, $\mathbf{Q}_{ap}$ is first partitioned into $K+1$ blocks as

$$\mathbf{Q}_{ap} = [Q_0, Q_1, \ldots, Q_K], \qquad (5.4)$$

each $Q_j$ $(j = 0, 1, \ldots, K)$ with size $N \times q \times K$. Then, similar to [9,10], $\mathbf{Q}_{ap}$ is further structured as follows

$$\mathbf{Q}_{ap} = \begin{pmatrix} Q_0 & & & \\ & Q_1 & & \\ & & \ddots & \\ & & & Q_K \end{pmatrix}. \qquad (5.5)$$

Each $Q_j$ $(j = 0, 1, \ldots, K)$ is further orthonormalized with each other. As such, proven in [9,10], the structured reduction by $\mathbf{Q}_{ap}$

$$\begin{aligned}
\widetilde{\mathbf{G}}_{ap} &= \mathbf{Q}_{ap}^T \mathbf{G}_{ap} \mathbf{Q}_{ap}, & \widetilde{\mathbf{C}}_{ap} &= \mathbf{Q}_{ap}^T \mathbf{C}_{ap} \mathbf{Q}_{ap}, \\
\widetilde{\mathbf{B}}_{ap} &= \mathbf{Q}_{ap}^T \mathbf{B}_{ap}, & \widetilde{\mathbf{L}}_{ap} &= \mathbf{Q}_{ap}^T \mathbf{L}_{ap}
\end{aligned} \qquad (5.6)$$

still preserves moments up to $q$-th order.

After being projected by $\mathbf{Q}_{ap}$, the reduced macromodel in time-domain can be solved by the Backward-Euler (BE) integration with the state-equation below at time-instant $t$ and time-step $h$

$$\left( \widetilde{\mathbf{G}}_{ap} + \frac{1}{h} \widetilde{\mathbf{C}}_{ap} \right) \widetilde{\mathbf{x}}_{ap}(t) = \frac{1}{h} \widetilde{\mathbf{C}}_{ap} \widetilde{\mathbf{x}}_{ap}(t - h) + \widetilde{\mathbf{B}}_{ap} \mathbf{I}_K(t)$$

$$\widetilde{\mathbf{y}}_{ap}(t) = \widetilde{\mathbf{L}}_{ap}^T \widetilde{\mathbf{x}}_{ap}(t) \qquad (5.7)$$

As shown in [9,10], the projection by $\mathbf{Q}_{ap}$ preserves the block matrix structure. As a result, $\widetilde{\mathbf{G}}_{ap}$ and $\widetilde{\mathbf{C}}_{ap}$ both have the same preserved lower-block triangular structure. It (5.7) can be efficiently solved by block-backward-substitution.

| ALGORITHM: SENSITIVITY BASED VIA ALLOCATION |
|---|
| 1 **Input**: $K$ principal input ports, $K$ principal output ports, maximum temperature bound $T_{max}$, maximum voltage-bounce bound $V_{max}$, signal-net-congestion bound $n_{max}$ and current-density bound $n_{min}$ |
| 2 *Construct* structured and parameterized macromodel; |
| 3 *Compute* nominal voltage($V$)/temperature($T$) and sensitivity $\mathbf{S}_V/\mathbf{S}_T$; |
| 4 *Check* $V_{max}$ and $T_{max}$ constraints for all tiles; |
| 5 *Increase* the via density $\mathbf{n}$ according to weighted sensitivity $\mathbf{S}$ in the range of $(n_{min}, n_{max})$; |
| 6 *Update* the structured and parameterized macromodel; |
| 7 *Repeat* from Step 3 until Step 4 is satisfied; |
| 8 **Output**:Via density vector $\mathbf{n}$ |

Fig. 5.1    Algorithm for the sensitivity-based via allocation with the use of macromodels.

Note that the reduced nominal response and its sensitivity are still separated at outputs:

$$\widetilde{\mathbf{y}}_{ap} = [\widetilde{\mathbf{y}}^{(0)}, \widetilde{\mathbf{y}}^{(1)}]^T = [\widetilde{\mathbf{y}}_0^{(0)}, \widetilde{\mathbf{y}}_1^{(1)}, \ldots, \widetilde{\mathbf{y}}_K^{(1)}]. \tag{5.8}$$

The according overall output response $\widetilde{\mathbf{y}}$, i.e., the integrity vector of voltage os temperature $(V/T)$ is

$$\widetilde{\mathbf{y}}(\mathbf{n}, t) = \widetilde{\mathbf{y}}^{(0)}(\mathbf{n}, t) + \widetilde{\mathbf{y}}^{(1)}(\mathbf{n}, t) \tag{5.9}$$

As discussed in Figure 5.1, such a structured and parameterized macromodel can be incorporated into a sensitivity-based via allocation.

The overall optimization flow to solve the problem formulation (5.3) is outlined in the Algorithm above. Its input is composed of two parts. The first is a principal system by (4.23) with the identified $K$ principal input and output ports. The second is the user–provided temperature bound $T_{max}$, voltage bounce bound $V_{max}$, signal-net congestion bound $n_{max}$, and current-density bound $n_{min}$. Then, a structured and parameterized macromodel is built once. Both nominal responses and sensitivities at $K$ principal input-ports for each perturbed allocation-pattern are solved at one time. If the integrity constraints are not satisfied for $K$ principal tracks, the vias density vector $\mathbf{n}$ is increased according to the sensitivity. This process repeats until the integrity constraints are satisfied. Details of this Algorithm can be found in [12].

The sensitivity vector $\mathbf{S}$ in Step 3 is a weighted-maximum of normalized voltage-sensitivity vector $\mathbf{S}_V$ and thermal-sensitivity $\mathbf{S}_T$:

$$\mathbf{S} = \mathbf{max}[\alpha \cdot \widetilde{\mathbf{y}}_{\mathbf{T}}^{(1)}/\|\widetilde{\mathbf{y}}_{\mathbf{T}}^{(1)}\|, \ \beta \cdot \widetilde{\mathbf{y}}_{\mathbf{V}}^{(1)}/\|\widetilde{\mathbf{y}}_{\mathbf{V}}^{(1)}\|] \in \mathbf{R}^{P_o \times 1} \tag{5.10}$$

where $\alpha$ and $\beta$ are weights for $\mathbf{S}_T$ and $\mathbf{S}_V$, and the maximum of the normalized sensitivity is selected at each track.

The sensitivity vector is iteratively updated to calculate the new via density vector by

$$\mathbf{n}^{(\text{iter+a})} = \mathbf{n}^{(\text{iter})} + \gamma^{(\text{iter})} \cdot \mathbf{S}^{(\text{iter})}, \tag{5.11}$$

till the integrity constraints are satisfied at all principal output ports. Note that $\gamma$ is an adaptive-controlled step size, which decreases geometrically by a factor of 0.99 as the iteration proceeds. Recall that at each step, $\mathbf{n}$ is constrained by the signal congestion induced bound $n_{max}$ and the current-density induced bound $n_{min}$.

The overall computational cost of Algorithm 5.1 is low. This can be illustrated in two folds. Firstly, the use of the principal input/output identification and the principal parameter clustering leads to an efficient evaluation of integrity at principal ports. Secondly, due to structure reduction, the nominal values and their sensitivities can be efficiently solved by the block-backward substitution of (5.7), and only the perturbed states are updated during each iteration. A formal complexity analysis of Algorithm 5.1 is illustrated as follows. The main computational cost of Algorithm 5.1 is from steps 2 and 3, the model order reduction and update of the integrity and sensitivity. A $q$-th order model order reduction by Krylov-subspace projection needs one LU decomposition, $q$ solves and the orthonormalization. Due to the lower-block triangular structure, the LU factorization is only applied for the nominal matrix in the diagonal. Moreover, there are only $K$ off-diagonal blocks for $K$ principal parameters. As such the total complexities of model order reduction are $O(qN^{1 \cdot y} + q(KN)^{1 \cdot x} + Nq^2)(x, y \in [0, 9])$. The computational cost of updating integrity and sensitivity is $O(q^3 + Kq^2)$. The model order reduction is done only once and the update is done iteratively until the optimization converges.

Experiments are implemented in C and MATLAB and run on a Sun-Fire-V250 workstation with 2G RAM. We call the separated allocation of thermal vias and power/ground vias the *sequential optimization*, and call our allocation of power/ground vias for both power and thermal integrity the *simultaneous optimization*. Moreover, the steady-state analysis is employed to calculate a static integrity [7,8]. We use the sequential optimization with the static integrity as the baseline, in comparison to the sequential optimization with the dynamic integrity and the simultaneous optimization with the dynamic integrity proposed in this work. The electrical, thermal constants and dimensions are the same with [12]. The targeted voltage violation $V_t$ is $0.2V$ and the targeted temperature $T_t$ is $52°C$. One modest 3D stacking is assumed with a 2-device-layer/2-dielectric-layer. Moreover, a 1-heat-sink and 2-P/G-plane were used in the example. In this section, we present the results of temperature reduction by via allocation, macromodeling, and the reduction in number of TSVs required while using the proposed algorithm.

Figure 5.2 further shows the steady-state temperature map across the bottom device layer. In this example, we assume that all thermal-power sources are located along one side of the device layer. The initial chip temperature at the bottom layer is $150°C$, and its temperature profile at the steady-state is shown in Figure 5.2(a). In contrast, the via-allocation results in a cooler temperature that closely approaches the targeted temperature

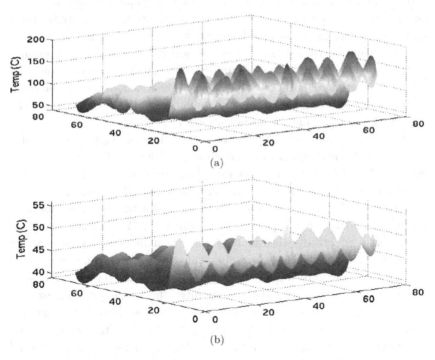

(a)

(b)

Fig. 5.2 Steady-state temperature maps of bottom device layer (a) before allocating via, and (b) after allocating via in different temperature scales.

Table 5.1   The complexity of the original circuits and the reduced circuits including: the size, number of input-ports, and number of output-ports.

| ckt | total tile# | reduced size (T,V) | input src# (T,V) | K-input (T,V) | output track# | K-output (T,V) |
|---|---|---|---|---|---|---|
| ckt1(2-layer) | 1.9K | (30,80) | (10,20) | (10,20) | $4^2$ | $(4^2,4^2)$ |
| ckt2(2-layer) | 6K | (15,48) | (100,200) | (5,8) | $4^3$ | (6,4) |
| ckt3(2-layer) | 12K | (80,160) | (300,600) | (10,16) | $4^4$ | (8,5) |
| ckt4(2-layer) | 27K | (96,180) | (1K,2K) | (12,18) | $4^4$ | (10,8) |
| ckt5(2-layer) | 52K | (96,220) | (1K,3K) | (12,20) | $4^5$ | (12,14) |

as shown in Figure 5.2(b). Clearly, even at steady-state the temperature is still spatially variant. An accurate measure of integrity is therefore needed to consider a time and space–averaged integrity at selected probing ports.

The details of the benchmark circuits are summarized in Table 5.1. The runtime and the number of allocated viass are compared in Table 5.2. In Table 5.2, columns 2–3 show the runtime and the number of allocated vias for the baseline, and columns 4–8 show the results for the optimizations using dynamic integrity. In detail, column 4 shows the runtime of transient analysis using macromodels without the port-compression, and column 5 shows the number of allocated vias under sequential optimization. Column 6 shows the runtime of transient analysis using macromodels with the port-compression, and columns 7-8 show the number of allocated vias under the sequential and simultaneous optimizations, respectively.

The use of macromodels reduces the computational cost to solve power and thermal integrities and their sensitivities. Compared to the macromodel without the port-compression, the macromodeling with the port-compression reduces the overall runtime by up to 16× with similar allocation results. Compared to steady-state analysis with full-matrix analysis, our macromodel with the port-compression has a 127× smaller runtime. Additionally, the steady-state analysis can not complete the largest example in a reasonable runtime. The maximum transient-waveform difference introduced by the macromodel is about 7% when compared to the exact transient waveform. We further compare the sequential thermal/power optimization with the simultaneous thermal/power optimization. Here both methods allocate vias with the use of dynamic integrity. Our simultaneous optimization reduces the via-cost by up to 34% when compared to the sequential optimization with static integrity, and by up to 22% when compared to the sequential optimization with dynamic integrity. This demonstrates that the

Table 5.2 Comparisons of via number and runtime for the sequential optimization with steady-state analysis, sequential optimization with transient analysis and simultaneous optimization with transient analysis. Two macromodels are used during the transient analysis. Macromodel-1 does not use the port-compression, and macromodel-2 uses the port-compression.

| ckt | Steady-state(direct) | | Transient(MACRO-1) | | Transient(MACRO-2) | | |
|---|---|---|---|---|---|---|---|
| | runtime (s) | total via # by seq-opt | runtime (s) | total via # by seq-opt | runtime (s) | total via # by seq-opt | total via # by sim-opt |
| ckt1(2-layer) | 5.4 | 178800 | 0.63 | 153800 (−13%) | 0.63 | 153800 (−13%) | 112800 (−36%) |
| ckt2(2-layer) | 29.7 | 184900 | 0.81 | 159600 (−13%) | 0.56 | 159600 (−13%) | 118200 (−36%) |
| ckt3(2-layer) | 182.2 | 218100 | 18.6 | 183800 (−16%) | 4.2 | 184200 (−15%) | 136200 (−38%) |
| ckt4(2-layer) | 1269.2 | 234800 | 165.7 | 199000 (−15%) | 10.3 | 199600 (−15%) | 145600 (−38%) |
| ckt5(2-layer) | NA | NA | NA | NA | 41.2 | 208600 (NA) | 154200 (NA) |

reusing of power/ground vias can reduce the via cost when compared to allocating the dummy thermal vias separately from the power/ground vias.

## 5.3 Clock-tree Design

Delay in 3D circuit varies with the temperature, hence, delay/skew optimization by insertion of TSVs are studied in this section. A 3D clock-tree shown in Figure 5.3 with TSVs for vertical interconnects is considered as a case study for determining the effect that different physical impacts have on electrical delay. The design of the 3D clock-tree is primarily involved with the reduction of delay difference at different clock-sinks, known as clock-skew [21, 125–131]. Compared to the clock-tree in 2D IC, the clock-tree in 3D IC experiences a much larger temperature and stress gradient both vertically and horizontally. As such, the traditional clock-tree design methods [125–127] without considering both temperature and stress gradients simultaneously will become inaccurate and unreliable. Additionally, such methods only conduct the clock network optimization based on linear or

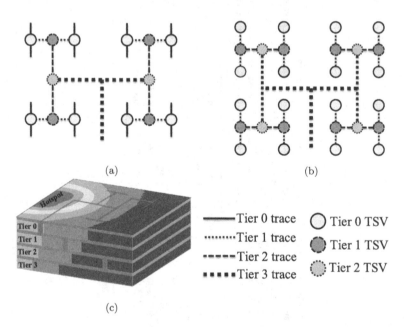

Fig. 5.3  3D clock-tree distribution network at different tiers: (a) Clock-tree with 14 TSV bundle locations in a H-tree; (b) Clock-tree with 28 TSV bundle locations in a H-tree; (c) Layer configuration under non-uniform temperature distribution.

partial or none electrical-thermal-mechanical coupling, ignoring the TSV physical model and hence, is not accurate. As such, there is no specific problem formulated for clock-tree design considering the reduction of both thermal and stress gradients in 3D IC. In a 3D clock-tree design, unlike 2D design, all the impacts discussed in the previous chapter need to be considered. A nonlinear problem is formulated to reduce the clock-skew.

### 5.3.1 *Problem Formulation*

By considering all the physical and electrical impacts, delay at clock-sink $i$ of a 3D clock-tree needs to be modeled as a nonlinear function $\Gamma$ of temperature and stress gradients. Thus, the delay at $i^{th}$ clock-sink can be modeled as $\delta D_i = f(\Gamma)$. *Clock-skew* **S** is defined as the maximal delay difference between two clock-sinks. Note that by adding dummy TSVs, one can balance the temperature and stress gradients and further reduce delay and skew. As such, we have the following problem formulation towards a reliable 3D clock-tree design under a temperature and stress gradient.

*Problem 1: For a pre-synthesized zero-skew 3D clock-tree with $N_S$ sinks, using signal TSVs for inter-tier connections, the clock-skew **S** needs to be estimated by considering the position and number of TSVs, i.e., temperature and stress gradients $\Gamma$, and by considering nonlinear electrical-thermal-mechanical coupling. A large number of dummy TSVs that can be inserted to minimize **S** under $\Gamma$.*

$$\mathbf{S} = \max : |\delta D_i - \delta D_j|, 0 \leq i, j \leq N_S \tag{5.12}$$

where $\delta D_i$ and $\delta D_j$ are delays from sinks $i$, $j$ respectively.

A nonlinear optimization is deployed to further minimize the clock-skew under non-uniform temperature and stress gradients. Note the pre-synthesized zero-skew 3D clock-tree is based on the work in [132] considering the wire length and driver but without electrical-thermal-mechanical coupling. To solve the above mentioned clock-skew reduction problem by insertion of dummy TSVs, reduction in temperature and stress gradients, their corresponding sensitivities, and skew sensitivity with dummy TSV density need to be studied first.

### 5.3.2 *Sensitivity based TSV Allocation*

As aforementioned, dummy TSVs are used for vertical interconnects without signal connection running through it. As dummy TSVs are filled with metal-material copper (Cu) which has a good thermal conductivity of $400W/m.K$, it can provide a vertical heat-dissipation path to reduce the temperature gradient. One can observe from (3.10) that the exerted mechanical stress decreases with temperature. Altogether, insertion of dummy TSVs can balance the density of TSV distribution and reduce temperature and stress gradients. However, insertion of dummy TSVs at improper locations can further worsen temperature and stress gradients. To solve the above mentioned clock-skew problem by insertion of dummy TSVs, reduction in temperature and stress gradients, their corresponding sensitivities, and skew sensitivity with dummy TSV density need to be studied.

### 5.3.2.1 *Reduction of Thermal Gradient*

For a chip level analysis, impact of a single dummy TSV is ineffective. Dummy TSVs are modeled in terms of local density $\eta$ as shown in Figure 5.4, where dummy TSVs occupy an area of $\eta A$ on a regular chip of area $A$. Considering the vertical heat-dissipation, the thermal conductivity

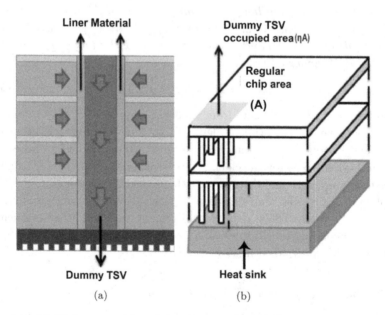

Fig. 5.4 (a) 3D heat-removal path by dummy TSVs; (b) 3D-view of dummy TSV insertion.

$\lambda$, given by

$$\lambda = (\eta + \delta\eta)\lambda_{TSV} + (1 - (\eta + \delta\eta))\lambda_0; \quad \delta n = \delta\eta A \quad (5.13)$$

where $\lambda_0$ is the initial thermal conductivity; $\eta$ is the initial TSV density; $\delta\eta$ is the change in TSV density; and $\delta n$ represents the change in the number of TSVs with respect to the initial number of TSVs $n$.

As such, the temperature gradient reduction with the change in TSV density $\delta\eta$ is given by

$$\delta T = T - T_0 = \frac{P \cdot l}{A\lambda_0} \cdot \frac{\delta\eta}{\frac{\lambda}{\lambda_{TSV} - \lambda_0} + \eta + \delta\eta} \quad (5.14)$$

where $P$ is the heat power flowing from chip to heat-sink; and $l$ is the length of heat-transfer path distance with a chip area of $A$. From (5.14), one needs to note that as $\delta\eta$ approaches or becomes larger than $\lambda/(\lambda_{TSV} - \lambda_0) + \eta$, the reduction in temperature due to dummy TSVs becomes saturated.

Further, the sensitivity of temperature gradient with TSV density is studied. In (5.14), it needs to be noted that after a certain dummy TSV density, the reduction in temperature gradient tends to saturate. Sensitivity of temperature gradient with respect to the dummy TSV density can be given by

$$\frac{\partial T}{\partial \eta} = \frac{P \cdot l}{A\lambda_0} \cdot \frac{\eta_0}{(\eta_0 + \eta)^2}; \quad \eta_0 = \frac{\lambda_0}{\lambda_{TSV} + \lambda_0}. \quad (5.15)$$

From (5.15), it can be clearly concluded that, when the dummy TSV density $\eta$ is smaller than its saturation value $\eta_0$, i.e., $\eta \ll \eta_0$, the sensitivity of temperature gradient with dummy TSV density remains almost constant; and as $\eta \gg \eta_0$ the sensitivity approaches zero, implying that the reduction of temperature tends to saturate. Thus, temperature sensitivity function with dependence on dummy TSV density can be used during the optimization of dummy TSV insertion.

To study the temperature gradient experimentally, a 4-layer stacked 3D IC is considered with each layer provided with same power density of $P$, which serves as the heat source and simulated in COMSOL. Study is performed for different values of thermal power densities with $P = \{6, 80, 115\}W/m^2$. In experiment, the temperature at each layer is collected without dummy TSVs initially. Note that the liner material of dummy TSV is $Si_3N_4$, of which the thickness is $200nm$. Based on the formed temperature distribution, dummy TSVs are inserted with density upper and lower bounds.

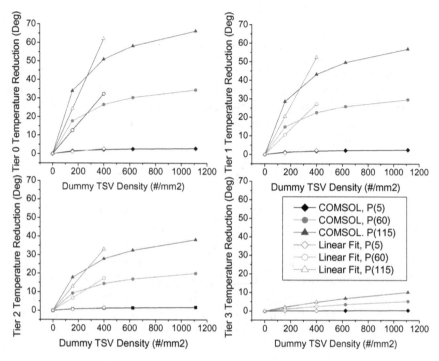

Fig. 5.5 Temperature gradient reduction in 4-layer 3D clock-tree with dummy TSVs under different power densities.

The temperature reduction in each layer with the insertion of dummy TSVs is presented in Figure 5.5. It is experimentally observed that the reduction in temperature initially increases but tends to saturate after a particular limit is reached. It can also be observed that the reduction in temperature for the bottom layer, i.e., layer-3, is less than those of other layers. As it is close to the heat-sink, the insertion of a dummy TSV does not make big difference compared with other layers. In addition, the maximum inserted dummy density observed is $400/mm^2$ with the maximum temperature reduction of nearly $50°C$ in layer-0 i.e. the top layer.

### 5.3.2.2 *Reduction of Stress Gradient*

To calculate the stress gradient variation, we consider a setup of a unit square having four TSVs at its four corners. All the transistors or devices inside the square will experience the stress from all the four TSVs. The stress contour from TSVs and its impact on each of transistors is shown in Figure 3.2(b). The stress on each of the transistors can be calculated using (3.10). What is more, it needs to be observed that there will be a reduction in stress gradient with the insertion of dummy TSVs at proper locations, and balances stress due to the reduction in temperature. The stress gradient reduction $\delta\sigma$ caused by TSV density difference $\delta\eta$ can be given as

$$\delta\sigma = -\frac{B\Delta\alpha\Delta T}{2}\left(\frac{R}{r}\right)^2\delta n = -\frac{B\Delta\alpha\Delta T}{2}\left(\frac{R}{r}\right)^2\delta\eta A \qquad (5.16)$$

where $\delta\eta$ represents change in TSV density due to insertion of $\delta n$ additional TSVs. When the density becomes more uniform, the stress gradient also becomes smaller.

Note that the stress and stress gradient depends on both TSV density and temperature gradient. The stress gradient at different temperature gradients and TSV densities are shown in Figure 5.6. For the purpose of simulation, two blocks A and B, in which TSV density can be varied and is constant respectively, are considered. As shown in Figure 5.6(a), block B having a constant TSV density of $400/mm^2$ is considered. Another block A has a temperature gradient of $180°C$, and its TSV density can be varied. The stress gradient between the blocks A and B is shown in Figure 5.6(b).

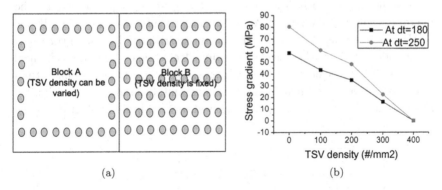

(a)                                                        (b)

Fig. 5.6    (a) Setup of TSV distributions for stress gradient calculation; (b) Variation of stress gradient reduction with TSV density and temperature gradient.

It can be observed that as TSV density is increased, the stress gradient tends to decrease. The stress gradient reduction for same setup but with a temperature gradient of $250°C$ is also plotted.

As given in (5.17), one can observe that the reduction in stress gradient depends on the TSV density. The sensitivity of stress gradient reduction with respect to TSV density is given by

$$\frac{\partial \sigma}{\partial \eta} = -\frac{B \Delta \alpha \Delta T A}{2} \left( \frac{R}{r} \right)^2. \tag{5.17}$$

The stress gradient sensitivity in (5.17) may look as if it was independent of TSV density, but that depends on the radius of TSV and area, which has an impact on TSV density. When the radius of the TSV becomes very small compared to the distance, the stress gradient tends to saturate early compared to TSVs with larger radius.

### 5.3.2.3   *Clock-skew Reduction*

The nonlinear electrical-thermal-mechanical coupling transforms the 3D clock-skew reduction problem into a nonlinear optimization problem. Here, a nonlinear programming–based algorithm is developed for the insertion of dummy TSVs to reduce the clock-skew. Before optimization is performed, sensitivity of clock-skew with respect to dummy TSV density needs to be discussed.

The nonlinear optimization of a 3D clock-tree is performed at the micro-architecture level by dividing each layer into $M \times N$ grids. If one TSV passes through a grid $g_i$, the delay contributed by that grid needs to be calculated by the developed coupled electrical-thermal-mechanical model in (3.18). Generally, temperature has much higher impact than stress. Moreover, the linear delay from horizontal metal wires and buffers are also considered in (5.18). The skew from $i^{th}$ individual grid can be given as

$$S_i = \begin{cases} S_c + [w_1 \delta T_i + w_2(\delta T_i)^2 + w_3(\delta T_i)^3] \\ \quad + [s_0 + s_1 \delta T_i + s_2(\delta T_i)^2] \delta \sigma_i : \quad 3D \quad signal \quad TSVs; \\ \\ z_0 + z_1 \delta T_i : \quad 2D \quad wires; \end{cases} \tag{5.18}$$

where

$$S_c = D_c + \left[ R_0\beta_1 + \alpha R_0 C_0 + S_{w1}R_{w1}S_T\beta_1 \right.$$

$$\left. + \frac{R_0\alpha(S_{w2}C_{w2} + S_L C_L)}{S_T} \right] T_0 + R_0\beta_2 T_0^2;$$

$$w_1 = 2R_0(\beta_2 + \alpha\beta_1)T_0 + \left[ R_0\beta_1 + \alpha R_0 C_0 + S_{w1}R_{w1}S_T\beta_1 \right.$$

$$\left. + \frac{R_0\alpha(S_{w2}C_{w2} + S_L C_L)}{S_T} \right];$$

$$w_2 = 2R_0\alpha\beta_2 T_0 + R_0(\beta_2 + \alpha\beta_1); \quad w_3 = R_0\beta_2;$$

$$s_0 = -\frac{mR_D}{S_D(1 + m\sigma_0)^2}[(S_D C_P + S_{w1}C_{w1} + S_T C_0 + S_L C_L + S_{w2}C_{w2})$$

$$+ S_T\beta_1 T_0 + S_T\beta_2 T_0^2];$$

$$s_1 = -\frac{mR_D}{S_D(1 + m\sigma_0)^2}[S_T\beta_1 + 2S_T\beta_2 T_0]; \quad s_2 = -\frac{mR_D}{S_D(1 + m\sigma_0)^2}S_T\beta_2;$$

$$z_0 = R_0 C_0; \quad z_1 = R_0 C_0\alpha$$

$$(5.19)$$

where $w_i$, $s_i$ are the skew coefficients in the presence of TSV; $z_0$, $z_1$ are the skew coefficients in the absence of TSV; $\delta T_i$ and $\delta\sigma_i$ represent the temperature and stress gradients in the $i^{th}$ grid respectively; and $S_c$ is the temperature and stress independent coefficient of clock-skew.

To optimize the clock-skew, dummy TSVs can be inserted to reduce the temperature and stress gradients. Here the impacts of dummy TSV insertion on temperature and stress gradients and corresponding sensitivities are presented, followed by clock-skew sensitivity to dummy TSV density and nonlinear optimization for clock-skew reduction.

**Clock-skew Sensitivity**  The sensitivity of the clock-skew with respect to dummy TSV density plays an important role during the optimization. From (5.18), the sensitivity of the clock-skew in $i^{th}$ grid can be derived as follows

$$\frac{\partial S_i}{\partial \eta_i} = (S_{T,T}^i + S_{T,\sigma}^i) \cdot \frac{\partial T}{\partial \eta} + (S_{\sigma,\sigma}^i + S_{\sigma,T}^i) \cdot \frac{\partial \sigma}{\partial \eta}; \tag{5.20}$$

with

$$S_{T,T}^i = w_1 + 2w_2\delta T_i + 3w_3(\delta T_i)^2;$$

$$S_{T,\sigma}^i = (s_1 + 2s_2\delta T_i)\delta\sigma_i; \tag{5.21}$$

$$S_{\sigma,\sigma}^i = s_0; \quad S_{\sigma,T}^i = s_0 + s_1\delta T_i + s_2(\delta T_i)^2.$$

Here, $S_{T,T}^i$ and $S_{T,\sigma}^i$ represent the temperature and temperature-stress gradient coefficients for temperature sensitivity in the $i^{th}$ grid; $S_{\sigma,\sigma}^i$ and $S_{\sigma,T}^i$ represent the stress and stress-temperature gradient coefficients for stress sensitivity in the $i^{th}$ grid; and $s_i$, $w_i$ are the skew coefficients in the presence of TSV and is given in (5.18).

The clock-skew sensitivity depends on the temperature and stress gradient sensitivities, which are eventually related to the dummy TSV density, and can be determined from (5.15) and (5.17). Based on the clock-skew sensitivity, the updated temperature and stress based on the new dummy TSV density $\eta_i$, in the $i^{th}$ grid can be given by

$$T_i^{new} = T_i + \gamma_T^i P_i \eta_i; \quad \sigma_i^{new} = \sigma_i + \gamma_\sigma^i \eta_i \tag{5.22}$$

where $\gamma_T^i$ and $\gamma_\sigma^i$ are the temperature and stress gradient sensitivities in the $i^{th}$ grid with respect to the dummy TSV density, and are given as $\partial T_i/\partial\eta_i$ and $\partial\sigma_i/\partial\eta_i$ determined from (5.15) and (5.17), respectively.

Moreover, $T_i$, $\sigma_i$ represent the temperature and stress in the $i^{th}$ grid; and $P_i$, $\eta_i$ represents the heat power density and TSV density respectively in the $i^{th}$ grid. Based on the updated values of temperature and stress and by updating the TSV density, the skew and its sensitivity values in (5.18) and (5.20) can be updated.

**Nonlinear Optimization**  Clock-tree branch $B_k$ is a set of grids that branch $k$, passes through, $B_k = \{g_i | \text{branch } k \text{ passes } g_i\}$, with $g_i$ representing the $i^{th}$ grid. Therefore, the clock-skew of a clock-tree branch is the sum of skews from all the grids it passes through

$$\mathbf{S} = \sum_{i \in B_k} S_i \tag{5.23}$$

where $S_i$ represents the skew from $i^{th}$ grid among set $B_k$. Based on the derived sensitivity and skew function, a nonlinear optimization can be performed as follows.

Substituting (5.18), (5.20) and (5.22) into (5.23), the clock-tree branch skew $\mathbf{S}$ converts to a quadratic function of inserted dummy TSV density $\eta_i$.

By considering the clock-skew from each grid, one can represent clock-skew into a matrix form as

$$\mathbf{S} = c + f^T \mathbf{x} + \frac{1}{2}\mathbf{x}^T H \mathbf{x} \tag{5.24}$$

with

$$c = S_c; \quad f = \begin{pmatrix} f_0 \\ f_1 \\ \vdots \\ f_{(M \times N)} \end{pmatrix}; \quad \mathbf{x} = \begin{pmatrix} \eta_0 \\ \eta_1 \\ \vdots \\ \eta_{(M \times N)} \end{pmatrix};$$

$$f_i = \begin{cases} (S^i_{T,T} + S^i_{T,\sigma})\gamma^i_T + (S^i_{\sigma,\sigma} + S^i_{\sigma,T})\gamma^i_\sigma : & if \quad i \in B_k; \\ 0 : & else; \end{cases}$$

$$H = \begin{pmatrix} H_{0,0} & H_{0,0} & \cdots & H_{0,N_s} \\ H_{1,0} & H_{1,1} & \cdots & H_{1,N_s} \\ \vdots & \vdots & \ddots & \vdots \\ H_{(M \times N),0} & H_{(M \times N),2} & \cdots & H_{(M \times N),N_s} \end{pmatrix}$$

$$H_{i,j} = \begin{cases} (6w_3\delta T_j\gamma^j_T + 2w_2\gamma^j_T + s_1\gamma^j_\sigma + 2s_2\delta T_j \\ \quad + 2s_2\delta^i_\sigma\gamma^j_T + 2s_2\delta T_i\gamma^j_\sigma)\gamma^i_T \\ \quad + (s_1\gamma^j_T + 2s_2\delta T_i\gamma^j_T)\gamma^i_\sigma : & if \quad i,j \in B_k; \\ \\ 0 : & else. \end{cases}$$

Here, $c$ represents the zero-order coefficient of clock-skew; $f$, $H$ represent the linear and nonlinear coefficients of clock-skew respectively; $\mathbf{x}$ represents the dummy TSV density vector; $N_s$ represents the total number of sinks; $M \times N$ is the total number of grids; $S^i_{T,T}$, $S^i_{T,\sigma}$, $S^i_{\sigma,\sigma}$, $S^i_{\sigma,T}$ are the coefficients of clock-skew sensitivity in the $i^{th}$ grid given in (5.21); $\gamma^i_T$ and $\gamma^i_\sigma$ are temperature and stress gradient sensitivities in the $i^{th}$ grid given by (5.22); $P_i$ represents the thermal power in the $i^{th}$ branch; $B_k$ represents the clock-tree branch; $S_c$ is the coefficient of clock-skew defined in (5.19); $\delta T_i$, $\delta \sigma_i$ are the temperature, stress gradients in $i^{th}$ grid; and $w_i$, $s_i$ are the coefficients of clock-skew and is given in (5.18). Note that each column in $f$ vector and $H$ matrix represents the clock-skew sensitivity.

Since clock-skew is the difference in the delay between two clock-sinks, the problem thus becomes minimizing the skew variance over all the clock-tree branches $B_k$, i.e., to minimize variance of $\mathbf{S}$ in (5.24)

$$\min : f(S) = \frac{1}{N_s - 1} \sum_{k=1}^{N_s} (\mathbf{S} - \bar{S})^2 \qquad (5.25)$$

where $\mathbf{S}$ represent the clock-skew for $B_k$ clock-tree branches; and $\bar{S}$ represents the average skew of $\mathbf{S}$ for $N_s$ sinks.

As such, $f(S)$ becomes a quadratic function of $\mathbf{x}$, given by

$$\bar{S} = \frac{1}{N_s} \sum_{k=1}^{C} \mathbf{S} = \bar{c} + \bar{f}^T \mathbf{x} + \frac{1}{2} \mathbf{x}^T \bar{H} \mathbf{x}. \qquad (5.26)$$

Where $c$ and $\bar{c}$ represent the zero-order coefficient of clock-skew and its mean; $f$ and $\bar{f}$ represent the linear coefficient vector of clock-skew and its mean; $H$ and $\bar{H}$ represent the nonlinear coefficient matrix of clock-skew and its mean.

Substituting (5.26) in (5.25), the original problem can be rewritten as one polynomial function by

*Problem 2:*

$$\min : f(\mathbf{x}) = \frac{1}{N_s - 1} \sum_{k=1}^{N_s} \left( \widehat{c}^2 + 2\widehat{c}\widehat{f}^T \mathbf{x} + \mathbf{x}^T (\widehat{f}\widehat{f}^T + \widehat{c}\widehat{H}_k)\mathbf{x} + \widehat{f}^T \mathbf{x}^T \mathbf{x} \widehat{H}_k \mathbf{x} \right.$$

$$\left. + \frac{1}{4} \mathbf{x}^T \widehat{H}_k \mathbf{x} \mathbf{x}^T \widehat{H} \mathbf{x} \right) \qquad (5.27)$$

where $\widehat{c} = c - \bar{c}$, $\widehat{f} = f - \bar{f}$ and $\widehat{H} = H - \bar{H}$. All these values represent the deviations from their respective means.

Though the clock-skew depends on the TSV density matrix $\mathbf{x}$, neither too many TSVs nor too few TSVs can be inserted, because of design constraints below

$$lb \leq \mathbf{x} \leq ub \qquad (5.28)$$

where lower bound $lb$ is determined by a foundry process such as minimum metal density, and upper bound $ub$ is determined by the maximum allowed overhead with respect to area and signal routing.

**Conjugate Gradient Solving**    Now the objective to minimize the clock-skew becomes minimizing (5.27). This can be done by finding the optimum value of the dummy TSV density vector $\mathbf{x}$ that also satisfies constraints in (5.28). The conjugate gradient method with line-search in [133] can be an efficient method to solve this nonlinear equation with given constraints.

To remove the inequalities in Problem 2, the Karush-Kuhn-Tucker (KKT) optimization method along with the Lagrange penalty factor, $\xi$, is used to reformulate original problem by

*Problem 3:*

$$\min : f^*(\mathbf{x}) = f(\mathbf{x}) + \xi h^2(\mathbf{x}) \tag{5.29}$$

with

$$h(\mathbf{x}) = \begin{cases} 0, & lb \leq \mathbf{x} \leq ub \\ \varphi \gg 0, & otherwise \end{cases} \tag{5.30}$$

where $f^*(\mathbf{x})$ is the objective function by considering boundary conditions, i.e., removing inequalities of $f(\mathbf{x})$; $\xi$ is the Lagrange penalty factor, determined as a stationary point of $f(\mathbf{x})$; and $\varphi$ is a weighting parameter.

The conjugate gradient method iteratively searches for the value of $\mathbf{x}$ that can minimize $f^*(\mathbf{x})$ along the search gradient vector, $g_k$, which points to the direction where the greatest rate of variation of objective function $f^*(\mathbf{x})$ lies. To achieve a converged solution, at iteration $k$, the search direction vector $d_k$, which indicates the direction where the objective variable has to be varied, moves in a negative gradient direction to minimize the variation. Thus, a new search direction vector $d_{k+1}$ can be obtained by linear addition of the previous search direction vector $d_k$ with the current negative search gradient vector $g_k$. Therefore, the next search direction vector becomes

$$d_{k+1} = -\nabla f^*(\mathbf{x}_k)^T + \frac{g_{k+1}^T g_{k+1}}{g_k^T g_k} d_k; \quad g_k = -\nabla f^*(\mathbf{x}_k). \tag{5.31}$$

Where $d_{k+1}$ is the search direction vector; and $g_k$, $g_k^T$ are search gradient vectors of $\mathbf{x}_k$ and its transpose, determined from the slope of the search vector.

Based on the search direction vector $d_k$, and gradient $g_k$, the optimal value for step-size $\alpha_k$ is decided on to minimize the function $f^*(\mathbf{x}_k + \alpha_k d_k)$. The new vector $\mathbf{x}$ can be updated based on the previous search direction vector $d_k$ as

$$\mathbf{x}_{k+1} = \mathbf{x}_k + \alpha_k d_k; \quad \alpha_k = \frac{g_k^T g_k}{g_k^T f(\mathbf{x}) g_k}. \tag{5.32}$$

This iterative search for the optimum value of TSV density vector $\mathbf{x}$ stops when the difference in successive approximations of $\mathbf{x}_k$ reaches a certain threshold. To perform with faster convergence and avoid the local minimum, the problem is solved by starting with some randomly generated approximation of $x_0$. Once the final value for $x$ is reached, it can satisfy (5.29) and density constraints based on the stress and temperature gradients, leading to the reduction of the clock-skew. The nonlinear optimization of clock-skew reduction for 3D clock-tree design by the insertion of dummy TSVs is presented below.

With the help of 3D-ACME [134], a 3D-IC thermal simulator based on Hotspot [135], the temperature distribution at each layer can be obtained. Moreover, average temperatures based on SPEC 2000 [136] benchmarks are taken to avoid an application specific temperature distribution [137].

As the clock-skew is the difference in delay between two sinks, the variation of clock-skew with TSVs of one 3D H-tree is shown in Figure 5.7. It can be noted from Figure 5.7, that the initial clock-skew without insertion of dummy TSVs is higher than the clock-skew after insertion of dummy TSVs. For example, for a row-grid 16 and column-grid 24, having dummy TSVs inserted results in a clock-skew reduction of nearly 17*ps*. To reduce the area overhead caused by dummy TSVs, the dummy TSV insertion density is limited to 7% of the total area.

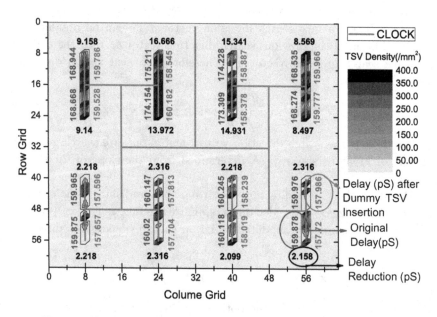

Fig. 5.7   Clock-skew reduction before and after insertion of dummy TSVs.

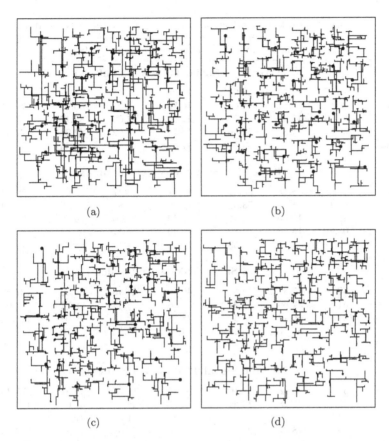

(a)        (b)

(c)        (d)

Fig. 5.8    3D clock-tree after insertion of dummy TSV (Black dots) with balanced clock-skews for: (a) Tier 0; (b) Tier 1; (c) Tier 2; (d) Tier 3.

Additionally, a 3D clock-tree from the IBM benchmark $r5$ [138] is studied. The 3D clock-tree after the insertion of dummy TSVs is shown in Figure 5.8 with TSVs in each layer indicated by solid dots. Dummy TSV insertion is performed under nonlinear optimization as discussed in this section. One can observe that a larger number of TSVs are inserted in the top layer, i.e., tier 0, compared to the other layers since the top layer is the farthest one from the heat-sink.

Finally, the comparison of clock-skew for different benchmarks before and after the insertion of dummy TSVs is given in Table 5.3 with a detailed summary, which shows the impact of the 3D electrical-thermal-mechanical coupled delay model and also the insertion of dummy TSVs to reduce

Table 5.3   3D clock-skew reduction by linear and nonlinear delay models

| Type | Orig | Lin | Impr% | Time (s) | Nonlin | Impr% | Time(s) |
|------|------|-----|-------|----------|--------|-------|---------|
| **HTree1 (14 Signal TSVs & 63 buffers)** | | | | | | | |
| T2 | 15.39 | 9.54 | 38.01 | 16.98 | 2.45 | 84.08 | 76.15 |
| T4 | 26.62 | 8.07 | 69.68 | 16.85 | 3.92 | 85.27 | 77.11 |
| T8 | 47.41 | 11.38 | 76.00 | 17.05 | 7.03 | 85.17 | 77.19 |
| T10 | 58.64 | 14.8 | 74.76 | 17.22 | 10.02 | 82.91 | 77.36 |
| *Mean* | — | — | 64.61 | 17.02 | — | 84.36 | 76.95 |
| **HTree1 (14 Signal TSVs & 63 buffers)** | | | | | | | |
| T2 | 23.55 | 8.15 | 65.39 | 17.41 | 3.31 | 85.94 | 79.02 |
| T4 | 44.13 | 11.61 | 73.69 | 17.51 | 4.81 | 89.10 | 79.59 |
| T8 | 82.00 | 14.60 | 82.20 | 17.40 | 9.00 | 89.02 | 79.98 |
| T10 | 103.70 | 15.50 | 82.80 | 17.49 | 9.51 | 90.83 | 79.32 |
| *Mean* | — | — | 76.58 | 17.45 | — | 88.72 | 79.48 |
| **r1 (45 Signal TSVs & 202 buffers)** | | | | | | | |
| T2 | 30.50 | 17.48 | 42.69 | 49.80 | 14.46 | 52.59 | 127.29 |
| T4 | 61.87 | 32.66 | 47.21 | 35.16 | 25.55 | 58.71 | 164.01 |
| T8 | 121.10 | 66.14 | 45.39 | 38.28 | 53.18 | 56.08 | 182.71 |
| T10 | 152.7 | 83.56 | 45.28 | 44.64 | 69.17 | 54.70 | 197.49 |
| *Mean* | — | — | 45.14 | 42.00 | — | 55.52 | 167.88 |
| **r2 (60 Signal TSVs & 365 buffers)** | | | | | | | |
| T2 | 35.13 | 23.48 | 33.16 | 160.80 | 19.33 | 44.97 | 475.36 |
| T4 | 69.75 | 46.28 | 33.65 | 123.36 | 34.17 | 51.00 | 485.19 |
| T8 | 134.70 | 86.89 | 35.49 | 127.80 | 71.01 | 47.28 | 490.58 |
| T10 | 169.00 | 118.89 | 29.65 | 167.16 | 88.08 | 47.88 | 513.43 |
| *Mean* | — | — | 32.99 | 144.84 | — | 55.52 | 491.14 |
| **r3 (75 Signal TSVs & 515 buffers)** | | | | | | | |
| T2 | 32.36 | 20.19 | 37.60 | 264.84 | 19.21 | 40.63 | 751.53 |
| T4 | 64.80 | 39.97 | 38.32 | 204.96 | 31.17 | 51.89 | 682.90 |
| T8 | 125.60 | 79.86 | 36.41 | 213.36 | 60.71 | 51.67 | 658.80 |
| T10 | 157.70 | 95.47 | 39.46 | 277.20 | 82.80 | 47.50 | 708.39 |
| *Mean* | — | — | 37.95 | 240.12 | — | 47.92 | 700.40 |
| **r4 (90 Signal TSVs & 604 buffers)** | | | | | | | |
| T2 | 31.68 | 17.97 | 43.28 | 254.16 | 16.53 | 47.83 | 912.39 |
| T4 | 64.57 | 36.21 | 43.92 | 278.64 | 26.13 | 59.53 | 844.46 |
| T8 | 126.80 | 70.61 | 44.31 | 280.80 | 64.50 | 49.13 | 898.21 |
| T10 | 159.80 | 88.81 | 44.42 | 393.12 | 76.83 | 51.92 | 909.25 |
| *Mean* | — | — | 43.95 | 301.44 | — | 52.10 | 891.08 |

(*Continued*)

Table 5.3   (*Continued*)

| Type | Orig | Lin | Impr% | Time (s) | Nonlin | Impr% | Time(s) |
|------|------|-----|-------|----------|--------|-------|---------|
| | | | **r5 (90 Signal TSVs & 1479 buffers)** | | | | |
| T2 | 35.00 | 21.00 | 40.00 | 798.72 | 18.90 | 46.00 | 1992.60 |
| T4 | 68.40 | 38.40 | 43.86 | 834.00 | 28.50 | 58.33 | 1843.77 |
| T8 | 131.00 | 74.60 | 43.05 | 870.24 | 63.80 | 51.30 | 1705.82 |
| T10 | 164.10 | 93.40 | 43.08 | 1126.20 | 79.43 | 51.60 | 1934.662 |
| *Mean* | — | — | 42.08 | 907.32 | — | 51.81 | 1869.21 |
| *Overall* | — | — | 49.10 | — | — | 61.30 | — |

gradient. Clock-skew values are reported in pico-seconds and runtime in seconds. The delay models with consideration of nonlinear electrical-thermal-mechanical-thermal impacts result in clock-skew reduction by 61.3% on average, listed under *nonlin* column, compared to clock-skew without insertion of dummy TSVs, listed under *Orig* column. Note that the reduced clock-skew by linear modeling is listed under *Lin* column with 49.1% clock-skew reduction compared to *Orig*. The runtime on average for nonlinear optimization is 611$s$ and 238$s$ for linear optimization.

## 5.4   Summary

Physical design for power/ground design and skew reduction in a 3D IC is studied in this chapter. Firstly, a TSV insertion problem for the power/ground design is formulated. With the use of sensitivity analysis, an efficient via allocation simultaneously driven by power and thermal integrity is developed. Experiments show that compared to sequential power and thermal optimization using static integrity, sequential optimization using dynamic integrity reduces non-signal vias by up to 18%, and simultaneous optimization using dynamic integrity further reduces non-signal vias by up to 45.5%. Further, considering a clock-tree design as an example, a dummy TSV insertion problem is formulated to optimize the clock-skew. The clock-tree skew is optimized by the nonlinear optimized dummy TSV insertion which is solved by the conjugate-gradient method. A number of 3D clock-tree benchmarks are used to verify the model and also the optimization. Simulation results have shown a reduction of clock-skew by 61.3% on average when the nonlinear electrical-thermal-mechanical delay model developed in the previous chapters is applied.

# Chapter 6

# Testing

## 6.1 Introduction

With more vertical through-silicon via (TSV) interconnects, 3D IC can provide high bandwidth with low power for memory-logic integration [13, 24]. However, 3D IC has limited yield due to the complexity of the TSV manufacturing process. The TSV may be shorted to substrate due to the pin hole, or open because of the micro-void or partial filling [139]. The reliability of the TSV is very important for a successful 3D IC. It is unacceptable to let the whole chip fail because of few broken TSVs. In order to increase the reliability of TSVs, various redundancy techniques are proposed. Redundant TSV insertion can be deployed to increase the yield when faulty TSVs are identified. Use of one spare TSV for each signal TSV largely increases the reliability, but it requires significantly increased area overhead. Therefore, one can imagine the great challenge of reducing redundant TSVs numbers without reducing reliability. An efficient TSV test methodology is thereby required to locate the defective TSV from a large volume of TSV interconnects.

Due to the stacking nature of 3D IC, pre-bond TSV test is needed for every layer before stacking. TSV testing is difficult due to the mismatch between TSV and probe pitch. Pre-bond data can only be collected from specially designed pads, as shown in Figure 6.1(a). The probe pad provides high alignment speed and contact stability. However, only a limited number of these pads are available for test. Moreover, post-bond test is required to detect the defects in the bonding process. After bonding, only the bottom die can communicate with the tester. As shown in Figure 6.1(b), the data has to transfer from the bottom die to the upper die through the TSV elevator as a functional TSV may not have enough I/O drivability [140]. As 3D

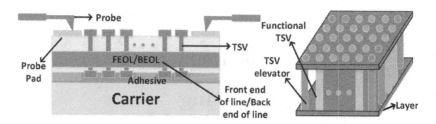

Fig. 6.1    (a) Probing on extra DFT pads in pre-bond test (b) functional TSV and TSV elevator in post-bond test.

stacking is mainly applied for a high-volume I/O circuit, one such type of chip can have a TSV density of $10,000/mm^2$ or more [141, 142]. As such, it poses a grand challenge on the test time for TSV pre-bond and post-bond test as the bandwidth of TSV test data becomes the primary bottleneck [143]. To reduce the test time, a more efficient method of TSV test data compression with maintained fault detection information is needed.

Compressive sensing is a promising technology for efficient signal acquisition, with reconstruction of data done by finding the most sparse solution to undetermined linear systems [144]. For the TSV test, signal (or test data) is sparse in nature as only a $1 - yield$ portion of data is nonzero if we compare the result to the expected outcome using XOR operation. This sparse signal can be compressed by multiplying a matrix to project data from high dimension to low dimension. Reconstruction can be achieved by finding the sparsest solution under determined equations. Error may come from an overly-high compression rate or failing to estimate the sparsity of the signal. In this chapter, a compressive sensing–based data compression for pre- and post- bond testing is presented.

## 6.2    3D IC Test

Here, we present the 3D IC test architecture for the pre-bond and post-bond TSV test with the probe pad and TSV elevator. Moreover, the problem of the output data compression is formulated based on this test architecture.

### 6.2.1    *System Architecture*

The on-chip TSV test vehicle architecture similar to that in [140] is considered. The adapted setup can support the pre-bond die test, post-bond

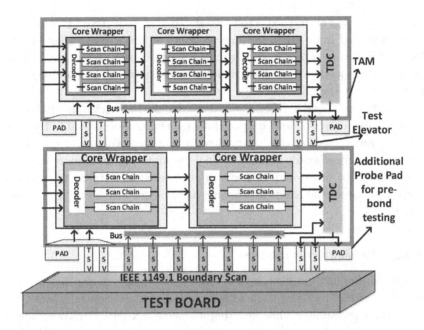

Fig. 6.2   TSV test vehicle with compressive sensing based testing circuits.

stack test and the board level interconnect test. The pre-bond test is mainly designed to detect TSV manufacturing faults; while the post-bond test is the IC test not only for the TSV interconnect test but also for the scan-chain and functional test. As shown in Figure 6.2, the test wrapper will provide the test access mechanism (TAM) and send/receive test signals to/from I/Os of external stack. Test data will be injected from the probe pad for the pre-bond test or from bottom die to upper die through the TSV elevator for the post-bond test. Decoder will decode the input test data and function as automatic test pattern generator (ATPG) to cover the sequential and combinational circuits. After the scan chain or from the bus, output data is compressed by the test data compaction (TDC) block and further transported down from the upper die to the bottom, eventually being collected by the tester or probe pad. The expected output data is known and can be sent to the die under test through the probe pad or TSV elevator and further stored in the test wrapper register.

For pre-bond TSV test, TSV groups are formed based on the pitch of probe head such that the probe can contact the whole TSV group at once. The faulty TSV interconnect can be detected by the difference between the received signal and expected signal. To test each TSV [141] is time consuming and impractical for large volumes of TSVs. The target of pre-bond TSV output compression is to minimize the output bits while be able to recover losslessly the original signal to save test time. The probe pad is used to collect the pre-bond TSV test data.

For the post-bond TSV test, as there are only TSV elevators and probe pads to provide input and output data for the non-bottom die, the test time is constrained by the bandwidth given a large volume of test data. Moreover, increasing the number of TSV elevators will incur more die area and reduce the functional TSV densities. Therefore, data compression is needed to reduce the bandwidth requirement of the TSV elevator and save test time.

### 6.2.2  *Problem Formulation*

The main proposed problem here is to fully utilize the compressive sensing to compress the TSV test data with high data compression rate under the given error probability while ensuring a lossless recovery. Here, we assume that the TSV failure probability is known as prior knowledge and the size of the TSV test data is large enough for the lossless signal recovery. As such, the TSV output test data compression problem can be formulated.

*Problem: Find minimum output bits OB to locate faulty TSV/IC index $E_{fault} \in R^N$*

$$Min. < OB = MLog2(Max(Y)) >$$
$$S.T.(i)\ X_r = (I + E_{fault})X_e$$
$$(ii)\ Y = \Phi(X_r - X_e)$$
$$(iii)\ \|E_{fault}\|_0 \leq K$$

(6.1)

where $X_r \in R^N$ and $X_e \in R^N$ denote the received and expected signals through TSV for pre-bond test or scan chain for post-bond test and $Y \in R^M$ is the output result. $\Phi \in R^{M \times N}$ is the compressive sensing matrix. $E_{fault} \in R^N$ are defective TSV indexes in the pre-bond test, while for the post-bond test, $E_{fault}$ represents the error bits introduced by faulty ICs. Sparsity $K$ represents the maximum number of non-zero values in $E_{fault}$, which can be estimated by the TSV failure probability or faulty IC probability. The minimum output bits thereby mean that when given the compressed result $Y$ and sensing matrix $\Phi$, we can losslessly recover $E_{fault}$ such that no TSV test data information is lost. By making use of sparsity of $E_{fault}$, an unique sparse solution can be found for the undetermined linear system [144, 145].

## 6.3 Compressive Sensing and Recovery of Testing Data

In this section, we present compressive sensing and the recovery of test data. We discuss the sparsity of the test data which is the foundation to perform compressive sensing. Moreover, orthogonal matching pursuit is introduced here to solve undetermined equations.

### 6.3.1 *Sparsity of Test Data*

Lossless data compaction such as length-run (LR) coding and Golomb coding (GLC) can be used for data compression. However, the complexity increases with the increase in data volume, which is quite impossible to be deployed in TSV test. Compressive sensing is to acquire and recover a sparse signal (or data) using least number of samples with the help of an incoherent projection basis. For a common TSV yield such as a 99% under pre-bond test, there are 99% interconnects functioning properly. The defects are sparse in this sense indicating that the difference between the received and expected outputs are sparse with only a 1% non-zero value at the locations of faulty TSV interconnects. Similar to the pre-bond test, the post-bond functional test output data can also be sparse by taking the difference between the received and expected data [146]. In order to know the sparsity, we need to estimate the TSV yield $Y_{pre}$ and IC fault-free probability $Y_{post}$. Figure 6.3 shows the difference between the expected and received test data by sending signal 1 through 1024 TSVs under test, where the difference is sparse.

Similar to [147, 148], we assume a uniform failure rate $p$ for a number of TSVs under test following binomial distribution. The overall probability of having $x$ defective TSVs is

$$Y_{pre} = C_{N_{TSV}}^x p^x (1-p)^{N_{TSV}-x} \tag{6.2}$$

where $N_{TSV}$ is the number of TSVs. The yield of pre-bond TSV can be calculated as $x = 0$. Similarly for the post-bond TSV test, the fault-free IC

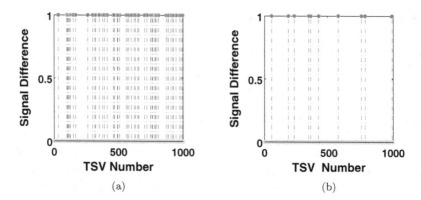

Fig. 6.3    (a) Sparse data for yield ($Y_{pre}$) of 95% (b) sparse data for yield ($Y_{pre}$) of 99%.

probability can be estimated as below using Poisson distribution [149].

$$Y_{post} = \frac{(N_{com} * P_{com})^{N_{fault}}}{N_{fault}!} e^{-N_{com}*P_{com}} \tag{6.3}$$

where $N_{faulty}$ is the number of faulty components. $N_{com}$ and $P_{com}$ are the number of components and the probability of them being faulty components respectively. The fault IC probability is $P_{post} = 1 - Y_{post}(N_{fault} = 0)$.

For post-bond, due to the application of ATPG and different functional test algorithms, the test output may not directly reflect the faulty component location. Therefore, the clustering effect is not considered. As the affected output bits due to faulty IC components depend on the testing algorithm, we assume the affected output signal (error bit probability) is proportional to faulty component probability. For pre-bond TSV test, the TSV testing data is directly related to the faulty TSV location. The clustering effect is considered and there exists a spatial correlation between defective TSVs. This indicates that the presence of a defective TSV increases the probability of a defective TSV nearby. Based on [147, 148], the probability of defect for the $i$-th TSV $P_i$ can be modeled as below.

$$P_i = P \left( 1 + \sum_{j=1}^{N_c} (1/d_{ic})^\alpha \right) \tag{6.4}$$

where $P$ is the single TSV failure rate, $d_{ic}$ is the distance between the $TSV_i$ and cluster center, and $\alpha$ is the clustering coefficients indicating cluster extent. A large $\alpha$ indicates higher clustering effect. In our simulations, we assume the cluster center is injected randomly and only forms a proportion of the total defective TSV number. The rest of the failure TSVs are generated with the combination of failure probability and clustering effect as mentioned in (6.4). The test data compression rate $N_C$ for both pre-bond and post-bond circuit can be defined as follows.

$$N_C = 1 - \frac{M * log_2 EC_{max}}{N} \tag{6.5}$$

where $EC_{max}$ is maximum value of the encoded number, and $N$ is the original data size. We select the maximum code number to ease the decoding after receiving them from the probe pad or tester.

### 6.3.2 Lossless Compression and Recovery

The lossless recovery can be formulated as a $L_0$ norm minimization problem given below

$$\begin{aligned} &\operatorname{argmin}_{x \in R^N} \quad \|E_{fault}\|_0 \\ &\text{subject to} \quad Y = \Phi E_{fault} \end{aligned} \tag{6.6}$$

where $E_{fault}$ is a N dimensional sparse signal $(E_{fault} \in R^N)$, $\Phi$ is the sensing matrix $(\Phi \in R^{M \times N})$, and $Y$ is the compressed data in low dimension $(Y \in R^M$ and $M \ll N)$.

Note the solution of $L_0$ norm is equivalent to $L_1$ norm with overwhelming probability [144]. This can be explained in a two-dimension example as Figure 6.4, which shows that $L_1$ norm provides the exact solution of $L_0$ problem; while $L_2$ gives a completely different solution.

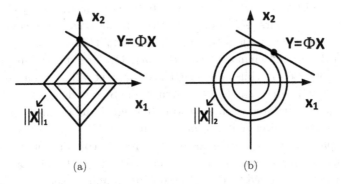

Fig. 6.4 (a) $L_1$ norm solution in 2D example (b) $L_2$ norm solution in 2D example.

To ensure a successful recovery without loss, the sensing (or sampling) matrix must satisfy the restricted isometry property (RIP) [144]. Here, we use a random Bernoulli matrix that can be implemented from a pseudo number generator in hardware with well recognized RIP property. The most attractive part of the compressive sensing is that it can sample the data based on its sparsity: the higher the sparsity, the better the compression rate. For the on-chip TSV test vehicles presented in the previous section, we can implement a counter to record non-zero values for lossless data compression and reconstruction. The minimum required sampled data [144] is

$$M = O(Klog(N/K)) \qquad (6.7)$$

where $M$ is the output of test data in measurements, $K$ is the sparsity of data or nonzero entries that can be estimated as (6.8), and $N$ is the total length of the signal. $N$ has a relationship with number of TSV $N_{TSV}$ by $N = N_{TSV} * N_{data}$, where $N_{data}$ is the number signal sent from each TSV. From (6.7), we can sample the signal based on its contents, which are the nonzero entries. As such, the sparsity $K$ can be estimated as below.

$$K = Ceil(N - NY_{pre}) \qquad (6.8)$$

where $Y_{pre}$ is the yield of the pre-bond TSV test. If we replace $Y_{pre}$ by $Y_{post}$, the sparsity $K$ for the post-bond TSV test can also be estimated.

### 6.3.3   *OMP Solver*

As discussed in Section 6.3.1, the $L_0$ norm solution can be applied to solve (6.6). We deploy the Orthogonal Matching Pursuit (OMP) solver for the $L_0$ norm solution, which is a heuristic solver based on the greedy algorithm to find the most sparse solution [150]. More details on signal recovery from OMP are provided in Algorithm 6.1. The residual is initialized as $E_{fault}$. Index set $\Lambda_0$ and chosen matrix $\Phi_0$ are empty. The largest correlated column is found from Step 4, while index and chosen matrix will be updated as Step 5. The new estimated signal is reconstructed in Step 6 via $L_2$ minimization. The residual is updated from the estimated signal and original signal. The iteration will stop after $K$ iterations. In summary, OMP performs two functions as follows. Firstly, it finds the most correlated column from the sensing matrix by comparing simple dot multiplication. Secondly, the largest correlated column is added to the selected column and by solving a $L_2$ norm minimization, the most fitted new signal is generated. This procedure will

---

**Algorithm 6.1** Orthogonal Matching Pursuit Algorithm

---

**Require:** An MxN Sensing Matrix $\Phi = [\varphi_1, \varphi_2, ..., \varphi_N]$, an M-dimensional data vector $Y$ and yield $Y_{yield}$

**Ensure:** An estimate $E_{faulty}$

1: Initialize the residual $r_0 = E_{faulty}$, the index set $\Lambda_0 = \emptyset$, $\Phi_0 = \emptyset$ and iteration counter $t = 1$.

2: Calculate sparsity $K = Ceil(N - NY_{yield})$

3: WHILE $(t \leq K)$

4: Find column index $\lambda_t$ of $\Phi$ correlates $Y$ most as below
   $\lambda_t = argmax_{j=1,...N} | < r_{t-1}, \varphi_j > |$,

5: Update column index set and matrix of chosen columns
   $\Lambda_t = \Lambda_{t-1} \cup \lambda_t$
   $\Phi_t = [\Phi_{t-1} \ \varphi_{\lambda_t}]$

6: Solve a least squares problem to obtain new signal
   $x_t = argmin|| Y\text{-}\Phi_t \ x ||_2$

7: Calculate the new approximation and residual
   $a_t = \Phi_t x_t, \ r_t = Y - a_t$

8: End While

9: $E_{faulty} = x_t$

---

repeat $K$ times to find the expected signal. Note that $K$ is the sparsity of the signal, which can be estimated from the $Y_{pre}$ and $Y_{post}$.

## 6.4  TSV Testing

In this section, we discuss the hardware implementation of compressive sensing and its applications in the pre-bond and post-bond TSV test.

### 6.4.1  *Testing Circuit*

To perform the data compression using the proposed algorithm, outputs from the scan chain and the probe pad are provided as inputs for the TDC block, as shown in Figure 6.2. The TDC block can be implemented using adders and XOR gates as shown in Figure 6.5. The scan chain output and the expected output from the probe pad are XORed to obtain the difference matrix, $E_{faulty}$, which is normally sparse, with 1 to denote the error/failure. The Bernoulli function is realized using a linear feedback shift register (LFSR) with $M$ measurements collected after performing $M$ shifts. Here, $M$ is the row number of sensing matrix $\Phi$. Furthermore, the Bernoulli matrix is multiplied with the difference matrix using AND gates, where bits are added using the adder and fed to the probe pad. This implementation

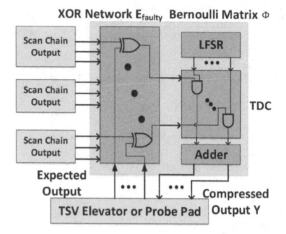

Fig. 6.5    Compressive sensing based testing circuit diagram for output data compression.

reduces the number of outputs from $N$ measurements to $M$ measurements. Note that the above mentioned TDC implementation is suitable for both the pre-bond and post-bond TSV tests.

We present the compressed output data of TSVs under different yields and the corresponding reconstructed data. The experiment is performed for 1024 number of TSVs with input 1 for each TSV to verify whether 1 is received. A faulty TSV is inserted based on (6.2). As an example, we collected 200 output measurements and plotted the adder output under different yields in Figure 6.6. From Figure 6.6(a), one can observe that the adder results are smaller for TSVs with 99% yield, compared to the adder results shown in Figure 6.6(b) for TSVs with yield 95%. It indicates that a higher yield data will require less bits for encoding, which results in a higher compression rate as in (6.5).

Furthermore, the minimum number of required measurements under different yields to achieve a lossless recovery is presented in Figure 6.7. As illustrated in Figure 6.7(a), the difference between the recovered and original data decreases dramatically as the measurements ($M$) increase, as expected from the compressive sensing theory [144], indicating a least number of measurements is required for lossless compression. As such, the higher the yield is, the lesser number of measurements ($M$) the required. For example, when the yield is as high as 99%, there are nearly 60 measurements good enough to fully recover the test data whereas nearly 190 measurements are needed to fully recover the test data for a yield of 95%. For lossless

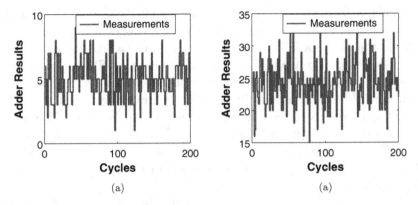

Fig. 6.6  (a) Compressed output for yield of 99% (b) compressed output for yield of 95%.

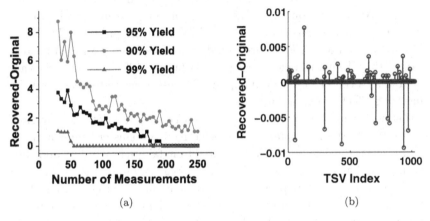

Fig. 6.7  (a) Maximum difference vs number of output (b) signal reconstruction with 60 measurements for yield of 99%.

reconstruction, the maximum difference between the recovered and original data should be sufficiently small. As Figure 6.7(b) shows, for 99% yield, the maximum difference between the recovered and original data is as small as $E^{-2}$ for 60 measurements to represent 1024 TSVs, corresponding to a 82.42% data compression rate $(1 - 60 \times 3/1024)$.

## 6.4.2  *Pre-bond TSV Test*

Short or open defects will lead to the receiving of incorrect data, which can be used to detect faulty TSVs after comparing to the expected data. In general, the size of probe heads is large compared to the size of TSVs. Hence, the probe head will contact a group of TSVs instead of each individual TSV. Among the group of TSVs, only a few can have the I/O driving ability, which can be used to input the test data. As multiple TSVs are connected to a probe head, it is important to differentiate each TSV at the receiving end. Similar to [151], a scan flip-flop (SF) controlled by a digital enable circuit is utilized to differentiate the outputs from the TSVs. This data output from the probe head is provided as the input to the XOR network of TDC (in Figure 6.5), and then the compressed output $Y$ can be collected. The original data $E_{fault}$ can be recovered from $Y$ based on the proposed Algorithm 6.1.

For the pre-bond and post-bond TSV tests, the proposed test data compression method is applied for ISCAS-85 benchmarks [152] in Verilog with the test pattern generated using Mintest [153], which provides 100% fault coverage. An 8-bit output signal after the scan test is assumed and error bit probability modeled based on (6.3) for TSVs. For the pre-bond TSV test, 4096, 16384 and 65536 TSVs are tested with a scan signal and a 3-bit test output [151]. To model defective TSV distribution, 10% defective TSVs are inserted as a center in a TSV map; and the rest is generated based on the clustering effect presented in (6.4).

Let us discuss the pre-bond test and its according test data compression. In Figure 6.8, the red square is the defective TSV cluster center and the black circle is the defective TSV generated from failure probability and clustering effect based on (6.4). The X-axis and Y-axis represent the location of TSV. The average failure probability is 20% and due to the clustering effect, the failure probability can be as high as 81.52% for the TSVs close to the center as Figure 6.8(c). Table 6.1 shows the output data compression rate for different clustering factor $\alpha$ and the number of TSVs. As the experiment set-up states, a 3-bit signal will be collected after scan test. A compression of nearly 89% is achieved for 4096 TSVs with a failure rate of 0.5%, but is reduced to nearly 66% with a failure rate of 1%. This indicates that more samples are needed for lossless reconstruction when the error probability increases. It also shows that, even with the clustering effect, if the yield has not changed, the compression rate will be maintained as almost the same.

In addition, as shown in Table 6.1, we compare our compression algorithm with length-run (LR) coding and Golomb coding (GLC) based compression algorithms. Note that Golomb coding is greatly affected by the tunable group size D. For example, if we consider 16384 TSV cases

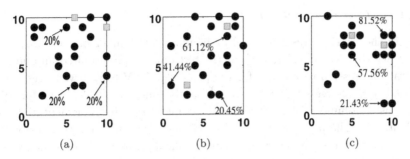

Fig. 6.8 (a) No clustering effect (b) clustering effect with $\alpha = 1$ (c) clustering effect with $\alpha = 2$.

Table 6.1    Test data compression in pre-bond test.

| TSV No. | Failure Prob. | Cluster | Proposed | LR Coding | GLC D = 8 | GLC D = 16 |
|---|---|---|---|---|---|---|
| 4096 | 0.5 % | $\alpha = 0$ | 89.45% | 81.80% | 76.83% | 79.98% |
| | | $\alpha = 1$ | 89.70% | 81.14% | 77.29% | 80.18% |
| | | $\alpha = 2$ | 89.32% | 81.52% | 76.68% | 79.76% |
| | 1% | $\alpha = 0$ | 65.29% | 50.99% | 35.57% | 44.38% |
| | | $\alpha = 1$ | 65.03% | 50.73% | 35.57% | 43.80% |
| | | $\alpha = 2$ | 66.48% | 51.52% | 37.16% | 45.85% |
| 16384 | 0.5% | $\alpha = 0$ | 89.16% | 80.37% | 75.95% | 79.11% |
| | | $\alpha = 1$ | 89.23% | 80.15% | 73.99% | 77.28% |
| | | $\alpha = 2$ | 89.35% | 80.47% | 75.59% | 78.89% |
| | 1% | $\alpha = 0$ | 64.80% | 49.42% | 34.08% | 43.49% |
| | | $\alpha = 1$ | 64.29% | 46.93% | 29.49% | 38.37% |
| | | $\alpha = 2$ | 64.86% | 50.62% | 34.51% | 43.84% |
| 65536 | 0.5% | $\alpha = 0$ | 89.17% | 79.77% | 73.48% | 76.86% |
| | | $\alpha = 1$ | 89.21% | 79.73% | 73.09% | 76.63% |
| | | $\alpha = 2$ | 89.24% | 79.38% | 73.49% | 76.86% |
| | 1% | $\alpha = 0$ | 65.32% | 50.03% | 34.27% | 43.35% |
| | | $\alpha = 1$ | 64.59% | 48.57% | 34.19% | 43.41% |
| | | $\alpha = 2$ | 64.88% | 47.08% | 36.68% | 40.21% |

with a failure probability of 1% and clustering factor ($\alpha$) of 1, our proposed algorithm can successfully compress 64.29% of samples whereas LR can only compress 46.93%; GLC with $D = 8, 16$ can only compress 29.49% and 38.37%, respectively. With the increase in $D$ value, compression rate increases, but at the cost of extra hardware and time.

### 6.4.3    *Post-bond TSV Test*

As in the previous discussion, after bonding the number of I/O pins, the bandwidth available [154, 155] for the test data is quite limited and hence data compression is required for the post-bond TSV test as well. The proposed compressive sensing–based test can be efficiently scaled even for the increased volume of data with its according hardware implementation in Figure 6.5. The post-bond TSV test refers to the functional test verifying the circuit operations according to its specifications under input conditions.

Figure 6.9(a) shows an example of a sequential circuit under the post-bond TSV test. There are 3 locations to be considered as being stuck at zero or one fault. The ATPG algorithm will determine a minimum set of vectors to cover sufficient faults. Figure 6.9(b) shows the input vector generated based on D-ATPG algorithm to detect the faults. D-ATPG use D value to indicate whether a circuit is good ($D = 1$) or faulty ($D = 0$). Expected value is always $D = 1$. As such, we can XOR the expected value and the actual result to obtain the difference $E_{fault}$, which is sparse have as we, discussed earlier. Through the proposed TDC hardware, the compressed output $Y$ is collected. The original output $E_{fault}$ can be recovered from $Y$ based on the proposed Algorithm 6.1.

For the post-bond TSV test, we assume 5% and 10% probabilities of faulty IC for an 8-bit output signal (signature), which mean 0.639% and 1.308% error probabilities for each bit. Similar to the pre-bond test, we compare our proposed compression algorithm with LR and GLC coding–based compression algorithms, and is presented in Table 6.2. It also shows that the data compression rate outperforms length-run coding and Golomb coding. The output compression for 5% error probability varies from 80.03% to 88.18%, but 74.27% to 68.26% for LR coding, and 76.42% to 72.49% for GLC with $D = 16$. For circuit $c7552$, our proposed algorithm has a 4.50%

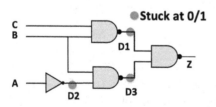

| A | B | C | Z | Exp. Z | Fault |
|---|---|---|---|---|---|
| 1 | 0 | 0 | $\overline{D1}$ | 0 | Stuck at 0 |
| 0 | 1 | 0 | D2 | 1 | |
| X | 0 | X | $\overline{D3}$ | 0 | |
| 1 | 1 | 1 | D1 | 1 | Stuck at 1 |
| 1 | 1 | 0 | $\overline{D2}$ | 0 | |
| 0 | 1 | 0 | D3 | 1 | |

Fig. 6.9   Example of a sequential circuit with input vector for D-ATPG algorithm.

Table 6.2   Test data compression in post-bond test.

| Error Prob. | Benchmark | Output (bits) | Proposed | LR Coding | GLC D = 8 | GLC D = 16 |
|---|---|---|---|---|---|---|
| 5% | c499 | 1696 | 88.15% | 78.22% | 73.01% | 76.42% |
| | c432 | 196 | 88.18% | 77.96% | 76.38% | 78.57% |
| | c1908 | 2475 | 86.89% | 73.69% | 72.83% | 76.34% |
| | c2670 | 6300 | 82.56% | 73.81% | 69.77% | 73.74% |
| | c3540 | 1870 | 87.76% | 80.55% | 75.75% | 78.72% |
| | c5315 | 4674 | 81.80% | 75.31% | 71.13% | 74.97% |
| | c6288 | 416 | 85.65% | 81.15% | 82.21% | 84.13% |
| | c7552 | 7992 | 80.30% | 74.27% | 68.26% | 72.49% |
| 10% | c499 | 1696 | 73.82% | 59.53% | 50.17% | 57.17% |
| | c432 | 196 | 82.19% | 68.67% | 63.06% | 66.94% |
| | c1908 | 2475 | 70.57% | 56.00% | 45.86% | 53.27% |
| | c2670 | 6300 | 61.42% | 55.17% | 44.25% | 51.80% |
| | c3540 | 1870 | 70.82% | 56.56% | 46.24% | 53.76% |
| | c5315 | 4674 | 63.71% | 52.38% | 40.88% | 49.14% |
| | c6288 | 416 | 76.06% | 58.51% | 49.83% | 56.49% |
| | c7552 | 7992 | 60.99% | 54.45% | 43.95% | 51.55% |

and 1.51% improvement compared to LR and GLC, respectively. However, as error probability increases, our proposed output data compression rate outperforms further by 6.55% and 9.45% compared to LR and GLC with $D = 16$ respectively.

## 6.5   Summary

With the advancement in the technology and increase in integration density, the number of TSV I/Os have increased in 3D integration. The 3D ICs with large number of TSVs having less number of I/O pads increased the complexity to perform testing. In this chapter, an output data compression for pre-bond and post-bond TSV tests via compressed sensing is discussed. By exploring the sparsity of test data, compressive sensing's encoding method can be easily implemented on-chip using XOR and AND networks, and the output bandwidth limitation is greatly alleviated by the output data compression. Simulation results for standard benchmarks have shown that a 89.70% pre-bond data compression rate can be achieved under 0.5% error probability; and a 88.18% post-bond data compression rate can be achieved with a 5% error probability.

# PART 3
# Thermal Management

# Chapter 7

# Power and Thermal System Model

## 7.1 Introduction

As 3D integration [5,156–158] emerged as an attractive alternative to alleviate the performance bottlenecks in 2D integration, high-throughput applications can now be executed on multi-processor system-on-chip (MPSoC) with a broader design space for exploration. By expanding the design space into the third dimension, 3D IC significantly reduces average wire length, wire delay, power consumption and footprint [159–161]. However, in 3D IC, due to a long heat-removal path from the silicon layers to the heat-sink at one side, the increased temperature can affect the cache accessing time and introduce a reliability concern. This effect might offset the benefits gained from the 3D integration. Additionally, with the increase in number of integrated cores and memory blocks, power management becomes equally essential. Therefore, a 3D multi-core cache-processor system needs to have an effective temperature and power management to maintain the temperature at a normal level.

Improper power management will have two consequences: increase in temperature and power density. Firstly, for a many-core microprocessor with DRAM, at high temperatures, leakage current will dominate the total power, which when coupled with temperature, forms a positive-feedback and results in thermal runaway failure [49,50]. Hence, from the device perspective, power models of core and DRAM with interconnects need to be studied first. Additionally, to avoid such failures, proper thermal management techniques are needed. Hence, thermal models need to be developed first.

Proper power and thermal management are mandatory to avoid thermal–induced aging, performance degradation and malfunctioning.

Previous dynamic thermal managements [118, 162–166, 166–169] in 2D microprocessors are mainly based on the task scheduling by thermal profile tuning, task migration and fetch toggling at runtime. Most of them assume a static demand fixed at runtime. In addition, most of the previous temperature management methods assume the presence of temperature sensors. The implementation of temperature sensors, however, is quite expensive in 3D. For thermal management of 3D circuits, most of the prior works are addressed during the design stage by utilizing the through-silicon vias (TSVs) [7, 9, 11, 12, 110, 170]. Until recently, it was found that the use of microfluidic channel arrays [171–174] is more effective to remove heat in 3D ICS by providing local thermal ground. In addition, liquid cooling exhibits a much higher thermal conductivity than air cooling by heat-sink, and is scalable to cope with the varying number of device layers [175]. All the above advantages have been well documented in the literature [171–176]. It is still unknown how an optimal thermal management scheme can be developed to control the flow-rate of microfluid. First, it needs to be runtime adjustable based on real-time demand. Second, it needs a robust controller with both prediction and correction capabilities. Third, it needs to have cost effective fine-grained management to improve the cooling efficiency. These properties are akin to the attributes of the latest closed-loop control scheme for the so-called cyber-physical systems [177–179].

For any of the above mentioned approaches, an accurate and computationally efficient thermal model is almost always needed to adapt and control the circuit temperature under a given power profile and cooling requirements. Numerical multi-physics simulators such as ANSYS [180] and COMSOL [181] are general and accurate. However, their long simulation cycle makes design phase and runtime thermal management inefficient. As a trade-off, analytical models were proposed to provide rough estimations to speed up the simulation. Early tools such as the well-accepted Hotspot [182] simulates both the transient and steady–state temperatures of heat-sink-cooled 2D/3D circuits. Results from Hotspot were shown to be accurate for 2D circuits. However, its modeling of 3D architecture was too simplistic. As microfluidic cooling was proposed as a scalable cooling solution with better cooling effect than heat-sink cooling for high performance 3D systems, various thermal models [183–186] were subsequently proposed for liquid-cooled structure, among which the latest model is 3D-ICE [185, 186]. As 3D-ICE ignores the entrance effect [186, 187] and adopts the conventional parameter correlation for macro-scale pipes [188], it tends to under-estimate the liquid cooling induced thermal gradient. Owing to its irregularity and profiling difficulty, thermal effects of TSVs in 3D-IC have not been properly investigated in previous thermal models.

In this chapter, we first present power models for core and DRAM blocks, followed by the power breakdown of a many-core microprocessor with 3D and 2.5D integrations. Later, a fast and accurate steady state thermal simulator for TSV-based 3D-ICs will be presented first. The proposed simulator models both heat-sink and microchannel-based cooling as the two major cooling methods found in literature. The anisotropic TSV thermal effects and entrance effect of microchannels are investigated and incorporated to address the deficiencies of Hotspot and 3D-ICE.

## 7.2   3D System Power Model

In a many-core memory-logic integrated system, the major contributors to the power are the core, memory and I/Os. We have already discussed about the power modeling of TSVs and TSIs in the previous chapters. Here, we discuss the power models of core and DRAM memory. In the later part of this section, we present the thermal runaway failure of the system caused by the positive feedback loop formed between leakage current and the memory.

### 7.2.1   *Core and DRAM Power Model*

The power dissipation of a microprocessor core $P_{\text{core\_total}}$ is the sum of its dynamic and leakage powers and can be given as

$$P_{\text{core\_total}} = \underbrace{\eta_a * C_{\text{core}} * V_{DD} * \Delta V * f}_{P_{\text{core\_dynamic}}} + \underbrace{V_{DD} * I_{\text{leakage}}}_{P_{\text{core\_leakage}}}; \quad (7.1)$$

with

$$I_{\text{leakage}} = \underbrace{A_s \cdot \frac{W_d}{L_d} \cdot v_T^2 (1 - e^{\frac{-V_{DS}}{v_T}}) \cdot e^{\frac{V_{GS} - V_{th}}{ns \cdot v_T}}}_{I_{\text{subthreshold}}} +$$

$$\underbrace{W_d \cdot L_d \cdot A_J (\frac{T_{oxr}}{T_{ox}})^{nt} \frac{V_g \cdot V_{aux}}{T_{ox}^2} e^{-B_J T_{ox}(a - b|V_{ox}|)(1 + c|V_{ox}|)}}_{I_{\text{gate\_leakage}}}; \quad (7.2)$$

where $v_T = k \cdot T/q$ is the thermal voltage; $P_{\text{core\_dynamic}}$ is the core dynamic power; $P_{\text{core\_leakage}}$ is the core leakage power; $\eta_a$ represents the activity factor; $C_{\text{core}}$ is the core load capacitance; $V_{DD}$ is the supply voltage with $\Delta V$ swing; $f$ is the clock frequency; $L_d$ and $W_d$ are the effective device channel length and width; $ns$ is the subthreshold swing coefficient; $A_s$, $A_J$, $B_J$, $a$, $b$ and $c$ are technology-dependent constants; $nt$ is a fitting parameter; $T_{ox}$ and $T_{oxr}$ are gate dielectric and reference oxide thickness; and $V_{aux}$ is auxiliary temperature–dependent function for the density of tunneling carriers.

In addition, consider DRAM as a data array of $n$ identical banks with each bank having a size $M$ and $B$ I/O channels per bank. The dynamic power of DRAM $P_{DRAM}$ can be given as

$$P_{DRAM} = n * B * C_{\text{channel}} * V_{DD}^2 * f \qquad (7.3)$$

where $C_{\text{channel}}$ is the channel capacitance; $n * B = N$ is the total number of channels; $V_{DD}$ is the supply voltage; and $f$ is the clock frequency.

Assuming that one DRAM memory bank with size $M$ has leakage current $I_{\text{leakage}}$ as given in (7.2), the total leakage power of DRAM $P_{Dleak}$ can be given as

$$P_{Dleak} = n * M * V_{DD} * I_{\text{leakage}}. \qquad (7.4)$$

It has been observed that the leakage current $I_{\text{leakage}}$ can vary significantly with temperature in an exponential fashion. This may have two consequences. Firstly, the leakage power can dominate the total power. Secondly, the leakage power may form a positive feedback loop with temperature that can lead to *thermal runaway failure*. For example, in high-performance computing with the increase in bandwidth $N = n * B$, i.e., the number of banks $n$, the leakage power of DRAM also goes up. As such, the increase in bandwidth by adding more I/O channels may worsen the electrical-thermal coupling to power and increase the probability of thermal runaway. Therefore, one may need to design high speed and low power I/O channels with each I/O channel supporting a high bandwidth, which can be satisfied by the use of 2.5D TSI T-line based I/O design.

### 7.2.2   *System Power Breakdown*

To evaluate the above mentioned power models and the thermal runaway failure, a gem5 [189] simulator is used. The multi-core system is set up as 16-core ×86 microprocessors (1GHz) operating in gem5, which is integrated with McPAT [190] to analyze the power of core, CACTI [191] for DRAM power and a modified thermal simulator [134] is employed to provide the thermal profiles for both 2.5D TSI and 3D TSV integrations. Each core has a 32KB L1 instruction and data cache, respectively. From the McPAT and

Table 7.1　System setup for data server system components.

| Components | Description | Value | Area Estimation |
|---|---|---|---|
| Core | Frequency | 1.0 GHz | $2.469\,mm^2$ |
| | L1 cache size | 32 KByte | |
| | Cache block size | 64 Byte | |
| DRAM | Number of banks | 4 | $32.025\,mm^2$ |
| | Bank size | 64 MByte | |
| | Page size | 8192 bits | |
| | Bus width | 128 bits | |
| TSV I/O | Number of channels | 64 | $19\,\mu m^2$ |
| | Length of interconnect | 50 um | |
| | Delay per channel | 22.50 ps | |
| TSI I/O | Number of channels | 64 | $10\,\mu m^2$ |
| | Length of interconnect | 1.5 mm | |
| | Delay per channel | 34.52 ps | |

CACTI simulation results, the area estimations are obtained and presented in Table 7.1. The total area is about $288\,mm^2$ for 2.5D integration and $64\,mm^2$ for 3D integration. As such, the thermal runaway failure can be examined in both 2.5D TSI and 3D TSV integrations.

What is more, the parameters of TSV/TSI I/Os are also shown in Table 7.1 based on the size of the core and DRAM. The number of channels for both 2.5D and 3D integrations is 64. The length of TSV I/O is $50\,\mu m$, which is the distance between adjacent layers; and the length of TSI I/O is $1.5\,mm$, which is the distance between two ICs.

The system power breakdown (core, memory, I/O) pie chart is shown in Figure 7.1. Power consumption for 2.5D integration executing the SPEC 2006 benchmark 401.bzip2 and Phoenix benchmark KMeans at $25°C$ and $120°C$ are shown in Figures 7.1(a) and (b) respectively. Power consumption for 3D integration executing SPEC2006 benchmark 401.bzip2 and Phoenix benchmark KMeans at $25°C$ and $120°C$ are shown in Figures 7.1(c) and (d) respectively.

One can make the following observations. The power of DRAM is the dominant factor, especially at high temperatures. For example, in Figure 7.1(b), with 2.5D integration at $120°C$, the DRAM power accounts for 65.33%, while the core power accounts for just 34.24%. It is because the DRAM leakage power will increase dramatically with the rise of temperature. Moreover, both the powers of TSI and TSV I/Os occupy a small portion of the pie chart. The TSI I/O power is less than 6% of the whole

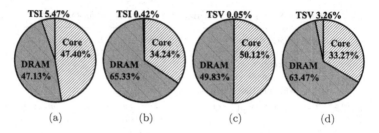

Fig. 7.1    Power breakdown of (a) 2.5D integration at $25°C$ (b) 2.5D integration at $120°C$ (c) 3D integration at $25°C$ (d) 3D integration at $120°C$.

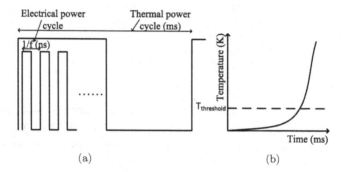

Fig. 7.2    (a) Thermal power and electrical power cycles; (b) thermal runaway and threshold temperature.

system power consumption in Figure 7.1(a) and (b), and is insensitive to the temperature. What is more, as shown in Figure 7.1(c) and (d), the TSV I/O power contributes less than 4% of the whole system power consumption. With rising temperature, TSV I/O power percentage will also increase.

### 7.2.2.1    Thermal Runaway Failure

The second kind of electrical-thermal coupling is between power and temperature. Under large bandwidth conditions, leakage power varying exponentially with the temperature dominates DRAM total power. What is even worse, leakage power and temperature forms a positive feedback loop resulting in thermal run away failure.

From Figure 7.2(a) it needs to be observed that the thermal power $P_{\text{thermal}}$ is sampled at $ms$-scale based on electrical power at $ns$-scale acting as thermal source. Thermal power $P_{\text{thermal}}$ is given by

$$P_{\text{thermal}} = P_{\text{core\_dynamic}} + P_{\text{leakage}} = P_{\text{core\_dynamic}} + a_2 * e^{a_1 \cdot T}; \qquad (7.5)$$

where $P_{\text{core\_dynamic}}$ is the dynamic power of core as in (7.1); and $a_1$ and $a_2$ are the coefficients describing the exponential dependence between leakage power and temperature $T$.

It can be observed from (7.5) that thermal leakage power $P_{\text{leakage}}$ increases with temperature. This increase in thermal power will form a positive feedback loop and increase temperature, resulting in a thermal reliability problem, which can be termed as thermal runaway failure. So, the leakage power is sensitive to the changes in temperature.

With each core serving as a thermal power source, interactions between $g$ adjacent cores in neighboring grids result in a flow of a heat flux through them, causing temperature to rise with time. The system thermal dynamics with heat-sink is given by

$$C_{TR}\frac{dT}{dt} = P_{\text{thermal}} - \sum_{j=1}^{g} \frac{T - T_j}{R_j}; \qquad (7.6)$$

where $C_{TR}$ is the thermal capacitance of a core, and $R_j$ is the thermal resistance path from $g$ chips to the heat-sink.

If the thermal source grows much faster than the heat removal ability of the heat-sink, temperature increases exponentially resulting in thermal runaway failure. Thermal runaway temperature $T_{threshold}$ is the temperature at which thermal runaway failure happens. As shown in Figure 7.2(b), $T_{threshold}$ represents the maximum temperature beyond which thermal runaway failure happens. Placing the heat-sink closer to processing cores can avoid thermal runaway failure. Since 2.5D integration has a much close heat-removal path to the heat-sink, the thermal-removal ability ($R_j$) has much better performance than that of 3D integration.

We evaluated the thermal runaway failure of high performance data servers by both 2.5D and 3D integrations. The experiment is based on both general purpose benchmarks and cloud-based benchmarks. Each cluster of cores execute different benchmarks. The number of memory layers for 3D integration is set to be flexible for the purpose of evaluation. The initial temperature is at room temperature ($25°C$). The size of the heat-sink is set as $4.0\,cm \times 4.0\,cm$ in 2.5D integration and $2.0\,cm \times 2.0\,cm$ in 3D integration, both with a heat-removal resistance set as $4.6\,K/W$.

The dynamic system temperature trend under 2.5D integration and 3D integration with different number of layers is shown in Figure 7.3. For 3D

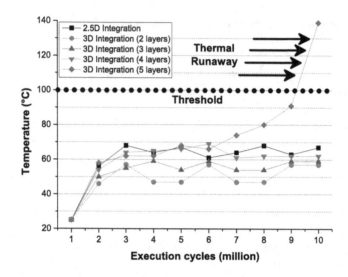

Fig. 7.3   Thermal runaway analysis of data server by 3D and 2.5D integrations.

integration, we vary the number of layers to see its heat dissipation performance from 2 to 5, i.e., 3 layers means it has one logic layer and two memory layers. When the temperature goes beyond the threshold temperature ($100°C$), meaning that the heat-sink cannot dissipate heat produced by the system, thereby forming a positive feedback loop with leakage current, there will be a dramatic increase in system temperature and cause thermal runaway failure. The temperature of 2.5D integration is maintained at between $50°C$ and $70°C$. In 3D integration, the temperature stayed stable for less than 4 layers. With 5 layers in 3D integration, rising trends in the temperature can be observed after 6 million execution cycles, and rises quickly beyond $100°C$ when thermal runaway happens. As such, a high performance data server formed by 3D integration is more liable to the risk of thermal runaway failure, if proper cooling technique is not applied. In this case, a 4–layers setting is the best from a thermal perspective.

## 7.3   3D System Thermal Model

As seen in the previous section, improper thermal management led to thermal runaway failure in 3D ICs, hence, as a first step, we perform the thermal modeling for 3D integrated systems. Figure 7.4 illustrates a schematic

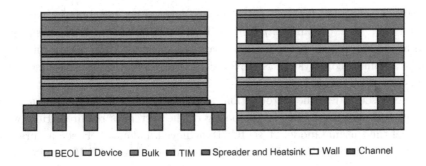

BEOL ▣ Device ▣ Bulk ▣ TIM ▣ Spreader and Heatsink □ Wall ▣ Channel

Fig. 7.4 Schematic of 3D integrated circuit with heat-sink and microfluidic cooling.

cross-sectional view of a typical 3D chip that was considered in most publications [172, 184, 185]. In 3D IC, multiple layers of circuit are stacked with fast interlayer connections. A circuit layer typically comprises of back end of line (BEOL), device and thinned bulk sublayers. Depending on the cooling demand, different cooling methods are applied. The schematics in Figure 7.4 show the cross-sectional views of 3D ICs with heat-sink cooling for low power circuits and microfluidic cooling for high performance 3D stacks. For heat-sink cooled 3D circuits, thermal interface material (TIM) is smeared in between circuit layers for better thermal connectivities. The generated heat is then dissipated into the surroundings through the spreader and heat-sink, probably with the aid of a fan. For microfluid–cooled 3D stacks, microchannels are etched at the back of dies before bonding. Those microchannels are assumed to be uniformly distributed with identical channel width and height. The channel walls are thus of the same width except possibly for the walls at the boundaries. Liquid coolant flows through the microchannels and carries away the generated heat. Without loss of generality, it is assumed that the coolant reservoir is large enough so that the inlet coolant is always at the ambient temperature, keeping in mind that the ambient temperature setting can always be modified to simulate for non-ideal cases where a higher inlet temperature is observed.

We assume an alpha-2 architecture for each computing core. Conventional 3D designs [7, 9, 12, 170] assume the use of the heat-sink together with the heat-spreader to dissipate heat, as shown in Figure 7.4(a). The heat-sink placed at the top of the device layers becomes the primary heat removal path to the ambient air. As shown in Figure 7.4(b), the microfluidic channels are fabricated in between silicon layers and hence provide the local thermal ground to the ambient. This can dramatically shorten the heat-removal path for each active device layer. Five benchmarks namely *bzip*, *eon*, *fma*, *gcc* and *gzip* from SPEC2000 [136] are randomly assigned to the microprocessors to simulate real life workload. More details on how this 3D multi-core system is explored for the simulation are discussed in this chapter.

As discussed in the introduction, to apply any of the thermal management techniques, an effective modeling of the system is required. Here we model the thermal model of 3D IC. To achieve a balance between accuracy and speed on estimating the steady–state temperature of 3D ICs, finite element approximation is adopted, where the 3D stack under study is divided into regular grid cells according to the user-defined row-column resolution. Thermal conductances between adjacent cells are then calculated by considering the effect of TSVs. At steady state, the thermal balance equation holds for every grid cell as

$$\sum (T_i - T_{i,adj}) * g_{i,adj} = P_i \tag{7.7}$$

where $T_i$ and $T_{i,adj}$ denote the temperatures of the $i^{th}$ grid cell and its adjacent grid cell, respectively. $g_{i,adj}$ represents the thermal conductance between two cells, and $P_i$ is the heat generated by the $i^{th}$ grid cell. The matrix equation of thermal equilibrium can be solved by the sparse solver KLU [192] with the provided power consumption.

### 7.3.1  *Microfluidic Channel Thermal Model*

Convective heat transfer between solid and fluid cells can be characterized by the convective conductance, defined as follows:

$$g_{Conv} = h * A_{conv} \tag{7.8}$$

where $h$ is the heat transfer coefficient and $A_{conv}$ is the area of interface where convection takes places. In our thermal model, the convective conductances of heat-sinks are considered as one of the configuration inputs in accordance with Hotspot [135]. Tabulated values of convective resistances for the heat-sink can be found from the manufacturer data sheet.

For microfluidic cooling through microchannels, $h$ is calculated as:

$$h = \frac{k_f Nu}{D_H} \tag{7.9}$$

where $k_f$ is the coolant's thermal conductivity. $D_H$ represents the hydraulic diameter and is given by (7.10) for rectangular channels.

$$D_H = \frac{4 \cdot A_{Ch}}{P_{Ch}} \tag{7.10}$$

where $A_{Ch}$ and $P_{Ch}$ denote the channel cross-sectional area and channel perimeter, respectively. The Nusselt number $Nu$ is usually obtained by empirical correlations to Reynolds number $Re$, Prandtl number $Pr$ and channel dimension in the literature. In our thermal model, the following correlation for microchannels [193] is used.

$$Nu_{PP} = 0.1165 \left(\frac{D_H}{Pitch}\right)^{0.81} \left(\frac{H_{Ch}}{W_{Ch}}\right)^{-0.79} Re^{0.62} Pr^{1/3} \tag{7.11}$$

$$Re = \frac{4 \cdot \rho \cdot Q}{\mu \cdot P}, \qquad Pr = \frac{SpHeat_f \cdot \mu}{f_f}$$

where $Pitch$ refers to the center-to-center distance of adjacent microchannels. $H_{Ch}$ and $W_{Ch}$ denote the height and width of microchannels, respectively. $\rho$ is fluid's density; $Q$ is channel's flow-rate; $\mu$ is fluid's dynamic viscosity; $P$ is perimeter of channel; and $SpHeat_f$ is the specific heat of fluid. Their values, $Nu_0 = 0.156$, $a = 0.67$, and $b = 0.3$, are taken from the experiment results done in [164].

In addition, it was noted that the heat transfer rate is significantly higher near the inlet of microchannels due to the thin developing thermal boundary layer [186,187], which was ignored by most of the previous works. To model the influence of such entrance effect, we make the following modification to the conventional macro-scale formulation [187].

$$Nu_{ST} = Nu_0 (Re * Pr)^{1/3} \left(\frac{D_H}{l}\right)^{\eta} \tag{7.12}$$

where $l$ denotes the distance from the channel inlet. The original values 1.86 and 1/3 of the coefficients $Nu_0$ and $\eta$ have been revised to 30 and 1.67 respectively, by model-fitting against COMSOL [181] for application on microfluid. The aggregated Nusselt number can thus be calculated as

$$Nu = Nu_{PP} + Nu_{ST} \tag{7.13}$$

### 7.3.2  Steady State Thermal Analysis

At steady–state, all the grid cells are under thermal equilibrium and (7.7) holds. For microfluidic cooling, the channel temperature is also held constant where the heat is transferred from upstream to downstream in the form of massive flow. Thus the following equation is satisfied by the liquid channel cells under thermal equilibrium.

$$\sum_{j\in\{walls\}} g_{ch,i,j}(T_{ch,i} - T_j) = (T_{ch,i} - T_{ch,i-1})QC_V \qquad (7.14)$$

where $T_{ch,i}$ and $T_{ch,i-1}$ indicate the temperatures of the $i^{th}$ channel cell and its upstream cell. $Q$ and $C_V$ are the coolants flow-rate and volumic specific heat, respectively. Based on (7.7) and (7.14), a matrix equation can be set up as follows irrespective of the cooling technique.

$$GT = P \qquad (7.15)$$

where $T$ is a column array for the unknown steady state temperature of each grid cell and $P$ is the heat generation column array. $G$ is an $N \times N$ coefficient matrix where $N$ denotes the total number of grid cells. Note that the thermal conductance is zero for two cells that are not in contact physically. Thus, $G$ is a sparse matrix with the number of non-zero elements linearly proportional to $N$. This sparse matrix equation can be efficiently solved by existing sparse solvers such as KLU [192] to obtain the steady-state temperatures of every grid cell.

### 7.3.2.1 *Software Implementation*

The discussed thermal model has been implemented in C program. The basic program flow is shown in Figure 7.5. The thermal model accepts a layer configuration file to set up the physical description of the 3D IC, including detailed sizing, cooling method, TSV setting, and power consumption, etc. After that, the model calculates the thermal conductances by considering the TSV effects. The $G$ matrix is then built by completing the column pointer array, row index array and numerical value array required by KLU. Finally, the $P$ array is created and the steady–state temperature is obtained by solving (7.15). In our program, the time spent on thermal conductance calculation is linearly dependent on the number of TSVs, which is an integer constant of limited value for a given circuit. The construction of $G$ and $P$ both have a time complexity of $O(m)$, where $m$ denotes the number of non-zero entries of the matrix. The KLU algorithm has a time complexity of $O(N + m)$, where $N$ is the total number of grid cells. Since $m$ is linearly proportional to N, our proposed thermal model has a time complexity of $O(N)$, which scales linearly with the total number

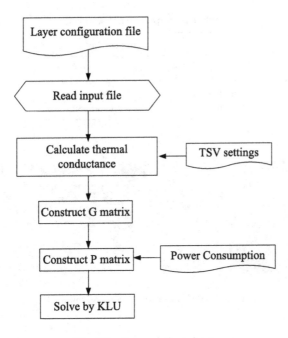

Fig. 7.5 Program flow chart.

of grid cells. The discussed thermal models are available for downloading via [134].

Based on the models proposed above, we verify the accuracy of the proposed thermal simulator and compare our results with existing works. For both cooling methods, a two-layer 3D stack is used for the test. The chip is set to a size of $1.8 \times 1.8 \, mm^2$ with $12 \, \mu m$ BEOL layers and $2 \, \mu m$ device layers. The thickness of the bulk layer is $48 \, \mu m$ for the top tier and $148 \, \mu m$ for the bottom tier. Ambient temperature is assumed to be $300 \, K$. The details of the cooling package setup is listed in Table 7.2. In the absence of physical prototypes, our thermal simulator is validated against the accurate commercial multi-physics solver COMSOL [181]. A simple $3 \times 3$ mosaic floorplan is adopted for the two circuit layers with customizable power densities. All the experiments are performed on a Linux server with Intel Xeon $3.47 GHz$ processor and $50 GB$ of memory.

### 7.3.2.2 *Validation of Proposed Model*

To verify the accuracy of the proposed thermal model on microfluidic cooling, we compare the simulation results with those of COMSOL and

Table 7.2    Experimental settings.

| Microfluid cooling setting | |
| --- | --- |
| Micro-channel size ($\mu m$) | $100 \times 100$ |
| Channel wall width | $100\,\mu m$ |
| Default flowrate | $9\,ml/min$ |
| Coolant type | Deionized water |

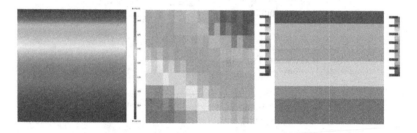

Fig. 7.6    Estimated thermal distribution comparison for top device layer.

3D-ICE 2.1. The same 2-layer 3D circuit is used except that it is cooled by the microfluidic channels between two circuit layers instead of the heat-sink. The same set of typical power densities is applied to the device layers. The maximum and minimum temperatures for the top device layer is reported in Table 7.3 with their corresponding percentage errors enclosed in brackets. It shows that our proposed thermal model is very accurate with a percentage error of less than 1.1% for this test case. In contrast, 3D-ICE exhibits large errors especially for the minimum temperatures due to the ignorance of the entrance effect, which causes the thermal gradient to be severely under–estimated. Figure 7.6 shows the temperature maps of the top device layer at a power density of $150W/cm^2$, plotted based on the results of three (COMSOL, 3D-ICE and proposed) simulators. It can be seen that the temperature distribution of our thermal model (right most) agrees very well with that of COMSOL while that of 3D-ICE deviates significantly.

Due to the predominant heat extraction through microchannels, the thermal effect of TSV is less pronounced compared with that found in heat-sink cooled 3D-ICs. For example, a reduction in maximum temperature of merely $0.28\,K$ is reported by our thermal model for the test bench at a power density of $250W/cm^2$ with TSV pattern C.

Table 7.3 Steady state temperature (K) comparison among COMSOL, HOTSPOT and proposed simulator.

| | PD | COMSOL | 3D-ICE | OURS |
|---|---|---|---|---|
| 50 | MAX | 314.3 | 317.4(1.0%) | 314.0(0.1%) |
| | MIN | 301.7 | 314.2(4.1%) | 302.4(0.2%) |
| 100 | MAX | 328.6 | 334.8(1.9%) | 328.0(0.2%) |
| | MIN | 303.4 | 328.4(8.2%) | 304.8(0.5%) |
| 150 | MAX | 342.9 | 352.2(2.7%) | 341.9(0.3%) |
| | MIN | 305.2 | 342.6(12.3%) | 307.2(0.7%) |
| 200 | MAX | 357.2 | 369.6(3.5%) | 355.8(0.4%) |
| | MIN | 306.9 | 356.8(16.3%) | 309.6(0.9%) |
| 250 | MAX | 375.1 | 387.0(4.3%) | 369.8(0.5%) |
| | MIN | 308.6 | 371.0(20.2%) | 312.0(1.1%) |

## 7.4   3D Cyber-physical System

The notion of a cyber-physical system in the context of thermal management implies a closed-loop control system that is driven by real-time demand and endowed with the prediction-and-correction capability and fine-grained adaptation. In this section, first we will present the system architecture of a cyber-physical system followed by the thermal management problem formulation by microfluidic cooling.

### 7.4.1   *System Architecture*

Figure 7.7 illustrates a schematic cross-sectional view of a typical 3D chip that was considered in most publications [172, 184, 185]. In 3D IC, multiple layers of circuit are stacked with fast interlayer connections. A circuit layer typically comprises of back end of line (BEOL), device and thinned bulk sublayers. Depending on the cooling demand, different cooling methods are applied. The schematic in Figure 7.7 shows the cross-sectional views of 3D ICs with heat-sink cooling for low power circuits and microfluidic cooling for high performance 3D stacks. For heat-sink cooled 3D circuits, thermal interface material (TIM) is smeared in between circuit layers for better thermal connectivities. The generated heat is then dissipated into ambient through the spreader and heat-sink probably with the aid of a fan. For microfluid cooled 3D stacks, microchannels are etched at the back of dies before bonding. Those microchannels are assumed to be uniformly distributed with identical channel width and height. The channel walls are

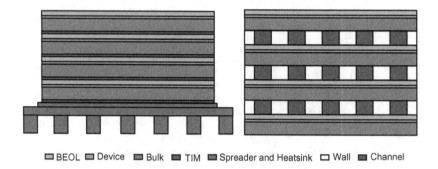

□ BEOL □ Device ▨ Bulk ■ TIM ▨ Spreader and Heatsink □ Wall ■ Channel

Fig. 7.7   Schematic of 3D integrated circuit with heat-sink and microfluidic cooling.

thus of the same width except possibly for the walls at the boundaries. Liquid coolant flows through the microchannels and carries away the generated heat. Without loss of generality, it is assumed that the coolant reservoir is large enough so that the inlet coolant is always at the ambient temperature, keeping in mind that the ambient temperature setting can always be modified to simulate for non-ideal cases where a higher inlet temperature is observed.

Similarly, to simulate the 3D IC with on-chip memory, we adapt the architecture presented in Figure 7.7. The heat-sink placed close to the bottom (or top) the device layers becomes the primary heat removal path to the ambient air. As shown in Figure 7.7(b), the microfluidic channels are fabricated in between silicon layers and hence provide the local thermal ground to the ambient. This can dramatically shorten the heat-removal path for each active device layer. For memory-logic integrated analysis, cores and caches on adjacent layers are arranged in a complementary manner to avoid highly concentrated heat generation. Each cache is divided into four banks, and one core can access one bank of cache at one time.

### 7.4.2   *Problem Formulation*

Given the 3D multi-core cache-processor system with microfluidic channels, our objective is to develop a thermal management system that can actively control the flow-rate of microfluidic channels to maintain the system temperature within a certain temperature range. As stated in the introduction, most of the previous dynamic thermal managements [118, 164–169] assume a static demand based on history, and often require temperature sensors to record the temperature. Moreover, the recent active microfluidic cooling methods [172–174] have the following limitations: (1) there is no real-time temperature demand estimation to adaptively update the needed flow-rate control; (2) there is no corrected prediction to account for the estimation error and variance of power consumption; and (3) there is no fine-grained flow-rate allocation to non-uniformly adjust the flow-rates to improve the cooling efficiency with minimum implementation cost. The solution for this thermal management problem will be discussed in the next chapter.

## 7.5    Summary

In this chapter, power models of microprocessor core and memory blocks are presented. To compare 3D and 2.5D integrations, system power breakdown is validated, which indicates 2.5D TSIs are thermal resilient compared to 3D TSVs. A thermal runaway failure is observed in 3D integration when more layers are stacked and no proper thermal management is carried out. To perform thermal management of 3D ICs with the aid of microfluids/TSVs, one needs to have an accurate steady–state thermal simulator for both heat-sink cooled and microfluidic cooled 3D ICs, capable of estimating the thermal effect of TSVs in fine-granularity and computationally efficient. The proposed thermal model considers the entrance effect of microchannels based on the most appropriate thermodynamics for microfluidic cooling. The performance of the presented simulator is in good concordance with the performance obtained from commercial software COMSOL, with a maximum error of 1.1% for microfluidic cooling.

# Chapter 8

# Microfluidic Based Cooling

## 8.1 Introduction

Due to the increased power density and accumulation of heat in top layers by vertical stacking, 3D ICs require effective thermal management. In this chapter, we present a novel scheme of temperature management for 3D multi-core cache-processor systems called cyber-physical temperature management based on the thermal models developed in the previous chapters. Under this scheme, the system temperature will be maintained at a desired level by adjusting the fluidic flow-rate according to the real-time temperature demand. Such a real-time temperature demand is determined by the future power consumption and thermal model of 3D IC with microfluidic channels. An Auto Regression (AR) predictor [194] is adopted to predict the future power consumption of on-chip components and the prediction is further corrected by Kalman filtering [195] to remove the error contributed by system variation and control. As such, the steady–state temperature, i.e., defined as the real-time temperature demand, can be estimated from the predicted future power consumption. With such an efficient determination of temperature demand on the thermal-time-constant scale, the flow-rate is then adjusted accordingly so that the system will not run out of the thermal threshold or over-cool. Moreover, the concept of "channel clusters" is introduced to provide a more effective fine-grained control of flow-rates for disparate groups of channels to reduce the number of hotspots. A smaller implementation cost is expected when compared to the use of temperature sensors. Lastly, a channel clustering algorithm is proposed to guide the grouping of microchannels based on the power density distribution. Experiment results show that with proper clustered microfluidic cooling, lower and smoother temperature distributions can be achieved under the same total flow rate constraint.

## 8.2   3D Cyber-physical Thermal Management

A many-core memory-logic integrated system with heat-sink cooled and microfluidic cooling mechanisms is shown in Figure 8.1. A cyber-physical system [177–179] has recently become one of the research focuses mainly due to the grand challenge faced in the real-time distributed control of a wireless sensor network. We extend the principle of a cyber-physical system for the thermal management of multi-core cache-processor systems as shown in Figure 8.2. The 3D system runtime thermal dynamics and power consumption is first estimated by the micro-architecture level simulators [135, 191, 196] under different workloads. The software-sensed power is collected and analyzed by a controller, which can predict the future power consumption, and at the same time, the estimation error and system variance are corrected. Based on the estimated real-time power, the future steady–state temperature, called temperature demand, can be obtained. The temperature demand here refers to the amount of heat to be removed at different functionary units. This demand is then passed to the flow-rate controller to adjust multiple flow-rates accordingly. A cyber-physical system

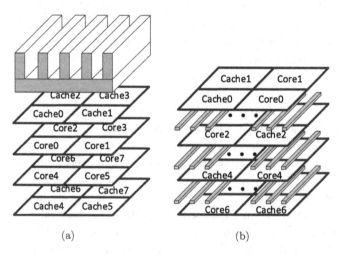

(a)                                      (b)

Fig. 8.1   (a) Conventional 3D multi-core cache-processor system with heat-sink; (b) 3D multi-core cache-processor system with microfluidic channels.

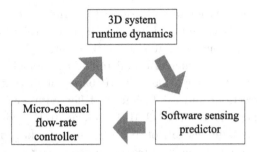

Fig. 8.2 Runtime microfluidic flow-rate control assisted by proactive temperature predictor with software-sensing. This close-loop control forms a cyber-physical resource management.

is formed by linking together the real-time power-state (software) sensing, temperature-demand prediction, and fine-grained flow-rate control in one closed-loop for thermal management.

One important feature in this cyber-physical thermal management is the granularity of flow-rate control. Since each functionary unit in the system works under different conditions, the power and temperature would be non-uniformly distributed across different layers. Therefore, different from previous works [172–174], non-uniform flow-rates are allowed in our system and they can be dynamically controlled to remove the heat in a more effective manner. Moreover, fine-grained flow-rate control allows graceful degradation instead of catastrophic failure when one microfluidic channel gets clogged. However, fine-grained control at every channel is not realistic for physical implementation due to the need for many pumps. Hence, a more practical approach using channel clusters is employed. The microfluidic channels can be divided into a number of groups based on the similarity of functionalities. The channels in the same group will have the same flow-rate. Each cluster can be powered by one micro-pump and its flow-rate is controllable through the pump.

To carry out thermal management on a 3D IC with on-chip memory, we adapt the architecture presented in Figure 8.1. We assume an alpha-2 architecture for each computing core and a set-associative SRAM structure for L2-cache. Conventional 3D designs [7, 9, 12, 170] assume the use of a heat-sink together with heat-spreader to dissipate heat, as shown in Figure 8.1(a). Five benchmarks namely *bzip*, *eon*, *fma*, *gcc* and *gzip* from SPEC2000 [136] are randomly assigned to the microprocessors to simulate real life workload. More details about how this 3D multi-core system is explored for the simulation of our proposed nonuniform flow rate microfluidic cooling scheme can be found in the later part of this chapter.

The above mentioned architecture is implemented in C/C++ with: (1) 3D thermal model including microfluidic channels; (2) real-time temperature demand calculation; (3) AR-based prediction and Kalman-filter-based correction; and (4) fine-grained flow-rate control. The power models are generated by CACTI [191] and Wattch [196]. The thermal network model is built inside HotSpot [135] by including the microfluidic channels, and is further analyzed by the sparse matrix solver from SuiteSparse [197]. Experiments are run on an Intel Dual-Core Server with 3GHz CPU and $2GB$ RAM. Table 8.1 summarizes the design parameters and their settings used in the following experiments. Other experimental settings are detailed below: Water is used as the coolant. The ambient temperature is set to be $45°C$ and initial temperature of the 3D system is set to be $50°C$. The temperature threshold is specified as $85°C$, maximum total flow rate is set to be $4896\,ml/min$, which is about one third of that cited in [174] in terms of per channel flow rate. For the cyber-physical controller, the $3^{rd}$ order AR prediction is used when implementing the cyber-physical temperature management. In addition, a 10% estimation error is assumed for the power trace generated by the power simulators. For all the case studies, the steady state power density varies between $2.98\,W/cm^2$ and $68.95\,W/cm^2$ across the entire 3D stack.

Table 8.1    Parameters of 3D multi-core system with microfluidic channels.

| Parameter | Value |
|---|---|
| Chip size | $32\,mm \times 32mm$ |
| Silicon layer thickness, $t_{Si}$ | $150\,\mu m$ |
| Inter-layer SiO$_2$ thickness, $t_{SiO_2}$ | $120\,\mu m$ |
| Microfluidic channel height, $h_{channel}$ | $100\,\mu m$ |
| Microfluidic channel width, $w_{channel}$ | $100\,\mu m$ |
| Microfluidic channel spacing | $200\,\mu m$ |
| Total Number of channels | 480 |
| Max total flow-rate | $4896\,ml/min$ |
| Max pressure drop of channels | $600\,mbar$ |
| Silicon thermal conductivity, $k_{Si}$ | $100\,W/(m\text{-}K)$ |
| Silicon specific heat | $1750\,KJ/(m^3\text{-}K)$ |
| SiO$_2$ thermal conductivity, $k_{SiO_2}$ | $1.4\,W/(m\text{-}K)$ |
| SiO$_2$ specific heat | $1750\,KJ/(m^3\text{-}K)$ |
| Max temperature allowed | $85°C$ |

## 8.2.1  *Real-time Temperature Demand Estimation*

The power trace at each functionary unit generated by the power estimator can have a large variance in time. Clock-gating can further introduce additional dynamic behaviour for the system power trace. As such, any thermal management that is based on static demand would be insufficient to follow the system power trace and be unable to maintain the temperature level accurately. To overcome this problem, a power trace library for all workloads can be first built in our cyber-physical thermal management system. For example, the benchmarks from SPEC2000 are used and they are characterized in Wattch simulator. Then, the power trace is tracked by periodically sampling the system power trace input and applying the update of both flow-rate and thermal conductance network. This period is called the *control-period*. The real-time here implies that the thermal power trace is not static but dynamically changing with time.

Moreover, it is found that the steady–state temperature with present power input is meaningful in indicating the demand for heat removal ability. When the system with microfluid reaches the steady state, the net gain of heat flux of each grid cell shall be zero. Assuming that there are $N$ grid cells and an additional node '$a$' for the ambient, the following equation can be established for any cell, $i$ $(i = 0, \ldots, N)$.

$$\sum_{j=0}^{j=N-1,a} g_{ij}T_i - g_{i0}T_0 - g_{i1}T_1 - \ldots - g_{i(N-1)}T_{n-1} = P_i + g_{ia}T_a. \quad (8.1)$$

As a result, $N$ equations can be set up for $n$ cells and the steady state matrix equation is obtained by

$$G_{steady} \cdot T_{steady} = P. \quad (8.2)$$

Hence, the matrix $T_{steady}$ for the steady–state temperature can be used as the real-time temperature demand for the next period. The challenge here is to solve (8.2) efficiently. Because the matrix $G_{steady}$ is sparse, symmetric and positive definite, this equation can be efficiently solved by Cholesky decomposition with the use of the sparse matrix solver from SuiteSparse [197]. As a result, the management time in our method remains small, at the scale of thermal time-constants $(ms)$.

## 8.2.2  *Prediction and Correction*

To calculate the real-time temperature demand for the next period, it is necessary to estimate the system power trace $P$ for the next period. As a result, one can perform thermal time-constant level temperature prediction when both the future power and thermal network models are provided. This is different from the traditional cycle-accurate thermal simulator.

Auto Regression, i.e., $AR(p)$, model [194] is a commonly used prediction method to correlate time series data. The basic idea of an $AR(p)$ model is to find the best polynomial fitting of a time varying series, $f(n)$, which is the system power trace in our problem. The $p$ coefficients, $c_1$-$c_p$, and $p$ previous thermal power values, $f(n-1), f(n-2), \ldots, f(n-p)$, are used to predict the future power state $f_p(n)$ by

$$f_p(n) = c_1 f(n-1) + c_2 f(n-2) + \cdots + c_p f(n-p) + \varepsilon(0, \sigma_p) \qquad (8.3)$$

where $\varepsilon(0, \sigma_p)$ denotes a random error with Gaussian distribution (zero mean and variance $\sigma_p$). The $AR(p)$ model can be applied to predict power in the next control-period and hence the real-time temperature demand.

If the electrical power of core and cache are deterministic, the system power trace can be exactly determined or predicted. However, in real-time applications, the workload can vary dynamically, the leakage current and power can fluctuate stochastically, and hence, both the electrical and thermal power can have an uncertain variability or error. To have an accurate and timely temperature regulation, a relatively accurate estimation of the future power and temperature is necessary to filter this variability. This can be resolved through a controller with prediction-and-correction by treating the uncertainty as a random process variable of a probability distribution.

The Kalman filter [195] is used for the correction of prediction error here. The purpose of the Kalman filter is to adjust the estimated result, $f_p(n)$ based on an additional physical measurement according to its noise level. In our cyber-physical management, the measured result $f_m(n)$ refers to the software-sensed power, which is first obtained from the cycle-accurate electrical power simulation and then further averaged to the thermal power at the scale of the thermal time constant.

Assuming that such a software-sensed physical measurement has a noise with variance $\sigma_m$ and the measurement result is $f_m(n)$, then the Kalman gain can be obtained by

$$K = \frac{\sigma_p}{\sigma_m + \sigma_p} \qquad (8.4)$$

Then, the corrected estimation can be calculated as

$$f_e(n) = f_p(n) + K \times (f_m(n) - f_p(n)) \qquad (8.5)$$

After this correction, a more accurate temperature demand is obtained and used for the flow-rate control as follows.

Based on the above discussion of real-time prediction and correction, we show the implementation of a 3D cyber-physical system. We first show the variation in the number of hotspots with control-period, followed by temperature prediction and correction.

Real-time: for real-time thermal management, the control period is dependent on the time-variance of the workload. Therefore, one needs to explore how to determine an optimal control period to capture the real-time behavior of the workload. For example, if the control-period is too long, it may be too late to recover from the temperature built-up from excessive temperature demand. As it takes time for the change of flow-rate to take effect, more hotspots can appear and the chip temperature may run away before the flow rate is increased. On the other hand, if the flow-rate is changed too frequently before the workload variation actually affects the temperature demand, the over-control will result in wasted power.

Figure 8.3 shows the temperature distributions of the *fma*, *gcc*, *gzip* and *bzip* benchmarks, each with three different control periods: 300*ms*, 1.5*s*, and 3*s*. During a 3*s* control-period, there are nine, one and zero times of controls performed for the control period of 300*ms*, 1.5*s* and 3*s*, respectively. Clearly, for the control-period of 3*s*, many hotspots can be identified

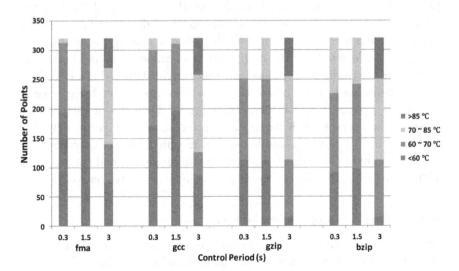

Fig. 8.3  Temperature distribution of multi-core system running four benchmarks with different control periods.

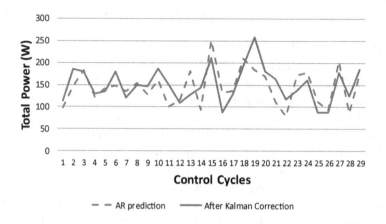

Fig. 8.4    Predicted total power with and without Kalman filter correction.

as no control is applied on time. In contrast, no hotspot is resulted for control-periods of $300ms$ and $1.5s$ because controls are applied frequently on time. However, over-cooling occurs in the benchmarks of $fma$, $gcc$ and $bzip$ when the control period is $1.5s$. Thus, if the workload varies with time, the small control-period is needed for real-time control in this case.

Prediction-and-correction: the prediction-and-correction in our cyber-physical thermal management is implemented by $AR(3)$ and the Kalman filter. At each control cycle, the $AR(3)$ model predicts the future power based on the power trace history. As the software sensing might still contain error or deviate from the real power trace value, Kalman filter based correction is introduced to reduce this randomly distributed error. We use the $gcc$ benchmark, and assume the software-sensed power trace error is 10% of its average power in one control cycle. Then the predicted total power without and with Kalman filter correction at each control cycle is presented in Figure 8.4. The difference in the predicted total power after the Kalman filter correction is discernible. Therefore, thermal management with corrected prediction is necessary for a more accurate control.

### 8.2.3    Clustering of Microchannels

Further, as controlling flow-rate for each channel is difficult, clustering of channels can reduce the complexity. The heat exchange paths of any silicon cell in the 3D grid can be classified into two categories: (a) silicon to channel heat transfer which contributes to microfluidic cooling; (b) silicon to silicon heat transfer which helps to spread the heat generation over the entire chip. The former can be controlled directly by varying the flow rate, while the latter depends only on the nearby thermal gradient due to the fixed silicon to silicon thermal conductance for a fixed architecture. However, the silicon to channel heat transfer is predominantly contributed by the larger temperature differences and highly conductive liquid cooling. An even temperature profile, as part of the goal of our proposed cooling method, will further reduce the silicon to silicon heat transfer to result in nearly zero thermal gradient. Therefore, in steady state, the heat generated by each silicon cell needs to be absorbed locally by the neighboring microchannels. In a nutshell, the cooling effort of microchannels should match the power distribution of the silicon cells. Hence, the power consumption level to be coped with on each channel makes a good criterion to divide the channel clusters.

As shown in Figure 8.1(b), every microchannel (denoted by the channel number $l_i$, $i = 0, 1, \ldots, M - 1$) is in contact with two columns of silicon cells that generate heat, one above and one below the channel. We define the two columns of silicon cells as the cooling region of channel $l_i$, denoted as $CR(l_i) = \{$silicon cell $j|$ $j$ is in contact with channel $l_i\}$.

Let the power consumption of a silicon cell $j$ be $P_{c_j}$. It is possible to define a cooling demand for each channel as a weighted sum of power dissipation based on the silicon cells in the region, i.e., $P_{sum,l_i} = \sum_{j \in CR(l_i)} (\alpha_j \cdot P_{c_j})$, where

$$\alpha_j = \begin{cases} 0.5 & \text{if cell } j \text{ is on an inner layer} \\ 1 & \text{if cell } j \text{ is on the top or bottom layer} \end{cases}$$

The factor $\alpha$ accounts for the advantage of cells situated at the inner layers that have twice the number of channels for cooling compared to those at the top or bottom layer. These cooling demands are then sorted in ascending order of magnitude into an array $P = [p_0, p_1, \ldots, p_{M-1}]$, where $p_i < p_j$ for $i < j$ and $p_i = P_{sum,l_i}$ for $i, l_i \in [0, M - 1]$. The channel indexes corresponding to the sorted power values in $P$ can be retrieved from an order array, $L = \{l_0, l_1, \ldots, l_{M-1}\}$, where $l_i = j$ if $p_i = P_{sum,j}$. The next step would be the segmentation of the sorted array $P$ to divide the $M$ channels into $N$ ($N < M$) clusters. Since channels in one cluster will share the same flow rate setting, the standard deviation of their $P_{sum,l_i}$ values should be as low as possible to achieve an uniform cooling under the same flow rate. This can be readily resolved by finding the natural separations, which are the largest differences between every pair of adjacent elements of array $P$. An array $D = [d_0, d_1, \ldots, d_{M-2}]$ can be obtained by computing the difference $d_i = p_{i+1} - p_i$ for $i = 0, 1, \ldots, M - 2$. The demarcations between clusters can then be found by searching for the $(N - 1)$ largest elements in $D$.

The number of clusters, $N$ can create subtle trade-offs between conflicting factors such as the thermal profile, total flow rate and overhead for a given power trace. The dilemma is, there is no easy way to predetermine an ideal $N$. Therefore, a divide–and–conquer approach is adopted in this work. By searching for the largest difference in $D$, the elements in the sorted array $P$ are divided into two clusters. If the largest difference is $d_j$, then all channels with indexes $l_i$ for $i \in [0, j]$ will be grouped into one cluster and the remaining channels with indexes $l_i$ for $i \in [j+1, M-1]$ will be grouped into another cluster. Let $r_i = p_{M-1} - p_i$ be the moderated cooling demand of the $i^{th}$ channel, and $\mu_R$ and $\sigma_R$ be the mean and standard deviation, respectively, of the modulated power values of all channels within the same cluster. If $\sigma_R > \eta \cdot \mu_R$, where $\eta$ is a parametric fraction, the cluster will be further divided into two clusters by the same approach. This process iterates until every cluster has their $\sigma_{R1} < \eta \cdot \mu_R$. The pseudo code of the clustering algorithm is shown in Algorithm 8.1.

The reason for using a biased $r_i$ value instead of the $p_i$ value directly for the computation of mean and standard deviation is to relax the uniformity criterion for clusters with lower power consumption while forcing a tighter homogeneity in power consumption for high power regions. This is because high power regions often have a low population with high variation in power values. If the clustering is not carried out with enough granularity, a severe mismatch between the cooling effort and cooling demand may occur in these regions and lead to potential thermal management failure. The control parameter $\eta$ provides the desired trade-off between cooling efficiency and overheads. If $\eta$ is set too high, the clustering may not be optimal; and if $\eta$ is set too low, excessive clustering may incur heavy overheads. In our simulation, the number of clusters is constrained to between two and six.

The total heat generation to be handled by each micro-channel in the five case studies is determined. In Cases 1–4, the five benchmarks (*bzip*, *eon*, *fma*, *gcc* and *gzip*) are randomly distributed to the 16 cores, while in Case 5, all the cores are assigned with an average power consumption, which is the mean value of the five benchmarks. The last case is presented to simulate the practical case where clustering is performed with an average power consumption, to cater to various runtime scenarios. The sorted power

---

**Algorithm 8.1** A divide and conquer algorithm for channel clustering

---

1: Function clustering $(L, \eta)$
2: Input: $L$: set of channels to be clustered; $\eta$: cluster control fraction
3: Output: $C$: set of channels in cluster;
4: $P_{c_i}$: power consumption of cell $i$;
5: $CR(l_i)$: cells in contact with channel $l_i \forall i \in [0, M-1]$
6: $M = n(L)$: number of channels in $L$;
7: **for** $(l_i = 0 \text{ to } M - 1 \text{ step } 1 \text{ do})$ **do**
8: $\quad P_{sum,l_i} = \sum_{j \in CR(l_i)}(\alpha P_{c_j})$;
9: **end for**
10: Sort $P_{sum,l_i} : P = \{p_i\}_{i=0}^{M-1}$, where $p_i \leq p_{i-1} \forall i \in [0, M-2]$;
11: Order $L = \{l_i\}_{i=0}^{M-1} : l_i \leftarrow j$ if $p_i = P_{sum,j}$;
12: Calculate $R = \{p_{M-1} - p_i\}_{i=0}^{M-1}$;
13: **for** $(i = 0 \text{ to } M - 2 \text{ step } 1 \text{ do})$ **do**
14: $\quad d_i = p_{i+1} - p_i$;
15: **end for**
16: $D = \{d_i\}_{i=0}^{M-2}$;
17: $t = argmax_j\{d_j\}$;
18: $C1 = \{l_i\}_{i=0}^{t}$; $C2 = \{l_i\}_{i=t+1}^{M-1}$;
19: Calculate $\mu_{R1}$ and $\sigma_{R1} \forall l_i \in C1$;
20: **if** $\sigma_{R1} < \eta \cdot \mu_{R1}$ **then**
21: $\quad$ return $C1$;
22: **else**
23: $\quad$ return clustering $(C1, \eta)$;
24: **end if**
25: Calculate $\mu_{R2}$ and $\sigma_{R2} \forall l_i \in C2$;
26: **if** $\sigma_{R2} < \eta \cdot \mu_{R2}$ **then**
27: $\quad$ return $C2$;
28: **else**
29: $\quad$ return clustering $(C2, \eta)$;
30: **end if**

---

values in ascending order of magnitude are charted in Figure 8.5. To be concise, each dot in the chart actually represents the power level of a channel bundle of five channels sharing the same cooling region.

The clustering results at different $\eta$ values are summarized in Table 8.2, where the number of channels in each cluster is listed in ascending order of cooling demand. As the value of $\eta$ decreases, the deviation in the total

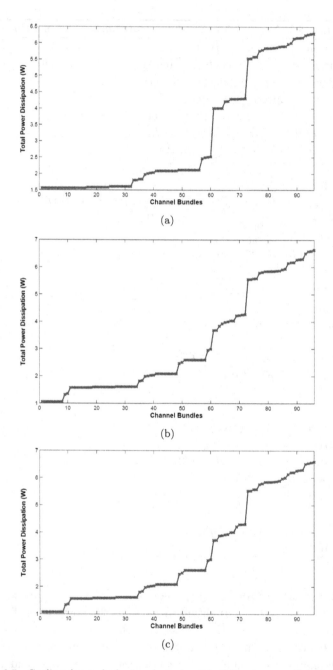

Fig. 8.5   Cooling demand of microchannels with different power distributions.

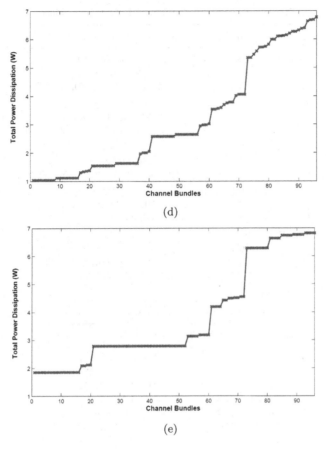

Fig. 8.5 (*Continued*)

power dissipation among channels in the cluster reduces. It can be seen that more clusters are defined with a stringent variation control while coarse clustering is resulted when $\eta$ is relaxed. To have the best cooling effect while not imposing excessive overheads, $\eta$ is set to be 20% for Cases 1–4 and 10% for Case 5.

**Uniform Cooling Effort vs Clustered Cooling Effort** With the above clustering results, the fixed total flow rate of $288ml/min$ are reallocated to each cluster, as shown in Table 8.3, where the flow rate per channel allocated to each cluster is listed in ascending order of cooling demand. Here

Table 8.2    Clustering results with different $\eta$ values.

| Case No. | $\eta$ | N | Number of Channels Per cluster |
|---|---|---|---|
| Case 1 | 20% | 3 | 300, 60, 120 |
|  | 10% | 5 | 300, 60, 20, 60, 40 |
|  | 5% | 7 | 280, 20, 60, 20, 60, 20, 20 |
| Case 2 | 20% | 3 | 360, 100, 20 |
|  | 10% | 6 | 300, 60, 20, 50, 30, 30 |
|  | 5% | 9 | 40, 200, 50, 40, 20, 20, 50, 30, 20 |
| Case 3 | 20% | 3 | 360, 100, 20 |
|  | 10% | 6 | 300, 60, 20, 50, 30, 30 |
|  | 5% | 9 | 40, 200, 50, 40, 20, 20, 50, 30, 20 |
| Case 4 | 20% | 3 | 360, 100, 20 |
|  | 10% | 6 | 200, 100, 60, 40, 60, 20 |
|  | 5% | 10 | 200, 100, 40, 20, 20, 10, 25, 25, 20 |
| Case 5 | 20% | 2 | 72, 24 |
|  | 10% | 4 | 300, 60, 40, 80 |
|  | 5% | 6 | 100, 300 ,60, 40, 20, 60 |

Table 8.3    Reallocation of cooling resources.

| Case No. | N | Flow rate per channel (ml/min) |
|---|---|---|
| Case 1 | 3 | 0.348, 0.798, 1.131 |
| Case 2 | 3 | 0.417, 1.126, 1.255 |
| Case 3 | 3 | 0.417, 1.128, 1.250 |
| Case 4 | 3 | 0.412, 1.139, 1.281 |
| Case 5 | 4 | 0.403, 0.696, 0.996, 1.068 |

the upper limit for flow rate per channel is specified as $1.5ml/min$, corresponding to an about $90KPa$ pressure drop, according to the method of calculation in [198].

Figure 8.6 depicts the top layer's steady state thermal maps in Case 1 under both uniform and non-uniform flow rate cooling by channel clustering. It is clear that with clustered microfluidic cooling, the overall temperature of the 3D chip is much lower. Numerical results for all the study cases are summarized in Table 8.4. Note that for Cases 1–4, the power distribution and channel clustering are ideally matched as the same power profile is used for both clustering and thermal simulation. While for Case 5, power profiles of Cases 1–4 are used for thermal simulation after clustering based on average power consumption. Only the maximum temperature is recorded because the minimum temperature is always close to $300K$ at the inlet area.

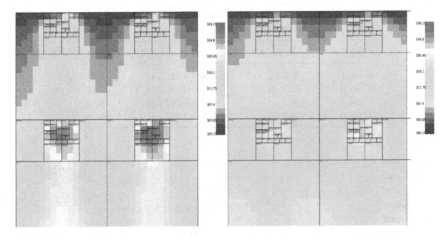

Fig. 8.6 Temperature distribution comparison: thermal maps of the top layer for Case 1 under uniform flow rate cooling and reallocated clustered flow rate cooling with the same total flow rate, coolant flows from north (upper) to south (lower) direction.

From the table, it is clear that the clustered cooling method is effective in reducing the peak temperature under both ideal and practical cases. On average, a drop of $6.22°C$ in the maximum temperature is observed.

### 8.2.4 *Allocation of Flow Rates*

The power profile of a 3D chip often accords with a Pareto distribution, i.e., there are a large number of low power areas and a few patches of high power points. As a result, the channel clustering algorithm will normally produce some clusters with many microchannels of low cooling demand and a few clusters with much fewer microchannels that require high cooling effort. If the cooling resource is constrained by the total available flow rate, the basic principle of its reallocation is just to move the flow rate saved from low-cooling-demand clusters to where it is needed the most. From the systemic point of view, the amount of heat removed is $(T_{out} - T_{in}) \cdot Q \cdot SpHeat_f$. If the flow rate is ideally adjusted, uniform temperature distribution can be achieved whereby the inlet and outlet temperature differences of microchannels are similar. In that case, the flow rate should be proportional to the amount of heat removed, which is equal to the steady state power dissipation. Quantitatively, the average cooling demand $\overline{P}(C_i)$ of any cluster $C_i$ can be obtained by calculating the mean of $P_{sum,l_j}$ for all channels

Table 8.4    Comparison of the maximum temperatures under uniform and clustered cooling.

| Case No. | Uniform (K) | Clustered (Case 5) (K) | $\delta T$ (K) |
|----------|-------------|------------------------|----------------|
| Case 1 | 333.62 | 326.40(327.48) | 6.68 |
| Case 2 | 332.02 | 326.64(326.02) | 5.69 |
| Case 3 | 333.63 | 327.45(327.50) | 6.16 |
| Case 4 | 333.94 | 327.49(327.68) | 6.36 |

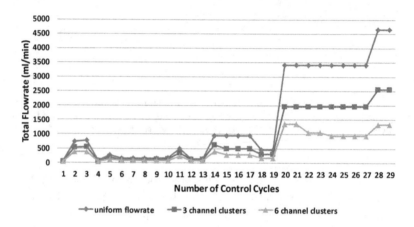

Fig. 8.7    Total flow-rate required at each control cycle (*fma* benchmark).

$l_j \in C_i$. Due to the predominant silicon to channel heat transfer, the following approximation is used to apportion the total flow rate to each cluster for the best cooling effect.

$$Q(C_0) : Q(C_1) : \cdots : Q(C_{N-1}) \approx \overline{P}(C_0) : \overline{P}(C_1) : \cdots : \overline{P}(C_{N-1}) \quad (8.6)$$

Thus, the flow rate per channel for the $i^{th}$ cluster is set to be

$$Q(C_i) = Q\text{total} \cdot \frac{\overline{P}(C_i)}{\sum_{j=0}^{N-1} n(C_j)\overline{P}(C_j)} \quad (8.7)$$

where $Q_{\text{total}}$ denotes the total available flow rate and $n(C_j)$ denotes the number of channels in Cluster $C_j$.

Figure 8.7 shows that the total flow-rate (summation of all channels) required decreases dramatically with fine-grained flow-rate control. The maximum saving of total flow-rate of a 6-channel-cluster control is about 72.1%, compared with that of uniform flow-rate control. The internal mechanism for this result is that: with fine-grained flow-rate control, the flow-rates have been optimally customized to local hotspots, and unnecessary flow-rate increments can be avoided. When there are more clusters, there is a better chance to encounter regions that are totally cool, and hence the corresponding flow-rates can be decreased. More importantly, the total flow-rate directly reflects the pumping power needed. Hence, our fine-grained flow-rate control has a much lower pumping power overhead. The fine-grained flow-rate control provides an additional design freedom to balance the pumping power and control circuit overheads.

### 8.2.5 *Minimization of Cooling Effort*

In most thermal management schemes, a maximum allowable temperature is set as the control target instead of a fixed total flow rate [174, 184]. As clustered microfluidic cooling is capable of producing an evenly distributed temperature profile, the cooling effort can be reduced and the total flow rate can be minimized effectively. In our proposed scheme, this is accomplished by incorporating the thermal modeling metric into the clustering algorithm. The procedure is described as follows:

- Step 1: Perform channel clustering according to the power information.
- Step 2: Simulate the steady state temperature of the 3D stack.
- Step 3: Record the current maximum steady state temperature of silicon cells as $T_{max}$.
- Step 4: Determine scaling factor.

$$\varphi = \frac{\overline{T_{max}(Ch)} - T_a}{\overline{T_{max}(Ch)} + (T_{th} - T_{max}) \cdot (1 - \frac{g_{S_{i,y}} + g_{S_{i,wall}}}{nr \cdot g_{S_{i,Ch}}}) - T_a} \qquad (8.8)$$

where $\overline{T_{max}(Ch)}$ denotes the average temperature of the channel sections that are in contact with the silicon cells of maximum temperature, and $T_{th}$ denotes the temperature threshold. $g_{S_{i,y}}$, $g_{S_{i,wall}}$ and $g_{S_{i,Ch}}$ denote the thermal conductances from silicon circuit cell to silicon circuit cell along the channel direction, from silicon circuit to channel wall cell, and from silicon circuit cell to channel cell, respectively.

- Step 5: update the flow rate settings to $Q_{new} = \varphi \cdot Q_{old}$.

The scale factor $\varphi$ provides a rough guide on the minimization of cooling effort. When flow rate is minimized, the maximum circuit temperature will approach the threshold. The temperature of channel cells in contact with the hottest circuit cell will also rise accordingly to remove the same amount of heat at low flow rate. Therefore, the flow rate is inversely proportional to the inlet-outlet temperature difference. In (8.8), the rise in temperature of channel cells in contact with the hottest spot is estimated as a scaled rise in temperature of the hottest spot. The use of thermal conductance at a high flow rate tends to overestimate the channel temperature while the inclusion of silicon circuit to wall flux tends to underestimate the channel temperature. This kind of offset makes the estimation relatively accurate.

As such, the total flow rate is reduced by $(1 - \varphi)$. However, since the overall steady state temperature of the 3D stack is raised, the thermal gradient will be amplified as well. A positive effect of such amplification is that a higher thermal gradient induces a bigger heat flux among the silicon cells to swiftly spread the temperature away from the hot spots. This has added a safety margin spontaneously into our minimization method to ensure that the maximum temperature will not exceed the thermal limit.

Based on the steady state simulation results, the total cooling effort can be further minimized by the proposed minimization steps. By finding the hottest point of the 3D stack under clustered microfluidic cooling and its nearby channel temperatures, the scaling factors $\varphi$ are calculated and the steady state thermal conditions are re-simulated after reducing every channel's flow rate by $(1 - \varphi)$. The maximum temperatures after minimization are reported in Table 8.5. It can be seen that the minimization method accurately predicts the flow rate demand under certain thermal constraints with a suitable safety margin in both ideal and practical cases. For the purpose of comparison, uniform flow rate thermal management is also implemented by a recursive decrease of flow rate until the maximum temperature of the 3D stack approaches the maximum temperature limit of $358.15K$ $(85°C)$. For a fair comparison, the maximum temperature after minimization is further pushed towards the thermal limit by reducing the flow rate proportionally. Some key findings are summarized in Table 8.6,

Table 8.5   Maximum temperature after minimization of cooling effort.

| Case No. | $\varphi$(Case 5) | Max Temperature (Case 5) (K) |
|---|---|---|
| Case 1 | 0.35 (0.38) | 354.06 (352.84) |
| Case 2 | 0.37 (0.36) | 352.52 (352.16) |
| Case 3 | 0.36 (0.38) | 354.3 (352.72) |
| Case 4 | 0.34 (0.36) | 356 (353.99) |

Table 8.6 Comparison of the cooling effort under thermal constraint.

| Case No. | | Case 1 | Case 2 | Case 3 | Case 4 |
|---|---|---|---|---|---|
| $T_{max}$(K) | Uniform | 357.28 | 357.73 | 357.27 | 358.01 |
| | Cluster | 357.85 | 357.43 | 358.01 | 358.1 |
| | Case 5 | 357.69 | 358.4 | 357.55 | 357.62 |
| $Q_{total}$ | Uniform | 119.95 | 112.58 | 120.01 | 117.72 |
| | Cluster | 91.93 | 94.7 | 94.97 | 93.42 |
| | Case 5 | 95.69 | 90.24 | 96.02 | 95.03 |
| Flow Rate Saving | | 21.80% | 17.90% | 20.40% | 19.96% |

where the total flow rate is measured in $ml/min$. It is shown that with the clustered cooling scheme, the total flow rate can be saved by up to 21.8% under the same thermal constraint.

## 8.3 Summary

In this chapter, thermal management of 3D ICs with microfluidic channels is presented. A cyber-physical thermal management by demand prediction-and-correction capability and fine-grained microfluid flow-rate adaptation is demonstrated. A novel channel clustering algorithm to guide the grouping of microfluidic channels for a customized fine-grained cooling based on power distribution and effort minimization for thermal management of 3D IC is illustrated. From the experiment results, it can be concluded that clustered microfluidic cooling guided by the presented channel clustering and flow rate allocation algorithms effectively reduces the peak temperature and thermal gradients.

# PART 4
# I/O Management

# Chapter 9

# Power I/O Management

## 9.1 Introduction

In the previous chapters, the modeling of interconnects and thermal management by TSVs and microfluid is studied. In a 3D integrated many-core microprocessor with large amounts of memory [8, 11–13, 16, 21, 55, 76, 104, 108, 132, 199–202], power management at the system level is one of the key issues to be handled along with thermal dissipation. Thermal dissipation in a 3D IC is discussed in the previous chapters. In this chapter, we focus on power management. I/Os utilized for power delivery are called power I/Os.

In a many-core microprocessor the power demands of cores will be different and varies with time, thus, providing an uniform voltage-level can result in high power density. To avoid a dark-silicon dilemma for many-core microprocessors, effective dynamic-voltage-scaling (DVS) [51–54] based power management has to be developed to provide cores with multi-level voltages at a scale of hundreds or thousands of cores. As such, supplying multi-level supply voltages with maintenance of low power density has become an emerging issue to address [32, 38, 203, 204].

From a physical hardware perspective, off-chip power converters may not be scalable for the surge in current demand by 3D many-core microprocessors due to long delivery latency, large delivery loss and severe delivery integrity [205]. On-chip power converters [32, 38, 52, 53, 203, 204, 206–208] are explored to provide prompt DVS power management with efficient power delivery. Since the chip area is quite limited for many-core microprocessors and the on-chip power converters may occupy considerable amount of area due to a non-scalable inductor, one power converter per-core based design cannot be deployed for the power management of many-core microprocessors. As such, there is a dire need to develop a reusing scenario that can fully utilize the on-chip power converters. What is more, by integrating a large number of cores on one chip, the remaining area is quite limited for on-chip power converters with a buck inductor. A single-inductor-multiple-output (SIMO) power converter [204, 209, 210] can be utilized to save area. One common single buck inductor is deployed to provide multi-level voltages in a time-multiplexed manner. The capability of the SIMO converter, however, still has limited scalability for many-core microprocessors. 3D integration introduces additional room for the integration of on-chip power converters. The work in [38] has demonstrated the possibility to design on-chip power converters integrated with 64-tile network-on-chip in 3D/2.5D manner. As such, it is meaningful to explore 3D designs that can provide effective demand-supply matching for DVS power management of large-scale cores and converters. Optimization techniques to overcome the thermal reliability issues as discussed in previous chapters can be implemented in 3D integration. An additional advantage of 3D integration is its short interconnect lengths, which is the key for power delivery.

From the cyber management perspective [19, 211–213], power management for the many-core microprocessor will not be same as the one for the traditional single-core microprocessor, because in a many-core microprocessor, there may exist multi-time-scale demands of supply voltages from different cores. Different power management schemes for many-core microprocessors are explored in [32, 38, 52–55, 203, 204, 206, 214–217], but the main challenge for a scalable DVS power management is still not resolved. In [32, 52, 54], voltage-frequency islands are utilized for power management of many-core microprocessors. However, as each core is statistically assigned with one fixed island, such a voltage/frequency assignment cannot be optimal with response to the time-varying characteristics of workloads. On the other hand, [53, 203] introduces the concept of a time-grained power management. However, such a power converter per-core based power management may not be scalable for large number of microprocessors. The recent work in [217] utilizes a controlled switch network to connect a set of cores to a set of power converters with space multiplexing but ignores the possibility of time multiplexing. A space-time multiplexing based power management in [55] addresses a dynamic power management with reconfigurable switch network to provide a demand-supply matching between many-core microprocessors and power converters. It has full flexibility in terms of power I/O connections as well as in reducing the number of power converters.

There exists a similarity between smart power management of a many-core microprocessor and a smart-grid, though at different time-scale with different workload behaviors. Thereby, the study of workload behavior with classification and also demand-response method can be leveraged from smart-grid management [218, 219] to deal with the large-scale on-chip demand-supply matching problem. In addition, workload balancing and peak-power reduction can also be addressed in the proposed approach.

In this chapter, we present a DVS power management for a many-core microprocessor, for small-scale and large-scale systems, followed by workload scheduling for workload balancing and peak-power reduction. The proposed power I/O management is verified by a system-level behavior model implemented in SystemC-AMS for an up to 64-core microprocessor. The physical design parameters are based on a $130nm$ CMOS process with TSV models. The power traces are generated from SPEC2000 benchmarks [136].

## 9.2 3D Power I/O Management

As aforementioned, power management of many-core microprocessor using on-chip power converters is more effective. The DVS power management with the use of SIMO power converter is presented here. Based on a 3D reconfigurable power switch network, space-time multiplexing (STM) based DVS power management is discussed for demand-supply matching between many-core microprocessors and multi-level on-chip power converters. The power switch network can be configured to perform space-time multiplexing between power converters and cores with connections by vertical through-silicon-vias (TSVs) in 3D, which can be formulated as two subproblems: resource allocation of power converters and workload scheduling. The objective of resource allocation is to achieve an optimal solution with the minimum number of power converters, while satisfying the constraints of both demands from different cores and hardware limitations from power converters. In order to solve the demand-supply matching problem for many-core microprocessors at a large-scale, integer-linear programming (ILP) and adaptive clustering of cores is deployed by learning and classifying the power-signature pattern of workloads. In general, similar workloads will be distributed to a number of cores for parallel computation i.e., thread-level parallelism. As such, those cores with similar workloads will show similar power-signature patterns and hence can be clustered together with a similar voltage-level. This is different from the single-core DVS that depends on the load current. What is more, power-signature of workloads with different patterns are initially classified by their magnitude levels as groups, such that power converters are allocated to be shared in the space between different groups, called *space-multiplexing*. In each group, power converters are further reused among different subgroups, formed based on their phases at different time instants, called *time-multiplexing*. Afterwards, the workload scheduling can be performed in a demand-response fashion, where workload is the amount of task performed on a core in one time-slot. The workloads on each of the allocated power converters are measured with available slacks determined. Based on the available slacks, the workload scheduling is performed without violating the workload priorities resulting in workload balancing and peak-power reduction.

### 9.2.1 *System Architecture*

The 3D many-core microprocessor system architecture shown in Figure 9.1 is utilised for power management and is basically composed of two tiers. The bottom tier is for power management, which includes arrays of power converters and power switches. Each power converter is of the SIMO type, capable of supplying multi-level voltages by one buck inductor (See Figure 9.2). The top tier includes array of many-core microprocessors. In between these

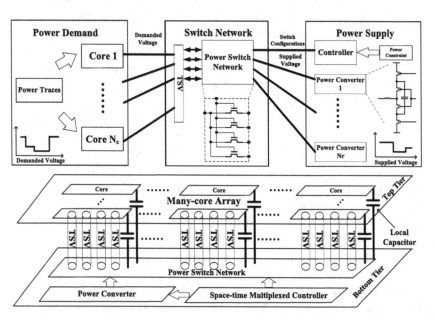

Fig. 9.1  3D reconfigurable power switch network for demand-supply matching between on-chip multi-output power converters and many-core microprocessors.

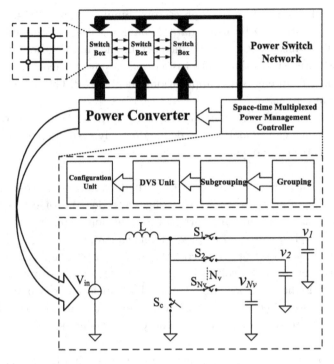

Fig. 9.2 Functional units of space-time multiplexing based power management for DVS with SIMO power converter.

two array-structured tiers, there are through-silicon-vias (TSVs), controlled by power switches, to connect power converters and cores. Note that with the use of through-silicon interposers (TSIs), a 2.5D integration of multi-core microprocessor and on-chip power converters is demonstrated with silicon prototype in [38]. Moreover, there is one local super-capacitor for each core, working as local power storage to supply voltage when a power converter is not available during multiplexing. The design and dimensions of TSVs are optimized for both speed of reconfigurability, the maximum driving current, and the thermal conductivity. One needs to note that 3D integration does not have to mean the traditional memory-logic integration. Similar to the recent work by IBM in [38], the 2.5D/3D integration can be utilized for on-chip power management efficiently, as well as scalability for large numbers of many-core microprocessors. What is more, the 3D architecture with space-time multiplexing proposed here has further provided the flexibility and also improved the efficiency when utilizing power converters for many-core microprocessors. Similar to [32, 53, 54], we evaluate the performance by using data-domain specified benchmark set like SPEC2000, which contains both memory-bound and cpu-bound applications with predictable data-patterns as well as power traces. Additionally, benchmarks from embedded applications such as MPEG4, JPEG decoder etc., can also be used.

To perform DVS power management for large-scale many-core microprocessors, one can model it as a demand-supply system composed of the following three components:

- *Power Demand*: a set of cores $C$ with demanded voltage-levels with set-size $N_c$. Each core $c_i$ has a voltage-level demand of $v_d(c_i)$ to meet the deadline of its running workload. In addition, $v_a(c_i)$ is the allocated voltage-level to core $c_i$ after power management.
- *Power Supply*: a set of power converters $R$ with set-size $N_r$. Each power converter outputs the voltage-level $v(r_i), v(r_i) \in V$, to supply the cores, where $V$ is the set of available voltage-levels before power management.
- *Power Switch Network*: a set of reconfigurable switch-boxes $SW$ with set-size $N_s$ to connect between $R$ and $C$ for demand-supply matching.

The power management circuit for DVS is shown in Figure 9.2. Initially, the voltage and current sensors sample voltage and current values from the cores as power profile. By tracking power profiles of cores, the demanded voltage-levels of cores for the next period of control can be tracked. The value for the next period of control is predicted based on the pre-stored training look-up-table. The data analytic of workloads can be performed to configure STM by learning and classifying power-signature patterns of workloads. Next, the DVS power management unit decides the optimal STM configuration that can match the demand of cores with supply from the minimum number of power converters. Figure 9.3 shows how to perform STM by on-chip SIMO power converters [209] utilizing a single inductor to provide multiple voltage-levels. The objective is to design SIMO power converters configured to satisfy the demand from cores. Based on the configuration of switches $(S_1, \ldots, S_{N_g})$, a corresponding voltage-level will be generated at the outputs of power converters to each group, divided in space, i.e., space-multiplexing. Power converters allocated to one group can be further reused by cores among subgroups divided in time, i.e., time-multiplexing. As such, one power converter is reused maximally to connect with one core at one allocated space-slot and time-slot.

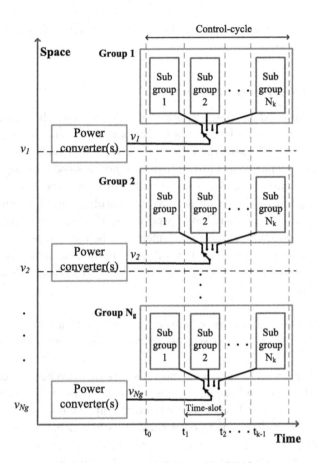

Fig. 9.3 Space and time multiplexing with on-chip SIMO power converters.

Note that in the traditional island-based [220] and SIMO-based [204, 209, 210] power management approaches, the connections between power converters and cores are assumed to be fixed, which is not feasible in providing the matched supply voltage-levels to many cores at the large-scale. The power management technique discussed in the next few sections is scalable for a large number of cores but with the assumption that the power management block including controller, switches, power converters and power delivery network are all in one layer separate from the layer of cores. As such, a different physical design may be required when routing a power/ground network to deliver to the layer of cores. Moreover, packaging a large number of power converters as well as cores may increase power density, and hence a thorough cooling design is required to satisfy thermal reliability concerns.

### 9.2.2   *Problem Formulation*

As aforementioned in the beginning of this chapter, the primary challenge here is to solve a large-scale DVS power management with matched demand-supply. Though there exists various workloads with different power-signature patterns, most of them can be classified by magnitude and phase when similar workloads are distributed to different cores. As such, if one can perform clustering of cores by learning and classifying power-signature patterns of distributed workloads, the complexity of demand-supply matched DVS power management can be reduced accordingly.

With the further consideration of the minimum number of power converters, one can formulate a resource (power converter) allocation subproblem as follows.

*Subproblem 1: Resource allocation problem is to decide the minimum number of power converters such that demands from cores can be satisfied.*

What is more, there may exist power slacks that can be utilized without violating the workload execution priority or deadline. One can vary/shift workloads on an over-loaded power converter at one time-slot to an other time-slot with under-loaded power converters in a demand-response fashion. As such, the peak-power can be reduced, and the workload can be balanced at power converters, which can be formulated as the second subproblem below after the first subproblem is done.

*Subproblem 2: Workload scheduling problem is to delay over-loaded workloads to under-loaded time-slots based on availability of slack and without violation of priority.*

To solve the above mentioned demand-supply matching by space-time multiplexing, an integer linear programming (ILP) based optimization for small-scale and adaptive clustering for large-scale problems are utilized. Before discussing more details of the algorithm, we redefine the resource allocation problem as follows:

*Resource allocation for STM:* there are $N_r$ power converters shared spatially among $N_c$ cores, connected by $N_s$ reconfigurable power switches with each power converter capable of switching among $N_v$ different voltage-levels at a fixed time-slot $H$ to supply multiple voltage-levels simultaneously.

## 9.3 ILP based Optimization

The space-time multiplexing (STM) problem was solved by an integer linear programming (ILP) in [55], with shown reformulated problem as a below.

*ILP Optimization of STM:* there are $N_r$ power converters shared spatially among $N_c$ cores, connected by $N_s$ reconfigurable power switches with each power converter capable of switching among $N_v$ different voltage-levels at a fixed time-slot $H$ to supply multiple voltage-levels simultaneously.

To perform ILP, first the constraints to be satisfied have to be defined. Due to physical hardware limitations, a power converter can be connected to a core only if the following constraints are met: (i) the maximal power converter inductance current does not exceed its maximal value, $I_L$; and (ii) the maximal core voltage-drop is within a specified value $\Delta V$ during multiplexing. Thus, one can formulate the following ILP optimization to determine the STM configuration.

One can have the following reformulation in the form of linear equations as

$$\text{min:} \quad \sum_{i=1}^{N_c} \sum_{j=1}^{N_r} \sum_{v=1}^{N_v} v_v \cdot x_{ij}^v$$

$$\text{s.t.:} \quad \text{(i)} \ \sum_{j=1}^{N_r} \sum_{v=1}^{N_v} x_{ij}^v = 1, \forall 1 \le i \le N_c$$

$$\text{(ii)} \ \sum_{j=1}^{N_r} \sum_{v=1}^{N_v} v_v \cdot x_{ij}^v \ge v_d(c_i), \forall 1 \le i \le N_c$$

$$\text{(iii)} \ \sum_{i=1}^{N_c} i_v \cdot x_{ij}^v \le I_L, \forall 1 \le j \le N_r, 1 \le v \le N_v$$

$$\text{(iv)} \ \sum_{i=1}^{N_c} \sum_{v=1}^{N_v} x_{ij}^v \le 1 + \frac{\Delta V \cdot C_L}{I_{max} H}, \forall 1 \le j \le N_r$$

$$\text{(v)} \ N_{min} \le \sum_{i=1}^{N_c} \sum_{v=1}^{N_v} x_{ij}^v \le N_{max}, \forall 1 \le j \le N_r.$$

$$(9.1)$$

In (9.1), the boolean variable $x_{ij}^v$ equals 1 if and only if the core $c_i \in C$ is supplied by power converter $r_j \in R$ with the voltage-level $v_v \in V$, as explained in (9.2).

$$x_{ij}^v = \begin{cases} 1 & c_i \text{ supplied by } r_j \text{ at voltage-level } v_v \\ 0 & \text{otherwise} \end{cases} . \qquad (9.2)$$

Therefore, the STM problem is now simplified to minimize (9.1) with the corresponding constraints being satisfied. This implies that the total voltage-levels allocated to cores are minimized. The constraints defined in (9.1) implies (i) each core is connected to at most one power converter at a particular time-slot; (ii) the allocated voltage-level must satisfy the demands of the core; (iii) the maximal inductance current does not exceed its maximal value $I_L$; (iv) the maximal core voltage-drop at any time instant $H$ for a core of capacitance $C$ does not exceed $\Delta V$; and (v) each power converter has minimum and maximum limits on the number of cores it could connect. The first two constraints ensure that adequate voltage is supplied to each core, the next two constraints ensure that inductance current does not exceed its limits, and core voltage drop is minimal, and the last constraint ensures that no power converter will be under– or over–utilized.

**Solution to Subproblem 1** The reformulated STM problem can be solved by a linear program *lp_solve* [221] deployed on one of the microprocessor cores with typical solving time ranging from microseconds [55], which is faster when compared to off-chip converter–based DVS management in the scale of seconds. Resource allocation can be performed using the above mentioned ILP optimization, however, the runtime increases exponentially with the number of cores. In the following, a power-signature learning based adaptive clustering is proposed to address the scalability problem for DVS power management of many-core microprocessors.

## 9.4　Space-time Multiplexing

As the number of cores increases, the complexity of solving ILP and the runtime may increase as well. As such, a scalable solution is required for resource allocation when dealing with a 3D many-core system. To perform resource allocation for large-scale systems in less time, learning and classification of the workload's power-signature pattern are employed to cluster cores by adaptive clustering, developed as discussed below. The resource allocation of power converters is then performed in both space– and time–based on the clustered cores. Solution for demand-supply matching with fewer power converters, i.e., subproblem 1 is discussed first followed by peak-power reduction and workload balancing by demand-response based workload scheduling. The main assumption here is based on the observation that similar workloads will be distributed to a number of cores for thread-level parallelism. As a result, they will have similar power-signature patterns and can be clustered together with the same voltage-level.

### 9.4.1 Adaptive Clustering

The adaptive clustering of cores is done by learning the similarity of power-signature patterns of workloads. The high-precision power profiles may not be necessary for clustering, and hence the first step of learning power-signature is to have a power-signature extract by envelope.

#### 9.4.1.1 Power-signature Extraction

Before proposing solutions for above mentioned problems, a few definitions are presented below.

*Control-cycle:* the amount of time required to finish all the allocated workloads in a group.

*Time-slot:* the amount of time required to finish all the allocated workloads in a subgroup.

Since it is impractical to perform power management with the use of continuous data of power profiles, one needs to extract the power-signature from power profiles. In the following, the procedure to obtain the power-signature pattern of one workload in one control-cycle by extracting the *envelope* from the power profile is presented. Based on the extracted peak-power envelope, or *power-signature*, one can build a workload behavior model to be utilized in the following resource allocation, as well as workload scheduling.

Assume that in one control-cycle $T^i$ for $i$-th group $g_i$, $g_i \in G$, having $N_k$ number of subgroups, each core is assigned with one workload. The relation

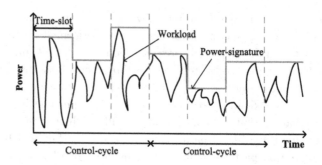

Fig. 9.4    Power-signature pattern extracted in each time-slot of control-cycle.

between control-cycle $T^i$ and time-slot $T^i_j$ is

$$T^i = \sum_{j=1}^{N_k} T^i_j. \tag{9.3}$$

As such, in one time-slot $T^i_j$, power-signature (peak-power envelope) $P_{s_z}$ is extracted for workload $p_z(t)$ from core $c_z$, $c_z \in C$ of one subgroup by

$$P_{s_z}(T^i_j) = max(p_z(t)). \tag{9.4}$$

This is repeated for the whole control-cycle $T^i$. Thus, the peaks are extracted and a peak envelope is formed. Power-signature extraction by forming peak-power envelope is shown in Figure 9.4. Power-signature is indicated by the blue line. In the example shown in Figure 9.4, the control-cycle is comprised of 3 time-slots, indicated by the red dotted line. The power-signature for core $c_i$ with power profile $p_i$ is denoted by $P_{s_i}$ ($P_{s_i} = max(p_i)$). Based on the knowledge of power signatures of workloads, classification of cores can be performed in two steps; namely grouping in space and subgrouping in time.

In order to allocate the voltage-level, load power has to be tracked and predicted. The prediction of power level is performed using the auto-regression (AR) algorithm [222]. At sampling interval $t$, based on the previous recorded load power values $p_i(t), p_i(t-1), p_i(t-2), \ldots, p_i(t-M)$ the transient power $p_i(t+1)$ needed for the next time instant can be predicted by

$$p_i(t+1) = \sum_{j=0}^{M} a_j p_i(t-j) + \epsilon \tag{9.5}$$

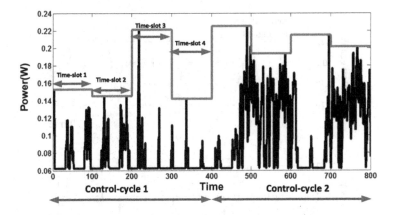

Fig. 9.5 Extraction of power-signature for workloads.

where $a_j$ is the auto-regression coefficient, $\epsilon$ is the prediction error and $M$ is the order of the prediction model. AR coefficients can be calculated based on the least-squares method. Prediction is performed in every control-cycle and accordingly, the predicted power is calculated, and the corresponding voltage-level is allocated. This makes the power converters allocation in a runtime fashion. It is assumed, ideally, that the driving capability of a power converter is independent of its voltage-level.

Figure 9.5 shows one example of the extracted power-signatures of workloads. The extracted power-signature is shown by the red line with a control-cycle of $400ns$ and time-slot of $100ns$. In two control cycles, there are a total of 8 power envelopes extracted for 8 time-slots.

To decide the demanded voltage-level under space-time multiplexing power management, the power-signature needs to be tracked and predicted. Here, power-signature tracking and prediction is performed based on (9.5). The order of prediction $M$ is set as 8 to guarantee the precision of prediction, and has a prediction error of 0.3% on average for SPEC2000 benchmarks. Based on the predicted power consumptions, the required voltage-level is looked up in the LUT. Figure 9.6 shows the power tracking and prediction of the core under benchmark *gcc*. One needs to observe from Figure 9.6(a) that the predicted power denoted by the red line using AR closely matches the actual power demand denoted by the blue line. Based on the predicted power-signature values and power-voltage pairs in the look-up-table, the supply voltage-levels to be allocated to cores are shown in Figure 9.6(b). For example, when the predicted power-signature

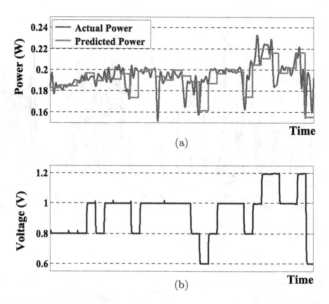

Fig. 9.6    Runtime power tracking and prediction for benchmark *gcc*: (a) power prediction; (b) voltage-level transition.

magnitude is between $0.17W$ and $0.19W$, a voltage-level of $0.8V$ needs to be supplied.

### 9.4.1.2   *Space Multiplexing: Grouping*

The cores with similar power-signature patterns in terms of magnitudes are grouped into one. For example, $z$-th group $g_z$, $g_z \in G$, of cores can be formed by the following criteria

$$g_z = \{c_i; v_d(c_i) = v_d(c_j) = v_z, \forall i, j = 1, \dots N_c, z \leq N_v\}. \tag{9.6}$$

Here, $v_z$, $v_z \in V$ represents the voltage-level; and $v_d(c_i)$ ,$v_d(c_i) \in V$ represents the voltage-level demand of core $c_i$, $c_i \in C$.

     Based on the power-signature magnitude levels, different groups are formed. Each group may contain a different number of cores which have similar power-signature magnitudes but may differ in power-signature phase. The number of cores in a group can change at different control-cycles. Note that the grouping process is based on the comparison of levels and hence has less complexity of computation.

### 9.4.1.3 *Time Multiplexing: Subgrouping*

For cores in the same group with similar magnitude levels, they can be further classified based on the power-signature phase, i.e., execution behavior with time. The study of power-signature phases cannot be simply classified based on power-signature magnitudes. Considering a set of power-signature patterns, subgroup $k_s$, $k_s \in K$, can be formed by the following criteria

$$k_s = \{c_i; (v_d(c_i) = v_d(c_j) = v_z)\&(P_{s_i} \sim Ps_j), \forall i, j = 1, ...N_c\}. \quad (9.7)$$

Here, $P_{s_i}$ represents power-signature of core $c_i$, $c_i \in C$, in one time-slot; $v_d(c_i)$, $v_d(c_i) \in V$, represents the demanded voltage-level of core $c_i$; and $v_z$, $v_z \in V$, represents the voltage-level allocated to group $g_z$.

To form subgroups based on the power-signature phases, the similarities between power-signature phases can be exploited. To find similarities between phases of power-signatures $P_{s_i}$ and $P_{s_j}$ in a group having between $N$ power-signatures in one control-cycle, correlation in term of a covariance matrix has to be evaluated as

$$X = \frac{1}{N} \sum_{i,j=1}^{N} (P_{s_i} - \overline{P_s})(P_{s_j} - \overline{P_s})^T \quad (9.8)$$

where $\overline{P_s}$ is the mean of all power-signatures ($\frac{1}{N} \sum_{i=1}^{N} (P_{s_i})$).

Based on the order of covariance matrix $X$, the number of subgroups $N_k$ can be analyzed by the singular-value-decomposition (SVD) of $X$ as

$$X = \mathbf{U} \times \mathbf{S} \times \mathbf{V}^{-1} \quad (9.9)$$

with

$$X = \begin{pmatrix} x_{1,1} & x_{1,2} & \cdots & x_{1,N} \\ x_{2,1} & x_{2,2} & \cdots & x_{2,N} \\ \vdots & \vdots & \ddots & \vdots \\ x_{N,1} & x_{N,2} & \cdots & x_{N,N} \end{pmatrix}; \quad \mathbf{U} = \begin{pmatrix} \mathbf{u}_{1,1} & \mathbf{u}_{1,2} & \cdots & \mathbf{u}_{1,N} \\ \mathbf{u}_{2,1} & \mathbf{u}_{2,2} & \cdots & \mathbf{u}_{2,N} \\ \vdots & \vdots & \ddots & \vdots \\ \mathbf{u}_{N,1} & \mathbf{u}_{N,2} & \cdots & \mathbf{u}_{N,N} \end{pmatrix};$$

$$\mathbf{S} = \begin{pmatrix} \mathbf{s}_{1,1} & 0 & \cdots & 0 \\ 0 & \mathbf{s}_{2,2} & \cdots & 0 \\ \vdots & \vdots & \ddots & \vdots \\ 0 & 0 & \cdots & \mathbf{s}_{N,N} \end{pmatrix}; \quad \mathbf{V} = \begin{pmatrix} \mathbf{v}_{1,1} & \mathbf{v}_{1,2} & \cdots & \mathbf{v}_{1,N} \\ \mathbf{v}_{2,1} & \mathbf{v}_{2,2} & \cdots & \mathbf{v}_{2,N} \\ \vdots & \vdots & \ddots & \vdots \\ \mathbf{v}_{N,1} & \mathbf{v}_{N,2} & \cdots & \mathbf{v}_{N,N} \end{pmatrix}.$$

$$(9.10)$$

Matrices $\mathbf{U}$ and $\mathbf{V}$ are orthogonal matrices with $\mathbf{S}$ as the diagonal matrix. One needs to note that SVD-based workload characterization is deployed in off-line learning of workload data and look-up-table (LUT) built online. As such, the runtime can be efficient for large-scale problems, and is reported in Table 9.2.

Based on the rank analysis of $\mathbf{S}$, the number of subgroups $N_k$ is decided. A new matrix can be formed with $N_k$ independent vectors, extracted from either of the orthogonal matrices. Let the newly formed matrix be $\mathbf{V}_k$, assuming it is extracted from $\mathbf{V}$. The product of $\mathbf{V}_k$ with the covariance matrix $X$ will result in a reduced matrix $X_k$, which forms basis of the clustering for subgrouping.

$$X_k = X \times \mathbf{V}_k \qquad (9.11)$$

with

$$X_k = \begin{pmatrix} x_{1,1} & x_{1,2} & \cdots & x_{1,N_k} \\ x_{2,1} & x_{2,2} & \cdots & x_{2,N_k} \\ \vdots & \vdots & \ddots & \vdots \\ x_{N,1} & x_{N,2} & \cdots & x_{N,N_k} \end{pmatrix} ; \quad \mathbf{V}_k = \begin{pmatrix} \mathbf{v}_{1,1} & \mathbf{v}_{1,2} & \cdots & \mathbf{v}_{1,N_k} \\ \mathbf{v}_{2,1} & \mathbf{v}_{2,2} & \cdots & \mathbf{v}_{2,N_k} \\ \vdots & \vdots & \ddots & \vdots \\ \mathbf{v}_{N,1} & \mathbf{v}_{N,2} & \cdots & \mathbf{v}_{N,N_k} \end{pmatrix} .$$

$$(9.12)$$

Based on the reduced matrix, the subgrouping can be performed by selecting the maximum value in each column and assigning it to the corresponding subgroup.

The control time reported in Table 9.2 is for total power management, which includes not only the converter switching time but also the time for performing workload characterization. Please note that there are two parts in workload characterization. The first part is the off-line SVD-based learning of workload data, which may consume a long time. The second part is the on-line look-up-table based clustering and prediction of workload data, which can be accomplished within a few nanoseconds. As such, the runtime can be efficient for a large-scale the whole process of grouping and subgrouping based on svd is presented in Algorithm 9.1.

### 9.4.1.4    *Allocation of Power Converters*

Once the groups and subgroups based on power-signature pattern learning are formed, the maximum workloads of one subgroup can be determined. As such, the minimum number of power converters is determined to supply to that subgroup of cores. This results in a feasible solution to solve

**Algorithm 9.1** Adaptive clustering based space-time multiplexing

---

**INPUT:** Power profile matrix $P$ with power profile vectors $p_i$

1. In one control-cycle, extract power-signatures $P_{s_i}$ from power profile vectors $P_{s_i} = max(p_i)$
2. Perform grouping of power-signatures by magnitude $g_z = \{c_i; |P_{s_i}| = v_z\}$
3. For each group $g_z$, compute the covariance matrix $X \in X^{N \times N}$
4. Perform SVD: $X = \mathbf{U} \times \mathbf{S} \times \mathbf{V}^{-1}$
5. Determine number of subgroups: $N_k = rank(\mathbf{S})$
6. Compute the first $N_k$ singular-value vectors $v_1, ..., v_{N_k}$ of $\mathbf{V}$
7. Let $\mathbf{V}_k = [v_1, ..., v_{N_k}] \in R^{N \times N_k}$ and $X_k = X \times \mathbf{V}_k$
8. Add $i$-th core to $j$-th subgroup if $X_K(i, j)$ is maximum in the $i$-th row
9. Form $P_k$ matrices within the group by finding corresponding indices in power profile matrix $P$
10. Perform the same subgrouping process for all $N_g$ groups

**OUTPUT:** New clustered matrices $P_{z,k}$ $(z = 1, ..., N_g; k = 1, ..., N_k)$

---

subproblem 1, which can be rephrased as below.

$$\text{min:} \quad \sum_{j=1}^{N_g} r_j$$

$$\text{s.t.:} \quad \text{(i)} \ v_a(c_i) \geq v_d(c_i), \forall c_i \in C$$

$$\text{(ii)} \ d(r_j) \leq N_{max}, \forall r_j \in R. \tag{9.13}$$

If one can determine the minimum number of power converters $r_j$ for each group, the total number of power converters for $N_g$ groups can be correspondingly minimized. Note that constraint (i) guarantees that the supplied voltage-level $v_a(c_i)$, $v_a(c_i) \in V$, from the power converter will satisfy the demanded voltage-level $v_d(c_i)$, $v_d(c_i) \in V$, from core $c_i$, $c_i \in C$. Moreover, constraint (ii) imposes the driving ability $d(r_j)$ of each power converter to be $N_{max}$, i.e., the maximum number of cores to drive. The driving ability varies with the supplied voltage-level: the higher the voltage-level, the lower the number of cores that one power converter could drive. With the increase in supplied voltage-level by a power converter, the load current increases, thereby reducing its driving capability.

Next, the minimization in a required total number of power converters which can be solved by grouping and subgrouping is discussed. By grouping, power converters can be shared in space among $N_g$ number of groups, and subgrouping makes sharing of power converters inside one group in time. Based on the driving ability $d^j$ of power converters in group $g_j$, $g_j \in G$, and the maximum number of cores among different subgroups $max(c_j)$, the maximum number of power converters allocated can be determined as

$$r_{g_j} = \frac{max(c_j)}{d^j}. \qquad (9.14)$$

As such, for the whole system, the total number of power converters needed will be $\sum_{j=1}^{N_g}(r_{g_j})$, which is the minimum number to satisfy the demand-supply matching. Therefore, the proposed learning of large number of power-signatures of workloads can classify cores into groups and subgroups to allocate the power converters, and hence reduce the complexity for large-scale problems, in contrast to the previously developed method by ILP. In summary, the large-scale demand-supply matching can be efficiently solved by the above-mentioned two-step clustering in every control-cycle. The obtained $r_{g_j}$ represents the minimum number of power converters needed to satisfy demand-supply matching i.e., solution to subproblem 1.

The procedure for the space-time multiplexing is further summarized in Algorithm 9.1. As an example, the formulation of groups and subgroups of cores by the learning of power-signature patterns with the reuse of on-chip SIMO power converters in space and time is illustrated in Figure 9.7 and described below.

(1) In the first control-cycle, cores with power-signatures: $P_{s_1}$, $P_{s_2}$, $P_{s_3}$, $P_{s_4}$ and $P_{s_5}$ are at one power magnitude level, and other cores are at a different power magnitude level. As such, one can assign them to two groups with voltage-levels $v_1$ for group 1 and $v_2$ for group 2 ($v_1 > v_2$). Thus, power converters can be shared in space. We assume the driving ability of the power converter supplying voltage-levels $v_1$, $v_2$ to be 1 and 2, respectively.

(2) In group 2, based on power-signature phase similarity, cores with power-signatures $P_{s_1}$, $P_{s_2}$, $P_{s_3}$ are clustered to form subgroup 1 and cores with power-signatures $P_{s_4}$, $P_{s_5}$ are clustered to form subgroup 2. To satisfy the demand of a group, a subgroup with more cores is selected, i.e., subgroup 1, having 3 cores. The corresponding number of power converters required is calculated based on (9.14), and 2 power converters ($r_1$, $r_2$) are allocated to group 2 under voltage-level of $v_1$ and driving ability of 2.

(3) In the first time-slot, power converters are connected to the cores in subgroup 1 and in the next time-slot, power converters are connected

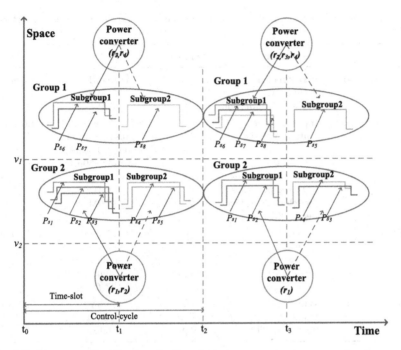

Fig. 9.7 Resource allocation by adaptive clustering: grouping by power-signature magnitude and subgrouping by power-signature phase.

to the cores in subgroup 2. Thus, power converters are shared in time by time-multiplexing.

(4) Similarly, for group 1, cores with power-signatures $P_{s_6}$ and $P_{s_7}$ are clustered to form subgroup 1; and the core with power-signature $P_{s_8}$ is allocated to subgroup 2 due to different power-signature phase. Considering subgroup 1 with 2 cores, the number of power converters is calculated and 2 power converters ($r_3$, $r_4$) are allocated under the voltage-level of $v_1$ and driving ability of 1.

(5) In the next control-cycle, due to changes in power-signature magnitude and phase, the core with power-signature $P_{s_5}$ is allocated to group 1, while other cores remain in the same group. Hence in group 2, subgroups 1 and 2 both will have 2 cores each. As such, 1 power converter supplying voltage-level $v_2$ can satisfy the demand.

(6) Whereas, in group 1, cores with power-signatures $P_{s_6}$, $P_{s_7}$ and $P_{s_8}$ form subgroup 1 due to the similarity in power-signature phase; and the core with power-signature $P_{s_5}$ is allocated to subgroup 2 due to the difference in phase.

Table 9.1 System settings of 3D many-core microprocessors, on-chip power converters, TSVs and power switches.

| Item | Description | Symbol | Value | Size |
|---|---|---|---|---|
| Microprocessor | Performance | N.A. | 410 DMIPS | $1.5\,mm^2$ |
| | Frequency | $f_c$ | 250MHz | |
| | Power Consumption | $P_c$ | 0.4W | |
| Power Converter | Input Voltage | $V_{in}$ | 2.4V | $1.6\,mm^2$ |
| | Output Voltage | $V_{out}$ | 0.6V, 0.8V, 1.0V, 1.2V | |
| | Load Current | $I_L$ | 120mA, 150mA, 220mA, 350mA | |
| | Number of Phases | N.A. | 2 | |
| | Inductor per Phase | L | 1nH | |
| | Switching Frequency | $f_s$ | 50–200MHz | |
| | Peak Efficiency | N.A. | 77% | |
| TSV | Length | $l$ | $25\mu m$ | $455\,\mu m^2$ |
| | Diameter | W | $5\mu m$ | |
| | Isolation Film | r | $120nm$ | |
| | Resistance | $R_{TSV}$ | $20m\Omega$ | |
| | Capacitance | $C_{TSV}$ | $37\,f\mathrm{F}$ | |
| Power Switch | Width | $w_s$ | 4mm | $520\,\mu m^2$ |
| | Length | $len$ | $130nm$ | |
| | Switching Time | N.A. | 300ns | |

(7) As such, in group 1, subgroup 1 will have 3 cores and subgroup 2 will have 1 core. To satisfy the demands, subgroup 1 having 3 cores is considered and the number of power converters needed will be 3. Instead of adding a new power converter, the power converter $r_2$ (previously allocated to group 2) can be used in group 1 to provide voltage-level $v_1$.

**ILP vs Adaptive Clustering**    The proposed system is validated by Matlab 7.12 and system-level models built from SystemC-AMS. Table 9.1 summarizes the system design specifications. All units are scaled or modeled at CMOS $130nm$ process. The specification of low-power MIPS microprocessor [223] is taken as the core model. Each core has a nominal frequency of $250MHz$ with maximal power consumption of $0.4W$. Benchmarks from SPEC2000 [136] are simulated by Wattch [196] simulator to generate power profiles. The extracted power-signatures from power profiles are used as workload models. Workloads are assigned to one core with a specified sequence, and a number of cores can have similar or same workloads. The typical control-cycle for power management is set to $400ns$.

A 2-phase multi-output power converter [207] is designed to generate four different voltage-levels. As the driving ability of a power converter depends on the supply voltage-level, driving abilities are set as 4, 3, 2 and 1 for voltage-levels $0.6V$, $0.8V$, $1.0V$ and $1.2V$ respectively. The look-up-table for power and voltage-levels is set as ($<0.17W$, $0.6V$), ($0.17W$-$0.19W$, $0.8V$), ($0.19$-$0.21W$, $1.0V$) and ($>0.21W$, $1.2V$). Moreover, the inductance value in the power converter is $1nH$ per phase to support the maximum current on the buck inductor. Such an inductor requires an area of $0.25mm^2$, occupying 30% of the area of the power converter. The local super-capacitor for each core is set as $1\mu F$ (max value) to support the time-multiplexing scheme between subgroups. The value of the local capacitor needs to be selected based on targeted application and available on-chip space. The design of an on-chip power converter thereby needs to consider the limitation of inductor and capacitor area, which are both placed in 3D integration and hence, the minimum area overhead to core area.

In addition, the vertical TSV [201] with the size of $455\mu m^2$ works as a connection between cores and power converters. According to the model in [76], it has a DC-resistance of $20m\Omega$. Considering the maximum current of $330mA$, the IR-drop of TSV is around $7mV$, which is quite small. Note that the capacitor of TSVs is in $fF$ scale and hence does not influence the load capacitance. What is more, for each TSV channel, one switch box is assigned with $N_r$ power switches to support the core-converter connection. The switch box offers a compact reconfigurable unit driven by the controller. The power switch inside each switch box occupies $520\mu m^2$ and is able to deliver the maximum core current with switching time of $300ns$. As such, the TSV coupling is also quite small to be considered under such a low-activity switching.

**Runtime and Power Saving Comparison** Experimental results using different sets of benchmarks from SPEC2000 [136] on different numbers of cores are presented in Table 9.2. Performance comparison is made for the non-STM method, ILP-based STM method and the adaptive clustering based STM method. Firstly, we compare the power saving of ILP and adaptive clustering when compared to non-STM power management. Considering the benchmarks in set 1 and set 2, there is a power saving of 29.01% and 51.82% by the ILP and 37.00% and 43.98% by adaptive clustering, respectively. The power saving depends on the workload deployed on the core. On average, the respective power savings are 34.68% and 40.38% with the ILP and the adaptive clustering based power management compared to

Table 9.2 Comparison of average power consumption and controller runtime for STM by ILP and STM by adaptive clustering.

| Number of Cores | Benchmarks | Power per Core (mW) | | | Power Saving (%) | | Controller Runtime (ms) | |
|---|---|---|---|---|---|---|---|---|
| | | ILP | Adaptive | Non-DVS Clustering | ILP | Adaptive Clustering | ILP | Adaptive Clustering |
| 4 | Set 1: art, eon, lucas, wupwise | 279.50 | 248.00 | 393.71 | 29.01% | 37.00% | 7.30 | 7.73 |
| | Set 2: apsi, gcc, gzip, mcf | 168.32 | 195.69 | 349.34 | 51.82% | 43.98% | 9.50 | 10.73 |
| | Set 3: facerec, galgel, twolf, crafty | 224.95 | 233.32 | 366.14 | 38.56% | 36.27% | 7.20 | 6.42 |
| | Set 4: vortex, parser, mgrid, sixtrack | 240.06 | 237.84 | 385.85 | 37.78% | 38.36% | 10.70 | 10.30 |
| 8 | Set 1 + Set 2 | 223.17 | 221.85 | 371.53 | 39.93% | 40.29% | 25.00 | 20.68 |
| | Set 1 + Set 3 | 252.24 | 240.66 | 379.93 | 33.61% | 36.65% | 27.10 | 20.17 |
| | Set 1 + Set 4 | 260.04 | 242.92 | 389.78 | 33.29% | 37.68% | 37.00 | 24.74 |
| | Set 2 + Set 3 | 195.34 | 196.64 | 357.74 | 45.40% | 45.03% | 21.70 | 20.31 |
| | Set 2 + Set 4 | 202.65 | 216.77 | 367.60 | 44.87% | 41.03% | 30.40 | 25.70 |
| | Set 3 + Set 4 | 231.71 | 235.78 | 376.00 | 38.38% | 37.29% | 29.80 | 25.23 |
| 16 | All Sets | 309.38 | 277.25 | 373.36 | 17.22% | 25.74% | 50.80 | 47.32 |
| 32 | All Sets | 319.93 | 290.63 | 374.14 | 14.49% | 22.32% | 336.30 | 97.35 |
| 64 | All Sets | 284.56 | 220.44 | 387.04 | 26.48% | 43.04% | 187500.00 | 120.29 |
| | Average | 245.53 | 235.22 | 374.81 | 34.68% | 40.38% | N.A. | N.A. |

the non-STM power management. What is more, the runtime comparison between the ILP method and the adaptive clustering method is observed as well. When applying benchmarks in set 1 and set 2 on an 8-core system, the runtime of ILP and adaptive clustering is $25ms$ and $20.68ms$ respectively. When on a 32-core system, the runtime of ILP becomes $336.30ms$, whereas adaptive clustering takes $97.35ms$, which is nearly linearly increased with cores. Furthermore, when the number of cores is increased to 64, the runtime of ILP is nearly 1000 times slower than the proposed adaptive clustering based power management. Please note that the control time report in Table 9.2 is for the total power management, which includes off-line SVD-based learning of workload data and the converter switching time along with the prediction and allocation of voltage-levels using LUT. The power converter switching time and LUT–based prediction consumes few $ns$.

**Adaptive Clustering** Adaptive clustering of cores by learning the similarity of power-signature patterns of workloads is described. In the previous ILP based resource allocation, complexity lies in searching the large solution space to satisfy the constraints involved, whereas in adaptive clustering, groups and subgroups are formed resulting in smaller search space, reduced complexity and less runtime.

Adaptive clustering performed for 32-core and 64-core microprocessors and their corresponding results are presented here. Input power traces are first divided into 4 groups based on their power magnitudes, then in each group, subgroups are formed based on their power phases. Based on the extracted power-signatures, cores with similar power-signature magnitudes are grouped and provided with the same voltage-levels. Power converters are allocated at different space (voltage-levels) for space-multiplexing. Inside each group, cores with similar power-signature phases are further subgrouped. Power converters are further allocated in different time for time-multiplexing. Based on the maximal number of cores among subgroups of a group, the number of power converters for each group are determined and allocated.

Figure 9.8 illustrates the adaptive clustering result of 32-cores in two consecutive control-cycles. Different filling patterns represent different groups or voltage-levels. Numbers on the downright-corner of cores represent different subgroups. For example, in the first control-cycle, the 12-th core is assigned to subgroup 4 with voltage-level 1 (group 1). And in the next control-cycle, it is assigned to subgroup 1 in the same voltage-level. Similarly, in the first control-cycle, 7-th core is assigned to subgroup 3 with

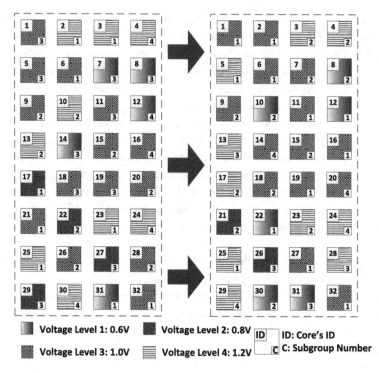

Fig. 9.8   Results of adaptive clustering for a 32-core microprocessor in two consecutive control-cycles.

voltage-level 1 (group 1) supplied by a voltage-level of $0.6V$. And in the next control-cycle, it is assigned to subgroup 2 with voltage-level 3 (group 3), supplied by a voltage-level of $1.0V$. This is how the voltage transition normally takes place with the aid of STM by adaptive clustering.

For the 64-core case, Table 9.3 summarizes the clustering results. The numbers in the table represent the core IDs. The runtime of the whole process is small and nearly $120ms$. One can observe that different groups and subgroups have different number of cores allocated and even some of the subgroups are left without any core, indicating that different power-signatures may have similar phases. For example, group 1 has been allocated with 14 cores, whereas group 2 has 11 cores, inside which subgroup 4 is empty but other subgroups of the group are allocated with cores. These unoccupied subgroups indicate the idleness of the power converter i.e., availability of slack. Considering group 1 in Table 9.3, the maximum number of cores in a subgroup among different subgroups is 6 and power converter

Table 9.3  Adaptive clustering result for 64-core microprocessor.

|         | Subgroup 1 | Subgroup 2 | Subgroup 3 | Subgroup 4 |
|---------|------------|------------|------------|------------|
| Group 1 | 31, 37, 52 58, 59, 63 | 33,43 | 7, 8, 14 | 12, 49, 54 |
| Group 2 | 17, 40, 41, 50 51, 56, 62 | 22, 42 | 27, 29 | N/A |
| Group 3 | 6, 21, 32 36, 39, 46 47, 64 | 9, 15, 16 20, 26, 28 35, 53, 55 | 1, 5 11, 18 19, 38 | N/A |
| Group 4 | 2, 3, 23, 25 34, 44, 45, 48 57, 60, 61 | 10, 13 | N/A | 4, 24, 30 |

driving ability 4, hence 2 power converters supplying a voltage-level of 0.6V is sufficient to drive cores in this group.

Next, the STM by adaptive clustering can lead to the reduction in number of power converters needed for demand-supply matching. When compared to two schemes, namely space-multiplexing (SM) and time-multiplexing (TM), STM takes advantage of both space-multiplexing (SM) and time-multiplexing (TM) with consideration of its driving ability. Table 9.4 shows the comparison of the number of power converters needed for a 32-core and 64-core case with the SM and TM schemes. One can observe a reduction of 55.00% (SM) and 35.71% (TM) in the number of power converters for a 32-core microprocessor, while 41.67% (SM) and 36.36% (TM) in number of power converters can be reduced for the case of 64-core. Therefore, STM by adaptive clustering can perform resource allocation with the minimum number of power converters to reduce the area overhead and also on-chip implementation cost.

### 9.4.2  Scheduling of Workloads

Once resource allocation is performed based on the learning of power-signature patterns, workload scheduling needs to be performed to reduce peak-power and achieve uniform workload balance. Recall that workload is the amount of work a core performs in one time-slot. A demand-response based workload scheduling will be developed towards uniform distribution with reduction in peaks on one power converter. Demand-response based workload scheduling is performed in two steps, namely, peak-power envelope extraction and peak reduction.

Table 9.4    Comparison of number of allocated power converters under different power management schemes.

|          |         | STM | SM | TM | STM/SM | STM/TM |
|----------|---------|-----|----|----|--------|--------|
| 32-core  | Group 1 | 1   | 2  | 3  | −50.00% | −66.67% |
|          | Group 2 | 1   | 2  | 2  | −50.00% | −50.00% |
|          | Group 3 | 3   | 7  | 5  | −57.14% | −40.00% |
|          | Group 4 | 4   | 9  | 4  | −55.56% | 0.00% |
|          | Total   | 9   | 20 | 14 | −55.00% | −35.71% |
| 64-core  | Group 1 | 2   | 4  | 6  | −50.00% | −66.67% |
|          | Group 2 | 3   | 4  | 7  | −25.00% | −57.14% |
|          | Group 3 | 5   | 12 | 9  | −58.33% | −44.44% |
|          | Group 4 | 11  | 16 | 11 | −31.25% | 0.00% |
|          | Total   | 21  | 36 | 33 | −41.67% | −36.36% |

### 9.4.2.1    *Slack Calculation*

Once the peak envelope of subgroup $k_j$, $k_j \in K$, is formed, it is compared with the threshold power $P_{th}(z)$ of group $g_z$ to determine *slack*, which is defined as the amount of extra workloads a power converter can handle without getting over-loaded at a time-slot, and is calculated as

$$s(z,j) = P_{th}(z) - P_s(T_j^z). \tag{9.15}$$

If the value of slack is negative then, the allocated power converter $r_j$, $r_j \in R$ is over-loaded and not capable of handling extra workload at that time-slot. After calculating the amount of slack, the workload on power converter $r_j$ can be rescheduled such that the priority is not violated. We call such a scheduling a *demand-response based workload scheduling*.

The procedure for scheduling by considering priorities $b_a$, $b_a \in B$, of workload $l_a$, $l_a \in L$, is described in Algorithm 9.2. It is deployed after resource allocation. Workload $l_a$, $l_a \in L$, assigned to subgroup $k_j$, $k_j \in K$, of group $g_z$, $g_z \in G$ is denoted as $l_a(z,j)$.

In the formulated algorithm, a set of voltage-levels $V$ is given as input along with workload priorities $B$ and calculated slacks $S$. It is aforementioned that a power converter can handle a workload $l_a(z,i)$, $l_a(z,i) \in L$, only if it has a positive slack $s(z,i)$, $s(z,i) \in S$, in subgroup $k_i$, $k_i \in K$, of group $g_z$, $g_z \in G$. When a power converter is over-loaded due to a larger workload or error in predicting power trace, the workload with lower priority will be shifted to another under-loaded power converter, as explained in Lines 2-8 of Algorithm 9.2. After workloads are scheduled (if needed) within the same group, workload scheduling between different groups is performed, and it is important that the allocated voltage-level after scheduling

---

**Algorithm 9.2** Demand-response based workload scheduling

---

**Require:** Initial set: voltage-levels $V$, workloads $L$, priorities $B$ and slack $S$

1: **if** $s(z,i) > 0$ **then**
2:    **while** $j > i$ **do**
3:       **if** $s(z,j) < 0, s(z,i) > 0$ && $\exists\, l_a(z,j)$ with $b_a{=}{=}1$ **then**
4:         $l_a(z,j) \rightarrow l_a(z,i)$
5:         $s(z,j) + +;$
6:         $s(z,i) - -;$
7:       **end if**
8:    **end while**
9:    **while** $v_y < v_z$ && $j > i$ **do**
10:      **if** $s(y,j) < 0, s(z,i) > 0$ && $\exists\, l_a(y,j)$ with $b_a{=}{=}1$ **then**
11:        $l_a(y,j) \rightarrow l_a(z,i)$
12:        $s(y,j) + +;$
13:        $s(z,i) - -;$
14:      **end if**
15:    **end while**
16:    **while** $j < i$ **do**
17:      **if** $s(z,j) < 0, s(z,i) > 0$ && $\exists\, l_a(z,j)$ with $b_a{=}{=}0$ **then**
18:        $l_a(z,j) \rightarrow l_a(z,i)$
19:        $s(z,j) + +;$
20:        $s(z,i) - -;$
21:      **end if**
22:    **end while**
23: **end if**

---

must satisfy its demanded voltage-level. The same is shown in Lines 9-15 of Algorithm 9.2. After high priority workloads are scheduled for over-loaded power converters, workloads with low priority on an over-loaded power converter can be scheduled to an under-loaded power converter within the group. Lines 17-22 of Algorithm 9.2 explain the scheduling of low priority workloads within a group. Low priority workloads can be delayed, ideally till the availability of slack. Though Algorithm 9.2 is defined for two levels of priority, it can be extended to multiple-levels.

Peak-power reduction can be shown by the shift in workloads on a power converter from one time-slot with negative slack to another time-slot having a positive slack. The example in Figure 9.9 shows the normalized

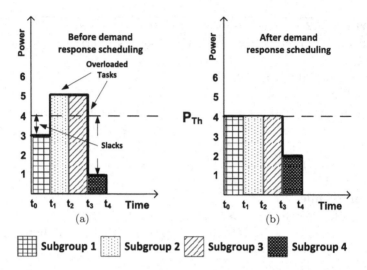

Fig. 9.9    (a) Workload before demand-response scheduling (b) Workload after demand-response scheduling with peak reduction and balancing.

peaks of four subgroups. Before performing demand-response based workload scheduling, subgroup 2 and subgroup 3 are over-loaded, and subgroups 1 and 4 have slacks for scheduling. The peak value in subgroup 2 and 3 is 5, which means that there are 5 peaks in those two subgroups. Peak-power reduction is then achieved with the comparison of the highest value in the subgroups before and after demand-response scheduling. After the demand-response scheduling, the peak value will be reduced to 4. So, a 20% peak-power reduction will be achieved. Workload balancing can be determined by calculating the standard deviation among cores between subgroups in a group. A larger standard deviation implies a less balanced workload.

### 9.4.2.2 *Scheduling of Workloads*

The aforementioned demand-response based workload scheduling can be deployed to solve subproblem 2, which can be reformulated as

$$\text{min:} \quad \sum_{j=1} \left| \sum_{z=1} s(z,j) \right| \tag{9.16}$$

$$\text{s.t.:} \quad P_s(T_j^z) < P_{th}(z).$$

The solution to this problem is to minimize the overall sum of slacks. This can be achieved by rescheduling workloads that overload the power converter. Based on the value of slack for a subgroup $k_j$, $k_j \in K$, if the slack is negative, then the workload on that subgroup needs to be delayed or advanced to an other time-slot. As such, the workloads are allocated to subgroups with highly negative slack, and the difference in slack is reduced. As a result, peak-power reduction and workload balancing can be achieved eventually.

Let us discuss the results of demand-response based workload scheduling of allocated power converters. The peak reduction is calculated as the difference in peak-power value before and after scheduling. The workload balancing is achieved by having a uniform number of workloads on a power converter. Workload balance is calculated by averaging the standard-deviation (SD) of workload on each power converter.

Firstly, based on the availability of slack and workload priority, demand-response based workload scheduling is performed. Table 9.5 shows the result of demand-response based workload scheduling (Algorithm 9.2) performed in addition to adaptive clustering based resource allocation for a 32-core case. For example, considering core 27, it is initially assigned to subgroup 3 of group 2; but after performing demand-response based workload scheduling, it is shifted to subgroup 2 of group 4. Note that the voltage supplied by group 4 is higher than that in group 2, which implies that the allocated voltage is higher than demand voltage. As such, this shifting of workload reduces the peak-power on power converter, thereby avoiding the over-loading of power converter. As this implementation involves comparison, not much runtime overhead is incurred.

Table 9.5  Demand-response based workload scheduling result for 32-core microprocessor.

| | Subgroup 1 | Subgroup 2 | Subgroup 3 | Subgroup 4 |
|---|---|---|---|---|
| **Workload distribution before workload scheduling (Algorithm 9.1)** | | | | |
| Group 1 | 31 | N/A | 7, 8, 14 | 12 |
| Group 2 | 17 | 22 | 27, 29 | N/A |
| Group 3 | 6, 21, 32 | 9, 15, 20 26, 28 | 1, 5, 11 18, 19 | 16 |
| Group 4 | 2, 3, 23, 25 | 10, 13 | N/A | 4, 24, 30 |
| **Workload distribution after workload scheduling (Algorithm 9.2)** | | | | |
| Group 1 | 31, 7 | N/A | 8, 14 | 12 |
| Group 2 | 17 | 22 | 29 | N/A |
| Group 3 | 6, 21, 32, 9 | 15, 20, 26, 28 | 1, 5, 11, 18 | 16, 19 |
| Group 4 | 3, 25 | 10, 13, 27 | 2, 23 | 4, 24, 30 |

What is more, based on (9.16), the number of power converters required to meet demand from cores can be calculated. After performing demand-response based workload scheduling, the maximum number of cores in a subgroup is eventually decreased. This leads to the reduction in the number of power converters required to drive the cores. For example, if STM by adaptive clustering is performed on a 32-core microprocessor, group 3 has a maximum of 5 cores among its subgroups driven by power converters that have a driving capacity of 2. To meet the demand, 3 power converters are allocated. However, when workload balancing is performed, the maximum number of cores among subgroups is reduced to 4, demanding only 2 converters. Thus, there is a reduction of 33.33% in power converters needed for group 2. The reduction in power converters after workload scheduling by DVS is summarized in Table 9.6.

Table 9.6   Comparison of number of allocated power converters with and without workload balancing.

|  |  | Adaptive Clustering | Workload Balancing | Reduction |
|---|---|---|---|---|
| 32-core | Group 1 | 1 | 1 | 0.00% |
|  | Group 2 | 1 | 1 | 0.00% |
|  | Group 3 | 3 | 2 | 33.33% |
|  | Group 4 | 4 | 3 | 25.00% |
|  | Total | 9 | 7 | 22.22% |
| 64-core | Group 1 | 2 | 1 | 50.00% |
|  | Group 2 | 3 | 2 | 33.33% |
|  | Group 3 | 5 | 3 | 40.00% |
|  | Group 4 | 11 | 6 | 45.45% |
|  | Total | 21 | 12 | 42.86% |

Table 9.7   Peak reduction and workload balancing by demand-response scheduling.

|  | Peak Reduction | Balance Before | Balance After |
|---|---|---|---|
| Group 1 | 33.33% | 0.91 | 0.58 (1.57X) |
| Group 2 | 50.00% | 1.09 | 0.75 (1.45X) |
| Group 3 | 33.33% | 0.93 | 0.17 (5.59X) |
| Group 4 | 45.45% | 0.51 | 0.36 (1.41X) |
| Average | 40.53% | 0.86 | 0.46 (2.50X) |

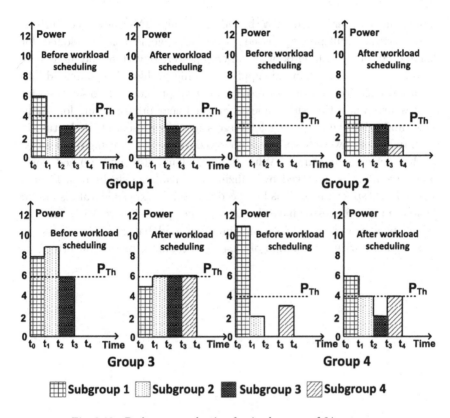

Fig. 9.10    Peak-power reduction for 4 subgroups of 64-core case.

For a 64-core microprocessor, the results of peak-power reduction is shown in Figure 9.10, where in group 4 the normalized peak-power value has been reduced from 11 to 6 with 45.45% peak-power reduction. The average standard deviation of workload on each power converter before and after scheduling are 0.86 and 0.46 respectively. Table 9.7 shows the summarized results of peak reduction and workload balancing by demand-response based workload scheduling for a 64-core microprocessor. One can observe an average of 40.53% peak-power reduction and 2.50× workload balancing.

## 9.5    Summary

Power management of a many-core microprocessor with fewer power converters is discussed in this chapter. A DVS based on-chip power management is performed with the aid of on-chip SIMO power converters. A space-time multiplexing based power management is developed for demand-supply matching between on-chip power converters and many-core microprocessors. The demand-supply matching problem is subdivided into resource allocation and workload scheduling subproblems. Two solutions for the resource allocation subproblem: integer linear programming for small-scale, and adaptive clustering for large-scale, are presented. Reduction in the number of power converters is observed as well while using STM based on-chip power management. What is more, demand-response based work-load scheduling is deployed by utilizing the available slacks to achieve a reduced peak-power as well as balanced workload. Experimental results for a 64-core microprocessor have shown that the space-time multiplexing can reduce peak-power by 40.53% and improve load balancing by 2.50× on average along with a reduction of 45.45% in required power converters.

# Chapter 10

# Signal I/O Management

## 10.1 Introduction

Existing many-core microprocessors with shared memory-logic integration [224] has limited bandwidth that is non-scalable for Exa-scale computing. With the increase in the number of integrated multi-core and memory blocks for a cloud server, the number of signal I/Os grew dramatically; hence there is an emerging need to develop high data-rate and low power I/O circuits [225, 226]. It is reported in [58] that I/Os consume nearly 50% of system dynamic power. Previous 2D wire-line communication by PCB trace of backplane [42, 102, 103] has a large latency, poor signal-to-noise ratio (SNR) and large channel loss. Though 3D integration by stacking several layers of dies vertically using through-silicon via (TSV) I/Os [13, 15, 16, 22, 199, 227–230] provides a scalable integration, accumulated heat in top layers and the complexity of heat removal techniques can limit the flexibility of the design [11, 12, 231]. As discussed in previous chapters, 2.5D integration by through-silicon interposer (TSI) in common substrate has a strong heat dissipation capability [24, 25]. With the design of a TSI based transmission line (T-line), one can further achieve high bandwidth and low power [3, 106] I/O communication without the area overhead, because the TSI is fabricated underneath the common substrate. Compared to 2D PCB trace for memory-logic integration, 2.5D TSI I/O has much higher energy efficiency in communication [106, 232]. Compared to 3D TSV for memory-logic integration, the 2.5D TSI I/O has much better thermal reliability [25, 50]. As such, the TSI T-line based I/O circuit designs have become of interest in the energy-efficient integration of multi-core microprocessor and main memory.

The use of the many-core microprocessor with shared memory for data-oriented commutation with large number of signal I/Os pose two main challenges, namely, bandwidth and I/O communication power management. As there are limited number of TSI I/O channels to access the shared memory, this calls for a channel reutilization under quality-of-service (QoS) [57] constraint. Memory controller [230, 233–237] is commonly employed as the bridge between cores and shared memory. A state-of-the-art memory controller [238] is mainly designed as an off-chip interface with a large number of I/O pins for memory-logic integration to be managed.

Note that the management of memory controllers can be classified into static and dynamic upon whether the requests are determined at run time or design time. A static memory controller such as [239] executes the controlling scheme based on the commands provided at the design time. A major drawback of the static memory controllers is their limited capability for only one already specified application at design time, thus limiting the utilization rate. Dynamic memory controllers [240] can perform scheduling at run time based on real-time requests. It is however non-scalable for many-core microprocessors to access one shared memory. For example, firm real-time controllers (FRT) are designed to meet the timing constraints [241], which are efficient only when all the requests are in the same kind due to less locality exploitation. Soft or no real-time memory controllers [234] maximize bandwidth, irrespective of satisfying timing requirements, which lacks memory efficiency with a better QoS. In this chapter, we focus on the reconfigurable dynamic memory controller for the reuse of I/O channels based on the memory access characteristics. As such, one can improve the utilization of the I/O channels with the good QoS.

On the other hand, previous I/O circuit designs such as [242–245] assume a constant and large output-voltage swing. For 2D wire-line communication, a large output-voltage swing is required to compensate the channel loss and noise of the PCB trace. But a large output-voltage swing with high data-rate can result in huge power consumption in communication. In [245], a single-ended low-swing voltage transmission scheme has been proposed with self-resetting logic repeaters (SRLRs) embedded inside the routers. Due to the use of single-ended signalling, the communication is very prone to the noise. Meanwhile, bit error rate (BER) requirements are not stringent for all applications. Therefore, the use of constant and large output-voltage swing for I/O communication may be an over design with low utilization. Thereby, in order to have an energy-efficient I/O communication, one needs to develop a dynamic output-voltage swing scaling to adaptively adjust the output-voltage swing level under the dynamic BER constraint [246, 247].

In this chapter, firstly, based on the memory-access characteristics, we introduce a reconfigurable dynamic memory controller for the reuse of 2.5D I/O channels such that one can improve the utilization rate with good QoS. To satisfy the demands from cores to access the shared memory with matched supply of 2.5D I/O channels, a QoS-constrained communication is achieved with reconfigurability of I/O channel connection by space-time multiplexing using crossbar-switches. Further, to improve the I/O communication efficiency, we make use of the conventional Q-learning to tune the output-voltage swing based on the workload characteristics. As conventional Q-learning has the convergence issue, an accelerated Q-learning is adapted to overcome the convergence problem. The proposed method can be extended to 2D and 3D integrations as well.

## 10.2 3D Bandwidth and Voltage-swing Management

Signal I/Os are utilized for the communication between cores and memory blocks. However, a 2.5D integration by TSI I/Os has less bandwidth utilization compared to a 3D integration by TSV I/Os. Additionally, the communication power is significant as well due to frequent communication between memory and microprocessor blocks.

### 10.2.1 *System Architecture*

Figure 10.1 shows the 2.5D architecture utilized for many-core memory-logic integration with the smart reconfigurable memory controller for TSI I/O channel management. It consists of 2 different dies on one common

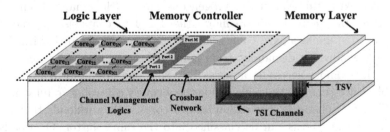

Fig. 10.1 Side-view of 2.5D many-core microprocessor and main memory integration by TSI I/O channels.

substrate with TSI I/Os. The die on the left comprises of many-core microprocessors, and the die on the right is the main memory. The core can access the main memory through a reconfigurable crossbar switch-network [248] to perform space-time multiplexing (STM) inside the memory controller, by configuring TSI I/O channel connections between many-core microprocessors and the shared main memory. The crossbar switch-network is suitable for such a many-core microprocessor with shared memory due to the following advantages: simple one-hop routing, and ease of implementing QoS policy. Besides, it can be easily reconfigured based on the data-pattern analysis. The switch-network can be reconfigured to connect with TSI I/O channels. A more detailed view of the proposed architecture is presented in Figure 10.2. In fact, one can model the 2.5D system architecture by a demand-supply system with the following three components:

- *I/O channel demander:* a set of microprocessor cores $C$ of set-size $N_c$ having demand for different bandwidth. Each core $c_i$ is to have a bandwidth demand of $B_d(c_i)$.
- *I/O channel supplier:* $N_{ch}$ TSI I/O channels which together can supply bandwidth $B_T$. The allocated bandwidth $B_a(c_i)$ for each core $c_i$ is supplied with TSI I/O channels by the reconfigurable memory controller.
- *I/O channel controller:* a set of $M$ memory ports with scheduler inside the reconfigurable memory-controller to map requests from $N_c$ cores to memory through $N_{ch}$ TSI I/O channels under demand-supply matching by flexible crossbar switch-network. Note that the TSI I/O channel management in memory controller can be implemented by the simple logics of address decoder, data queue, request queue and scheduler.

### 10.2.2  *Problem Formulation*

As there exists time-varying heterogeneous bandwidth demands from cores to access the shared main memory, the TSI I/O channels may not be fully utilized under a fixed connection. To manage I/O channels with time-varying bandwidth demanded from many cores, the main idea here is to learn the memory-access data-patterns so that one can perform a reconfigurable space-time multiplexing inside the memory controller to fully utilize the TSI I/O channels. For example, two cores with workloads that have similar data-patterns can be classified into the same cluster. Overloading of I/Os due to large bandwidth demands (by clustering the cores) are mitigated by assigning priority inside a cluster. Based on the classified clusters,

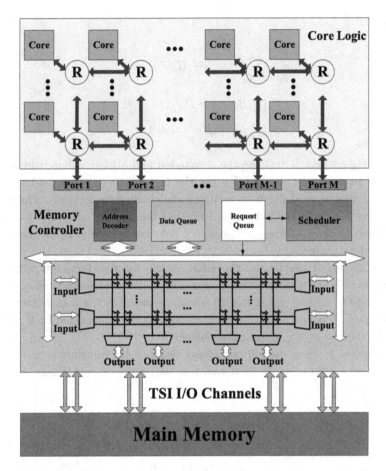

Fig. 10.2 Memory-access data-pattern aware reconfigurable memory controller with space-time multiplexing.

each with a demanded signature bandwidth, one can further schedule the cores to occupy the I/O channels inside one cluster based on the priority. This is how the complexity for a large-scale demand-supply matched signal I/O management can be reduced to implement inside the smart memory controller. One can formulate a space-time multiplexing design problem as follows:

*Problem 1: A reconfigurable memory controller can perform space-time multiplexing that adaptively allocates $N_{ch}$ TSI I/Os to $N_c$ cores* *such that:*

$$\text{min:} \quad \sum_{i=1}^{N_c} |B_a(c_i) - B_d(c_i)|$$

$$\text{s.t.:} \quad \text{(i)} \ B_a(c_i) \geq B_d(c_i); \tag{10.1}$$

$$\text{(ii)} \ Q_i \leq E_R \quad \forall i = 1, 2, \ldots, N_{ch}$$

where $B_d(c_i)$ and $B_a(c_i)$ are the demanded and allocated bandwidths for core $c_i$ respectively; $Q_i$ and $E_R$ are the number of request at $i$-th channel and maximum requests capacity of one channel respectively. The main objective is to reuse I/O channels such that the demand and supply can be matched. In the following, we present a solution by studying the memory-access data patterns.

As discussed previously, the BER at the receiver increases with the decrease of I/O communication power due to channel loss and noise. Thus, one needs to find an optimal output-voltage swing at the transmitter such that balanced power reduction is obtained with the maintained BER, which can be defined as the following problem.

*Problem 2: Tune the output-voltage swing at the transmitter to achieve low power at the cost of BER based on the I/O communication channel characteristics.*

$$Opt. Pw_i, \ BER_i$$

$$S.T.(i) \ Pw_i \ \leq \ Pw_T \tag{10.2}$$

$$(ii) \ BER_i \ \leq \ BER_T$$

where $Pw_i$ and $BER_i$ denote the I/O communication power and BER under the $i$-th output-voltage swing level $V_i$. Note that the BER and power are both functions of the output-voltage swing. $Pw_T$ and $BER_T$ represents the targeted I/O communication power and BER of one TSI I/O channel under the normal operation. With the increase of output-voltage swing, the I/O communication power increases, and BER decreases, and vice-versa. As such, the output-voltage swing level $V_i$ needs to be adaptively tuned to optimize the I/O communication power and BER simultaneously. Here, a self-adaptive tuning of the output-voltage swing at transmitter is performed based on the conventional Q-learning and further accelerated Q-learning discussed later in this chapter.

## 10.3 Signal I/O Bandwidth Management

In this section, we study the memory-access data-pattern characteristics first, followed by the classification of the memory-access data-pattern to allocate the signal I/Os.

### 10.3.1 *Memory-access Data-pattern*

The analysis of memory-access data-pattern with the according QoS analysis, which will be utilized for space-time multiplexing, is presented here. In this chapter, we define a control-cycle with a period of $T_c$, and a control-cycle is divided into time-slots with a period of $T_s$. Within each control-cycle $T_c$, I/O channels are allocated; while at each time-slot $T_s$, cores are allocated with I/O channels.

#### 10.3.1.1 *LLC MPKI Pattern*

There exists many approaches to describe the memory-access data-pattern of workloads. Last-level cache (LLC) misses-per-kilo-instructions (MPKI) is an important metric to indicate the communication intensiveness between cores and memory. A higher LLC MPKI indicates a larger bandwidth requirement. A DRAM row buffer hit-rate is another metric to show the spatial locality of the workload. Since the focus is mainly on logic-memory communication, a LLC MPKI pattern is considered for analyzing the memory-access data-pattern. Note that the memory access number can be inferred by LLC MPKI [249].

Memory-access number $\mathbf{M}(c_i)$ for a core $c_i$ within a control-cycle $T_c$ is the sum of access numbers at each time-slots of $T_s$, given as

$$\mathbf{M}(c_i)|_{T_c} = \sum_{t=1}^{T_c/T_s} \mathbf{M}(c_i)|_t. \tag{10.3}$$

The bandwidth demand $B_d(c_i)$ of core $c_i$ is related to the memory-access number $\mathbf{M}(c_i)$ by

$$B_d(c_i) = \frac{\mathbf{M}(c_i) * L_c}{T_c} \tag{10.4}$$

where $L_c$ is the last-level cache line size.

Based on the memory access number, one can classify the communication traffic pattern of cores as clusters indicated by the similarity in demanded bandwidth. What is more, inside one cluster, the arrival time of requests and the number of memory access requests can be different among cores. As such, one can further define priority $R_i$, indicating core $c_i$'s ability to occupy the allocated I/O channels at time-slot $T_s$. The priority $R_i$ will be raised when core $c_i$ has an early arrival request or more memory access requests to be served if multiple requests arrive at same time. The core assigned with the highest priority will be allocated with the I/O channel for accessing memory.

The classification of memory-access data-pattern based on the bandwidth demands (LLC and number of memory requests) is shown in Figure 10.3. We classify benchmarks as high, medium and low memory-access benchmarks. For example, a core running a *lbm* benchmark having a LLC MPKI of 59 (represented by vertical bars) and 233k memory requests can be categorized under high memory-access, whereas a core running a *libquantum* benchmark will be classified under low memory-access benchmarks due to low LLC MPKI of 0.24 and 1776 memory requests. One needs to observe that the LLC and number of memory access requests are proportional.

Further, the process of classifying data-patterns with similar memory-access numbers is illustrated here. From Figure 10.4(a), one can observe that *lbm* and *gobmk* have similar bandwidth signatures and high memory-access demands compared to *libquantum* and *mcf*. Thus *lbm* and *gobmk* can form a cluster (space multiplexing). In Figure 10.4(b), we present the

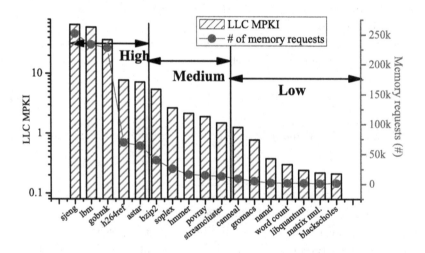

Fig. 10.3   Analysis of memory-access data pattern.

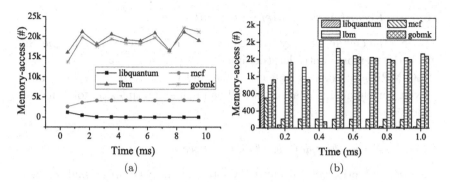

Fig. 10.4    SPEC 2006 memory-access data-patterns: (a) memory access for execution time of $10ms$; (b) memory access for one control-cycle of $1ms$.

memory-access number within one control-cycle. A variation in the number of memory access demands of cores in one cluster (*lbm* and *gobmk*) can be observed, based on which priorities can be assigned to occupying I/O channels in a time multiplexed manner.

### 10.3.1.2    *Quality of Service*

The system performance is sensitive to long-latency memory requests because instructions dependent on the long latency load cannot proceed until the load completes. As the main memory access latency is much higher than cache access, the last-level cache miss will inevitably stall the core executions. More memory access served indicates a better performance. Hence, the memory controller must balance the accesses from different cores with good quality-of-service (QoS) maintenance mechanisms [250].

A higher *QoS* value indicates better performance. Hence, QoS is considered as the metric to determine the matching between demand and the supply, given by

$$QoS = \frac{\sum_{i=1}^{N_c} r_i \mathbf{M}(c_i)}{\sum_{i=1}^{N_c} \mathbf{M}(c_i)} \tag{10.5}$$

where $r_i$ is the processed ratio of requests for core $c_i$. Here QoS is defined as a ratio of total number of requests processed to the total number of access requests. A higher number of requests processed indicates a higher QoS and better performance.

## 10.3.2   Reconfigurable Memory Controller

To solve the demand-supply matching problem by space-time multiplexing inside the memory controller, the time-varying memory-access data-pattern (LLC MPKI) of cores is first extracted and then utilized in space multiplexing for channel allocation as well as in time multiplexing for time-slot allocation.

### 10.3.2.1   Space Multiplexing: Channel Allocation

Here, the adaptive allocation of I/Os to the cores by space multiplexing is discussed. The demand here is the bandwidth requirement from the cores running workloads and the supply is the I/Os to support the required access to memory. To deal with the large number of cores, the cores can be classified based on the memory-access data-pattern (LLC MPKI).

*Clustering* here can be defined as the process of grouping cores with similar demanded signature bandwidths. Each cluster is allocated with cores and connected to one of the ports of the memory controller through on-chip routers. Inside the memory controller, cores are virtually partitioned and allocated to each port of the reconfigurable memory controller which can meet the demanded bandwidth. The number of cores for different clusters may be different. Note that the configuration of virtual clusters can change adaptively. If the demand does not vary, the core is expected to remain in same cluster.

For example, $z$-th cluster $G_z$ at port $p_z$ is formed by

$$G_z = \{c_i | B_d(c_i) \leq B(p_z); c_i \in C, z = 1, 2, \ldots, M\} \tag{10.6}$$

where $B_d(c_i)$ is the bandwidth demanded from core $c_i$; and $B(p_z)$ is the bandwidth allocated to the port $p_z$ of the memory controller. Such a clustering by the magnitude of the memory-access is called space multiplexing.

Once the clusters are formed, cores in one cluster can be further assigned with priorities to access TSI I/O channels in a time-multiplexed manner to avoid overloading of TSI I/O channels.

### 10.3.2.2   Time Multiplexing: Time-slot Allocation

The time-slot allocation for time multiplexing can be described as follows. Memory-access requests from different cores can arrive at different time-instants and can have different memory-access numbers. The core with earliest request will access the I/O channel first, while other cores are stalled.

The core with an early request arrival time will be assigned high priority defined as

$$R_i = H \; if \; t_a(c_i) < t_a(c_j). \tag{10.7}$$

where $H$ indicates high priority and $t_a(c_i)$ indicates the arrival time of the memory-access request from core $c_i$; and $R_i$ indicates the priority for core $c_i$. In the case of multiple requests with of same arrival time, the priority is assigned to the core with more the memory access requests. This is how time multiplexing is performed inside the cluster based on the priority assigned to the core. It needs to be noted that the priority can change due to the change in the memory-access data-pattern (LLC MPKI), and so does the multiplexing of I/Os.

### 10.3.2.3 *Space-Time Multiplexing based I/O Bandwidth Management*

The space-time multiplexing based I/O management inside the memory controller is given in Algorithm 10.1.

Firstly, a set of ports $P$ are labeled in the ascending order of bandwidth it can provide i.e., $B(p_z) > B(p_{z-1})$. Based on the demanded bandwidth, cores with similar bandwidth demands are grouped into one cluster and connected to one port with required TSI I/Os. As such, $B(p_z) = B_a(c_i) > B_d(c_i)$. This is discussed in Lines 1–4 of Algorithm 10.1. Once the cores are

---

**Algorithm 10.1** Proposed I/O Management

---

**Require:** Set of cores $C$, ports $P$, bandwidth demands $B_d(c_i)$
  1: **for** $i = 1 : N_c$ **do**
  2:     $G_z = \{c_i | B(p_{z-1}) < B_d(c_i) \leq B(p_z); z = 1, 2, \ldots, M\}$
  3: **end for**
  4: channel allocation for each cluster
  5: **for** $z = 1 : M$ **do**
  6:     With all cores inside $G_z$
  7:     **for** $t = 1 : T_c/T_s$ **do**
  8:       **if** $t_a(c_i) = \min(t_a|_{G_z})$ **then**
  9:         $R_i = H$
10:       **else if** $t_a(c_i) = t_a(c_j)$ **then**
11:         $R_i = H$ if $\mathbf{M}(c_i) > \mathbf{M}(c_j)$
12:       **end if**
13:     **end for**
14: **end for**
**Ensure:** cluster $G$, allocated I/O channels

---

clustered, priority based scheduling will be performed at each time-slot. The requests from cores will be processed on a first-come-first-serve principle. At each time-slot, if there is only one request, the corresponding core will be served first (Lines 8–9). If there is more than one request, then the core with more requests is assigned higher priority and served (Line 11). Here 'served' implies that the TSI I/O channel is occupied by the core with memory access demand. This process is repeated for every control-cycle $T_c$. Thus, I/O channel management can be performed in a space-time multiplexed manner to improve I/O utilization and QoS. With the above mentioned space-time multiplexing based I/O management, demand-supply matching is achieved, indicating a better utilization rate of TSI I/Os. This implementation may pose some challenges due to the extra control logic required.

Figure 10.5 shows the memory-access data-pattern by LLC MPKI from different cores. Figure 10.5(a) illustrates core bandwidth demand $B_d(c_i)$, where $i = 1, 2, 3, 4$. At control-cycle 1, cores $c_1$, $c_3$ and $c_4$ are allocated with bandwidth $B(p_1)$, while core $c_2$ is allocated with bandwidth $B(p_2)$. As such, cores with different workloads can be clustered based on their magnitude in space, for space multiplexing.

Further within the cluster consisting of cores $c_1$, $c_3$ and $c_4$, all cores will compete for a channel at each time-slot. To avoid overloading of I/O channels, a priority based time multiplexing is implemented. Figure 10.5(b) shows the allocation of channels based on the priority at a time-slot. The memory access request from one core is shown by a rectangular box. At time-slot 1, since there is only one request from $c_4$, the I/O channel will be allocated to it. For the next time-slot, there are simultaneous requests from

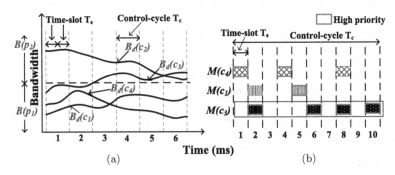

Fig. 10.5    Illustration of LLC MPKI memory-access data-pattern: (a) bandwidth; (b) priority.

both $c_1$ and $c_3$, so priority needs to be assigned to access memory through the I/O channel. Here, when multiple requests arrive at the same time, priority is decided based on the total number of memory access requests, which is 2 for $c_1$ and 4 for $c_3$. Hence, core $c_3$ is assigned a higher priority and occupies the I/O channel. The core assigned with high priority is shown with red outline. Thus, the cores are further classified based on the phase or priority in time, for time multiplexing. We set the control-cycle $T_c$ and time-slot $T_s$ as $1ms$ and $0.1ms$ respectively.

To evaluate the I/O bandwidth management, the gem5 simulator [189] is utilized for many-core microprocessors and the DrSim simulator [251] is employed for shared main memory. The proposed space-time multiplexing I/O management using a smart reconfigurable memory controller is implemented inside the DrSim simulator. The TSI I/O channel model is based on [24] and the length is $1.5mm$. The crossbar switch-network inside the memory controller is estimated $1mm^2$ area and $1GHz$ frequency for a 64-core system under $32nm$ design [248]. The benchmarks are selected from SPEC 2006 [252] which have high memory-access demand, PARSEC [253] with medium memory-access demand and Phoenix [254] with less memory-access demand. Furthermore, we assume a baseline system which has fixed core-to-memory connections, while the proposed system employs the reconfigurable connections. A reconfigurable crossbar switch with STM logic may incur small power and latency overhead similar to [224]. Additional details about system architecture is presented in Table 10.1.

**Adaptive Clustering Analysis** Here, the adaptive clustering analysis of cores based on the signatures of demanded bandwidth is discussed. For a 16-core case, the assumption is that the total number of I/O channels is 8 and the number of ports is 4, and each core is assigned randomly with a benchmark. For the baseline system with fixed connections, each port is assigned with four cores and connected to 2 I/O channels. For the proposed system, each port connects to 4, 2, 1 and 1 I/O channels respectively to meet

Table 10.1    System parameters.

| | |
|---|---|
| Processor core | 1GHz x86 |
| L1 I-cache | 32kB private, 64B cache line |
| L1 D-cache | 32kB private, 64B cache line |
| L2 cache | 256kB private, 64B cache line |
| Main memory | 4GB capacity, 800MHz DDR3-1066 channel, x8 DRAM chips, 8 banks per channel |

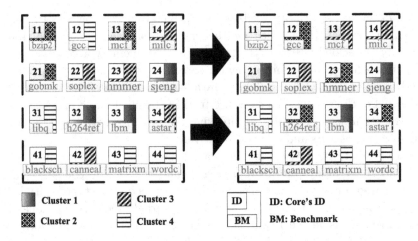

Fig. 10.6    Adaptive clustering of cores at two consecutive control-cycles.

the memory access demands from high traffic to low traffic benchmarks. A similar setup is assumed for the 64-core case as well. Figure 10.6 illustrates the adaptive clustering results for the 16-core case for 1 control-cycle ($5ms$ to $6ms$). Different filling patterns represent different clusters. Cluster 1 handles high-traffic workloads, cluster 2 is used for middle-traffic workloads, and cluster 3 and cluster 4 are allocated with low-traffic workloads. For example, at $5ms$, core 12 assigned with *gcc* benchmark is assigned to low traffic cluster 4, but is assigned to high traffic cluster 2 in the next control-cycle due to time varying memory-access characteristics.

Figure 10.6 illustrates the adaptive clustering results of 16-core case at two consecutive control-cycles ($5ms$ to $6ms$). Different filling patterns represent different clusters. Cluster 1 handles high-traffic workloads, cluster 2 is used for middle-traffic workloads, and cluster 3 and cluster 4 are allocated with low-traffic workloads. For example, at $5ms$, core 12 running *gcc* benchmark is assigned to low traffic cluster 4, but is assigned to middle traffic cluster 2 in the next control-cycle due to time varying memory-access characteristics. Similarly, Figure 10.7 shows the variation in the number of cores allocated to a cluster with time. For example, at $3ms$, cluster 4 is allocated with 18 cores, but at $4ms$ is allocated with 20 cores.

**Bandwidth Balancing**    Bandwidth balancing across all ports can improve the I/O channel utilization efficiency. We use requests per channel to measure the traffic flow at each control-cycle and calculate the standard deviation for all four ports. The standard deviation shows how much

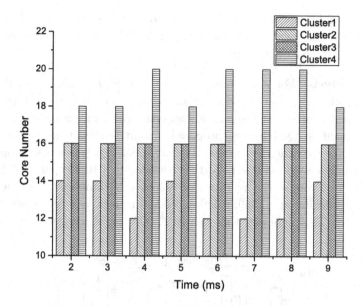

Fig. 10.7 Variation in number cores allocated to a cluster with control-cycles.

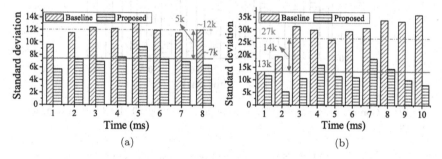

Fig. 10.8 Bandwidth balance across TSI I/O channels with: (a) 16-core; (b) 64-core under randomly distributed benchmarks of SPEC 2006, Parsec and Phoenix.

variation there is from the average value. A low standard deviation indicates more bandwidth balancing. Bandwidth balancing for a 16-core and 64-core microprocessor, with randomly distributed SPEC 2006, Parsec and Phoenix benchmarks to cores, is shown in Figure 10.8. For example, in a 64-core microprocessor, at $5ms$, the baseline system showed a standard deviation of 26k memory requests, while the proposed system just requires 11.5k with 55.90% bandwidth balancing improvement. The average of standard deviation for the baseline and proposed systems is 27k and 13k memory requests, indicating a 14k reduction in deviation. On average the proposed system can improve the bandwidth balancing by 58.25% under the 16-core case and 58.85% under the 64-core case.

**QoS Analysis** For a 16-core microprocessor with SPEC 2006, Parsec and Phoenix benchmarks randomly distributed on cores, the comparison of QoS for the proposed system and baseline architecture is presented first. Improvement in QoS by the proposed system is shown in Figure 10.9. The average QoS for the baseline system is 0.472, whereas the proposed STM achieves a QoS of 0.561. Further considering a $10ms$ span (100 time-slots) as an example, the average QoS achieved by proposed, baseline and time multiplexing are 0.589, 0.470 and 0.529 respectively, as presented in Table 10.2. This indicates that nearly 11.90% more requests are served,

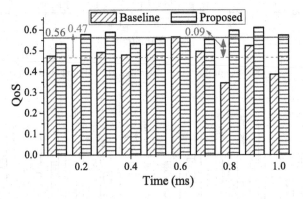

Fig. 10.9   Communication QoS efficiency improvement by STM I/O management for 16-core.

Table 10.2   Number of requests served and achieved QoS

| Method | Requests Served | QoS |
|---|---|---|
| **For 16-core microprocessor** | | |
| Baseline | 448578 | 0.473 |
| Time multiplexing | 523840 | 0.528 |
| Proposed | 580399 | 0.589 |
| **For 64-core microprocessor** | | |
| Baseline | 2092083 | 0.461 |
| Time multiplexing | 2340843 | 0.514 |
| Proposed | 2622005 | 0.577 |

and an improvement in QoS by the proposed method compared to baseline. Additionally, the number of requests served by the memory controller with the method is also given in Table 10.2. In the 64-core case the proposed system achieves a QoS of 0.577, compared to a QoS of 0.461 and 0.514 achieved by baseline and time multiplexing techniques respectively. Use of space multiplexing alone may result in heavy congestion at the memory controller, and hence is not compared.

## 10.4 Signal I/O Voltage-swing Management

Before performing the voltage-swing management, we present the trend of BER, communication power versus voltage-swing levels for different benchmarks. Figure 10.10 shows the trade-off between BER (blue circle) and the I/O communication power (black rectangle). With the increase in output-voltage swing, BER decreases at the cost of I/O communication power, which is validated through 4 benchmarks. For example, for

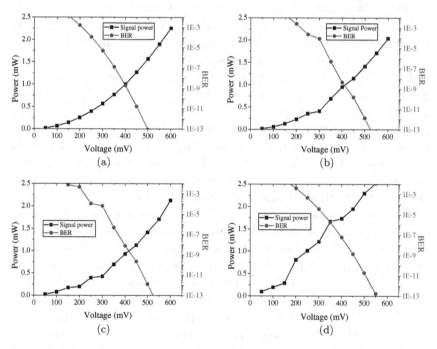

Fig. 10.10  Trade-off between BER and power under various benchmarks: (a) *FFT* benchmark; (b) *Filetransfer* benchmark; (c) *bzip2* benchmark; (d) *gzip* benchmark.

the *filetransfer* benchmark, at an output voltage-swing of $350mV$, the I/O communication power is $0.693mW$ with a BER of $4.85E-7$; whereas at an output-voltage swing of $400mV$, BER goes down to $4.47E-9$ at the cost of increased communication power by $0.953mW$. This observation shows that there is a balance point where we can use less power with the BER guaranteed. Furthermore, as the sensitivity of power and BER with voltage-swing level are different for each benchmark, adaptive tuning of the I/O voltage swing based on conventional Q-learning and further accelerated Q-learning can help to achieve an optimal trade-off between the communication power and BER for different benchmarks respectively.

On-line machine learning based power management has been recently practiced [26, 255–259]. For example, Q-learning [260, 261] can be utilized to find an optimal action-selection policy from a set of states. A Q-learning based management is applied to adjust the level of output-voltage swing at the transmitter (associated with cores) such that one can achieve a reduced power under a specified BER requirement. However, the conventional Q-learning is limited by the learning rate with slow convergence [262]. In [263], a large range of learning rates is dynamically applied to increase the convergence rate. However, its complexity increases with the range of learning rates. A reinforcement Q-learning can improve the policy decision by using prior knowledge of the system with a Markov decision process (MDP) [256]. In [256], reinforcement Q-learning with accelerated convergence is applied to a multimedia system to achieve a trade-off between distortion rate and complexity.

Firstly, one conventional Q-learning based I/O management is applied to adjust the level of output-voltage swing at transmitter of 2.5D TSI I/Os, such that one can achieve a reduced power under specified BER requirement. To overcome the slow convergence issue in the conventional Q-learning algorithm, an online reinforcement Q-learning [256], termed as an accelerated Q-learning based management approach, is further developed in this paper to adaptively adjust the output-voltage swing levels of 2.5D TSI I/Os. Instead of transmitting data under a fixed large voltage-swing, an on-line reinforcement Q-learning based approach is applied to select the output-voltage swing at the transmitter. Based on historical data, the voltage-swing adjustment is formulated as a MDP problem solved by model-free reinforcement learning under the constraints of both a power budget and BER. To accelerate the adjustment convergence, a prediction of BER and power as virtual experience is applied to the reinforcement Q-learning algorithm. One corresponding 2.5D TSI I/O is designed in the

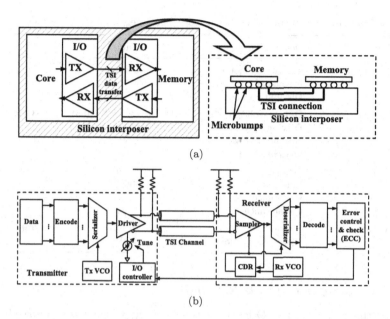

(a)

(b)

Fig. 10.11 (a) Core-memory integration by 2.5D TSI I/O interconnect and its cross sectional view; (b) adaptive tuning I/O based on error checking and correction.

$65nm$ CMOS process for a multi-level output-voltage swing with balanced power and BER.

Figure 10.11(a) shows the block diagram of memory-logic integration by the 2.5D TSI I/O. Each of the core and memory blocks comprises of a transmitter as well as receiver to enable a full duplex communication. To further reduce the I/O communication power between logic and memory blocks, a self-adaptive voltage-swing adjustment is required. The current-mode logic (CML) buffer is shown in Figure 10.11(b) with a tunable tail-current (based on the output of I/O controller) to adaptively tune the output-voltage swing. By adjusting the I/O output-voltage swing, the I/O communication power can be reduced with improved energy efficiency compared to the previous designs [242–244] that utilize the fixed full output-voltage swing. However, BER increases when the output-voltage swing decreases. Hence, a trade-off needs to be maintained between the I/O communication power and BER, which requires an optimized on-line management. The geometry and technology details are presented in Table 10.3. To solve problem 2 of voltage-swing tuning, we make use of online Q-learning.

Table 10.3   System settings for memory-logic integration with TSI I/O.

| Item | Description | Value | Size |
|------|-------------|-------|------|
| Microprocessor | Technology node | $65nm$ | $0.3mm^2$ |
| | Frequency | $500MHz$ | |
| | Dissipation power | $15mW$ | |
| I/O controller | Output-voltage swing | $0.1V$, $0.15V$, $0.2V$, $0.3V$ | $0.03mm^2$ |
| | Driving current | $2mA$, $3mA$, $4mA$, $5mA$ | |
| | Number of levels | 4 | |
| | Switching time | $0.4ns$ | |
| TSI | Length | $3mm$ | $3mm^2$ |
| | Inductance | $300pH$ | |
| | Resistance | $5\Omega$ | |
| | Capacitance | $60fF$ | |
| Memory | SRAM | 16 KB | $0.2mm^2$ |
| | Power dissipation | $6mW$ | |

## 10.4.1   *Reinforcement Q-learning*

Here, the basics of the Q-learning theory is first presented followed by conventional Q-learning, modeling of Markov decision process (MDP) and the according accelerated Q-learning algorithm for adaptive voltage-swing tuning.

### 10.4.1.1   *Q-learning Theory*

Machine learning algorithms such as Q-learning theory [264], and reinforcement-based Q-learning [256], are generally practised to find an optimal action-selection policy from the set of states $S$. Both these algorithms evaluate *state* and *action* pairs from the previous inputs. To solve (10.2) using basic Q-learning and further by reinforcement Q-learning algorithms, the I/O communication power $Pw$ and BER $BER$ are considered as the state vector; and the tuning of output-voltage swing level $V_i$ as the action. State vector $S$ can be given as

$$S = \langle Pw, \ BER \rangle.$$

Before describing the basic Q-learning and reinforcement-based Q-learning algorithms, a few terminologies are presented below:

- State $S$: set of states indicating the value of system variable(s). We consider the communication power and BER the state vectors.

- Action $A$: set of actions indicating the change of state. We consider the change of output-voltage swing level as the action.
- State transition probability $P\ (s_i, a_k, s_j)$: probability indicating whether to take an optimal action $a_k$ based on Q-learning or perform random action.
- Reward $R(s_i, a_k, s_j)$: evaluation value of action $a_k$ to change the state from $s_i$ to $s_j$, which is dependent on historical $BER$ and power $Pw$.
- Q-value $Q(s_i, a_k)$: set of accumulated Q-values to measure the benefits of taking action $a_k$ at state $s_i$.
- Expected Q-value $\hat{\mathbf{Q}}(s_i, a_k)$: set of values to measure the expected benefits of taking action $a_k$ at state $s_i$.
- Policy: process of state change under sequence of action.

In order to obtain the state-action pairs and form a look-up-table (LUT), the input samples (voltage-levels) are trained and the corresponding communication power and BER are noted. A sample LUT can be formed as follows:

| Action | State | |
|---|---|---|
| $(Voltage swing)$ | $Power$ | $BER$ |
| $a_1(V_1)$ | $Pw_1$ | $BER_1$ |
| $\vdots$ | $\vdots$ | $\vdots$ |

The input samples are collected at regular time intervals, called control cycle, at scale of $ns$. A control cycle can be defined as the minimum time required for the state transition. The duration of control cycle is based on the speed of I/O controller circuit. The next state variable needs to be predicted with an action for the next input sample. This can be done by calculating a reward function to achieve an optimally estimated value based on the state vectors, given by

$$R = f(Pw, BER). \tag{10.8}$$

Here, reward $R$ is the function of communication power $Pw$ and BER value $BER$ as the state vectors. The relation between state variables and reward value is presented later. The next state and the current state can be the same depending on the workload characteristic. The reward $R$ forms a part of the expected Q-value, which decides the direction of state transition. The optimal estimation is chosen among the set of states to satisfy the required criteria by taking the corresponding action selected from the formed LUT.

The expected Q-value is calculated as

$$\hat{Q}(s_i, \ a_k) = Q(s_i, \ a_k)(1 - \alpha) + \alpha(R + \gamma E). \qquad (10.9)$$

Here $\alpha$, $\gamma$ denotes the learning rate and discount factor respectively. The optimal estimation $E$ of state $s_i$ can be calculated as follows

$$E = min\{\hat{Q}(s_i, \ a_k)\}, \ \ k = 0, \dots, M. \qquad (10.10)$$

Here, $\hat{Q}(s_i, a_k)$ represents the expected Q-value after taking action $a_k$; $M$ denotes the number of possible actions available at state $s_i$. The optimal estimate can be *min* or *max* depending on the reward function.

The 2.5D adaptive TSI I/O verification is performed in Cadence Virtuoso (Ultrasim-Verilog) and Matlab. An 8-core MIPS microprocessor with 8-bank of SRAM memory is designed with GF $65nm$ CMOS. The 2.5D TSI T-line is of length $3mm$ and $10\mu m$ width, driven by the CML buffer. The power traces are measured from Cadence Virtuoso and control cycle is set as $1ns$, larger than the switching time of the I/O controller. The I/O management controller is based on the basic Q-learning and reinforcement Q-learning output to balance the I/O communication power and BER at receiver. The look-up-table (LUT) is designed with I/O communication power and BER, stored in LUT for adaptively tuning the output-voltage swing. A LUT is set up as follows: $(100mV, 6.27E - 2mW, 7.03E - 2BER)$, $(150mV, 1.41E - 1mW, 1.35E - 2BER)$, $(200mV, 2.51E - 1mW, 1.61E - 3BER)$, and $(300mV, 5.64mE - 1mW, 4.93E - 6BER)$. This LUT is almost robust, since this depends on the characteristics of the circuit rather than the application. The learning rate $\alpha$ and discount factor $\gamma$ are set as 0.5 and 0.9 respectively. An auto-regression (AR) of order 8 is used for load current (or I/O communication power) prediction. The error between the predicted and actual values is less than 0.3%. The circuit is designed in Cadence with the according technology PDK. The overall I/O performance can provide a minimum of $76mV$ peak-to-peak signal swing with $4Gb/s$ bandwidth, and the power consumption is only $12.5mW$ when implemented in $65nm$ CMOS process. The adaptive self-tuning of output-voltage swing may come with a little area overhead of $0.03mm^2$ for additional control circuits and a latency of 100-200$ps$.

### 10.4.1.2   System Model

The prediction of power and BER of an I/O communication channel and its models are discussed here. The first component of the state vector is I/O communication power. The system power model includes the I/O communication power of driver and the TSI T-line. Both are functions of the output-voltage swing $V_i$. For the CML based driver with TSI T-line [60], I/O communication power is given by

$$Pw_i = V_i \cdot \left( I_t + \frac{\eta * V_{dd} * s}{R_D + Z_{diff}} * f \right). \qquad (10.11)$$

Here $I_t$ is the driver tail current; $s$ is the duration of signal pulse; $\eta$ is the activity factor; $R_D$ is the resistance of driver; and $Z_{diff}$ is the characteristic impedance of the TSI T-line.

The tail current $I_t$ at the current control cycle is set by analog design and can be obtained by measurement. The tail current for the next control cycle can be predicted by auto-regression (AR) as

$$I_t(k+1) = \sum_{i=0}^{M-1} w_i I_t(k-i) + \xi. \qquad (10.12)$$

Here $I_t(k+1)$ denotes the predicted tail current at the $(k+1)$-th control cycle; $w_i$ represents the auto-regression coefficient; $\xi$ is the prediction error and $M$ represents the order of the AR prediction. Based on the predicted tail current, the I/O communication power for the next control cycle is calculated as $Pw_i(k+1)$.

The second component of the state vector is the BER of I/O communication, which is feedback from the receiver. BER depends on the output-voltage swing, external noise, channel noise and some other parameters [265]. In a wire-line communication system [266], the BER can be estimated with the dependence on the output-voltage swing as

$$BER_i = \frac{1}{2} erfc \left( \frac{V_i}{\sqrt{2}\sigma_v} \right). \qquad (10.13)$$

Here $erfc$ is the complementary error function; $V_i$ refers to the $i$-th output-voltage swing level and $\sigma_v$ is the standard deviation of the noise.

As such, BER can be obtained from the ECC at the receiver by counting errors during a period. During the learning process, based on the obtained BER at ECC under one output-voltage swing, the standard deviation of noise, $\sigma_v$ is estimated from (10.13).

### 10.4.1.3 *Adaptive I/O Voltage-swing Tuning*

The self-adaptive tuning of the output-voltage swing at the CML buffer is performed based on conventional Q-learning. The I/O communication power $Pw_i$ and BER $BER_i$ corresponding to the output-voltage swing level

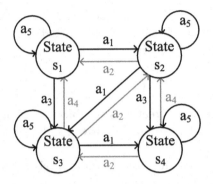

Fig. 10.12    State transition based on conventional Q-learning.

---

**Algorithm 10.2** Conventional Q-learning based adaptive tuning of output-voltage swing

---

**Require:** Communication power trace $P_i$, BER feedback from receiver and look-up-table (LUT)

**Ensure:** Adaptive tuning of output-voltage swing $V_i$

1: Predict tail current: $I_t(k+1) = \sum_{i=0}^{N-1} w_i I_t(k-i) + \xi$
2: Calculate corresponding communication power and BER
3: Reward: $R_w(s_i, a_k, s_{i+1}) = b_1 \Delta(Pw_i) + b_2 \Delta(BER_i)$
4: $\hat{\mathbf{Q}}(s_i, a_k) \leftarrow Q(s_i, a_k)(1 - \alpha) + \alpha(R_w + \gamma E)$
5: Optimal value estimate: $E = min\{\hat{\mathbf{Q}}(s_i, a_k)\}, \; j = 0, \ldots, M$
6: By adjusting tail current using control bits, tune corresponding $V_i$

---

$V_i$ are considered the components of state vectors $S$ and output-voltage swing level as the action variable.

Figure 10.12 illustrates an example of a state diagram depicting the change of states by the conventional Q-learning algorithm. In this example, four states and five actions are considered. When the system is in state $s_1$ ($s_2$) and action $a_1$ ($a_2$) is selected, then the system changes to state $s_2$ ($s_1$); when the action $a_3$ is chosen, the state changes to $s_3$ ($s_4$) and remains in the same state when action $a_5$ is chosen. Similarly, other state transitions also happen. One needs to note that the state transition depends on the action chosen at the current state.

A self-adaptive output-voltage swing tuning by conventional Q-learning is presented in Algorithm 10.2. LUT is formed with output-voltage swing as action and the I/O communication power and BER as the state vectors. The tail current of the CML buffer is firstly predicted by auto-regression, as given in Line 1 of Algorithm 10.2.

Based on the predicted tail current, the corresponding I/O communication power and the BER are calculated. Further, using the present I/O communication power and BER values, reward $R_w$ is calculated. Since we have two factors, we consider the weighted sum of I/O communication power and BER. The reward function is given as follows

$$R_w(s_i,\ a_k,\ s_{i+1}) = b_1\Delta(Pw_i) + b_2\Delta(BER_i) \qquad (10.14)$$

here $b_1$ and $b_2$ denote the weighted coefficients for normalized rewards of the communication power $\Delta(Pw_i)$ and BER $\Delta(BER_i)$. This calculation of reward is given in Line 3 of Algorithm 10.2. After calculating the reward, the expected Q-value is calculated from Line 4 of Algorithm 10.2, and the optimal action is selected based on Q-values, as in Line 5 of Algorithm 10.2. This is how adaptive tuning is performed by implementing the conventional Q-learn algorithm. As a summary, the whole flow of adaptive tuning by the conventional Q-learning algorithm is shown in Figure 10.13.

An example of how adaptive tuning of output-voltage swing by conventional Q-learning is shown in Figure 10.14. Initially, the voltage-swing for

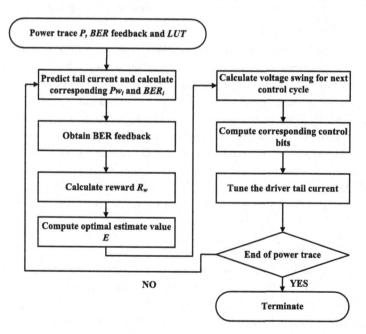

Fig. 10.13   Flowchart showing conventional Q-learning based self-adaptive voltage swing tuning.

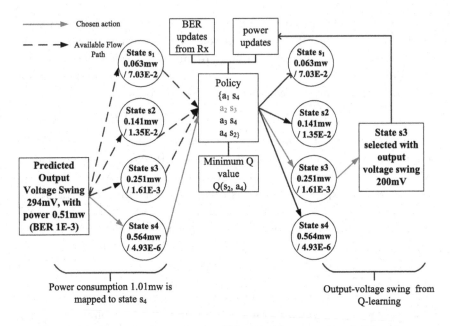

Fig. 10.14   Example of one adaptive I/O control by conventional Q-learning.

the next control-cycle is predicted as $294mV$. As it is impractical to have LUT with continuous values, we map the incoming value to the closest value in LUT. Thus, the voltage-swing $294mV$ is fit to the state $s_4$ because of the closest voltage level. Based on the predicted value and the current value, reward is calculated as given in (10.14). Further, the corresponding voltage-swing level is obtained based on the calculated reward and optimal estimate function. This results in setting the voltage-swing level (action) to $200mV$ with $0.25mW$ communication power i.e., state $s_3$. It needs to be noted that the reduction in voltage swing level is due to the tolerance to BER. Further, the BER and I/O communication power are updated.

### 10.4.1.4   *Accelerated Q-learning*

The conventional Q-learning algorithm [261] converges to the optimal after unlimited or large number of iterations that may be too slow for convergence [262]. To overcome this convergence issue, an accelerated Q-learning for adaptive tuning of output-voltage swing can be performed to achieve low power and faster convergence. Here, an example to illustrate the difference between conventional Q-learning and accelerated Q-learning is presented,

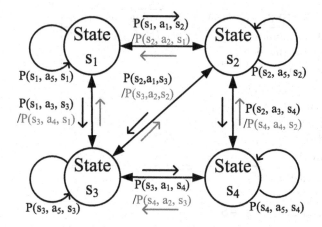

Fig. 10.15    State transition based on accelerated Q-learning.

followed by the modeling of a MDP and the according reinforcement Q-learning algorithm for the adaptive tuning.

One example with a 4-state is shown in Figure 10.15. For state $s_1$, action $a_1$ can change its state to state $s_2$ with probability $P(s_1, a_1, s_2)$; For state $s_2$, action $a_2$ can change its state to state $s_1$ with probability $P(s_2, a_2, s_1)$, whereas action $a_5$ causes no change in state, whose probability is given as $P(s_1, a_5, s_2)$. The state transition probability $P$ is given as a decaying function. The probability under the decaying function is given by $P = 1 - 1/(log(N_{s_i} + 2)$ with $N_{s_i}$ denoting the number of visits to state $s_i$. The transition probability based action will ensure the visit to all states in the starting period. This will calculate the Q-value to every available state accordingly. After this, Q-value based action will dominate and the optimal action with the smallest Q-value will be selected.

To find the optimal value of MDP, probability based action selection, an accelerated Q-learning algorithm can be utilized to evaluate the pair of state and action as the Q-value. The conventional Q-learning algorithm converges to the optimal value after an unlimited number of iterations [262]. Accelerated Q-learning [255] can be utilized to find an optimal solution with a faster convergence based on the next predicted state and the according transition probability with an initialized random action at the first few states. Initial random actions make the system explore the environment faster, and more easily find optimal states. To achieve a faster convergence, the transition probability is utilized to select the action instead of directly selecting the next state. Further, the reinforcement Q-learning can

---

**Algorithm 10.3:** Accelerated Q learning algorithm

---

**Input:** Communication power $Pw$, $BER$ feedback
**Output:** Output-voltage
**function** Init()
    $1 \rightarrow P(s_i, a_k, s_{i+1})$
    Reward $R(s_i, a_k, s_{i+1}) = L$
    $v_{predict} \rightarrow V_{s_i}$
    Selection()
**end function**

---

**function** Selection()
    **for** $k = 1 : n$
        $V_{s_i}, BER_i \rightarrow s_i \in S$
        $\hat{Q}(s_i, a_k) \leftarrow (1 - \alpha) * Q(s_i, a_k) +$
        $\alpha * (R(s_i, a_k, s_{i+1}) + \gamma * \min(Q(s_i, a_k))$
        **If** $P(s_i, a_k, s_{i+1}) > rand(0, 1)$
            $a_k \leftarrow rand(A)$
        **else**
            $a_k \leftarrow \min(\hat{Q}(s_i, a_k))$
        **end if**
        Update()
    **end for**
**end function**

---

**function** Update()
    *Reward:* $R(s_i, a_k, s_{i+1}) = b_1 \Delta(P_i) + b_2 \Delta(BER_i)$
    Update Policy $(s_i, a_i)$, based on new Q
    $\forall\, s_i \in S$ {
        $a_k \leftarrow rand(A)$
        $Q(s_i, a_k) = \hat{Q}(s_i, a_k)$
        $P(s_i, a_k, s_{i+1}) = 1 - \frac{1}{log(N_{s_i} + 2)}$
    }
**end function**

---

be utilized to find the optimal point for the modeled MDP to solve the discussed adaptive tuning problem. Accelerated Q-learning is presented in Algorithm 10.3.

The first phase is the initialization to form a LUT with states and corresponding actions. In addition, the transition probability $P$ for all the states is set as 1 and the reward is set to a maximum value $L$. This process of initialization is presented as $Init()$ of Algorithm 10.3.

A prediction of the next state is performed to obtain the corresponding action. In the action selection phase, given by $Selection()$, the Q-value for the state and action pair is found iteratively, where the expected Q-value is given by

$$\hat{Q}(s_i, \ a_k) = (1 - \alpha) * Q(s_i, \ a_k) + \alpha * delta \qquad (10.15a)$$

$$delta = R(s_i, a_k, \ s_{i+1}) + \gamma * \min_{a \in A}(Q(s_i, \ a_k)). \qquad (10.15b)$$

where $Q(s_i, a_k)$ represents the accumulated Q-value and $\hat{Q}(s_i, a_k)$ represents the expected Q-value after taking action.

In each iteration, the action is selected either based on the transition probability or based on the minimum Q-value (or policy). If the transition probability is larger than the threshold, a random action is selected; otherwise, the action policy with the minimum Q-value is selected. The random action will happen in the first few rounds to explore the design space. As the learning process continues, the policy action with the calculated Q-value will dominate and become more accurate to use. As such, a higher probability exists that the action $a_k$ with the minimum Q-value will be selected. The policy action with the minimum Q-value can be described as below

$$a_k \leftarrow \min(\hat{Q}(s_i, \ a_k)). \qquad (10.16)$$

Lastly, the $Update()$ phase is activated at the end of each iteration of the $Selection()$ function. The reward is defined as the weighted value of $BER$ and $Pw$ and updated as given in (10.14).

At the end of $Update()$, each state will be randomly visited and the Q-value will be updated accordingly as given in (10.15). The transition probability $P(s_i, a_k, s_{i+1})$ is also updated as $N_{s_i}$ (increases with the number of visits to state $s_i$) after each iteration.

Note that with the prediction of states $s_i$ as given in function $Init()$ and $Update()$ and the transition probability, the convergence to the optimal solution is accelerated [256]. This is done at the end of each round with the random action $a_k$ to visit the state $s_i$. In brief, the whole flow of adaptive tuning by accelerated Q-learning is presented as a flowchart in Figure 10.16.

The LUT can be implemented online with the corresponding control bits calculated and fed back to the CML buffer to tune the DAC current

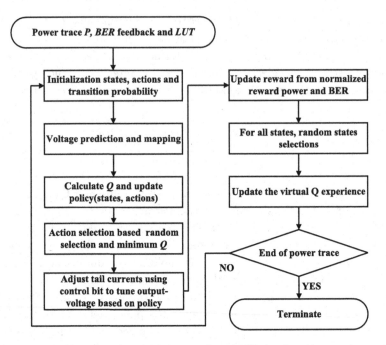

Fig. 10.16    Flowchart of accelerated Q-learning based self-adaptive voltage swing tuning.

of the CML buffer. Note that LUT can be implemented in the hardware with multiple AND/OR partial matching logic circuits instead of read only memory (ROM). This LUT based implementation has higher speed and lower power consumption compared to the ROM.

An example of one adaptive tuning by accelerated Q-learning is shown in Figure 10.17. Initially, the voltage-swing for the next control-cycle is predicted as $294mV$. The voltage-swing $294mV$ with power $0.51mW$ fits to the state $s_4$ because of the nearest voltage level. The action will be selected based on the probability check as in *Selection*() of Algorithm 2. As shown in Figure 10.17, the solid red line and red dotted line indicate the policy-based action selection and probability-based action selection, respectively. For the probability-based action selection (red dotted line), a random action is selected to visit any of the available states. This takes place with high probability at the starting period to ensure a visit to all the available states, where its Q-value will be updated accordingly. Afterwards, the action with minimum Q-value will dominate and be considered as the optimal action. As Figure 10.17 shows, action $a_4$ is eventually selected, leading to the voltage level associated with state $s_2$. Afterwards, the new

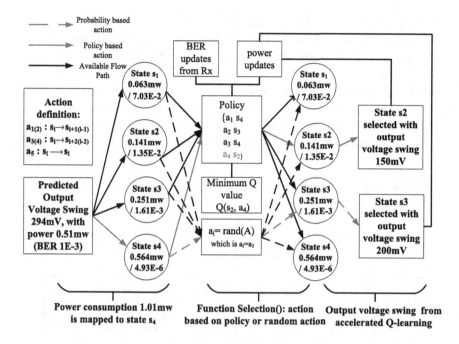

Fig. 10.17 Example of one adaptive I/O control by accelerated Q-learning.

voltage-swing is assigned based on the state $s_2$ as $0.141mW$. As such, the driver tail-current is tuned to have the output voltage-swing as $150mV$. Lastly, the reward $R(s_4, a_4, s_2)$ will be updated based on the feedback of BER and power obtained at Rx. One needs to note that as shown in Figure 10.14, conventional Q-learning selects state $s_3$ due to the action based state change, however, it may converge to state $s_2$ after a few iterations.

### 10.4.1.5 *Comparison of Adaptive Tuning by Conventional and Accelerated Q-learning*

The I/O communication power saving is verified for various SPEC benchmarks [136] in Table 10.4 by the self-adaptive tuning using: no Q-learning (normal); conventional Q-learning without reinforcement (conv); and the accelerated Q-learning (acc). It shows that the accelerated Q-learning algorithm is more power efficient, with faster convergence. For example, for *bzip2* benchmark, the I/O communication power without the adaptive tuning is $0.267mW$, which is reduced to $0.224mW$ when adaptively tuned by conventional Q-learning; and is further reduced to $0.210mW$ with accelerated Q-learning. On average, the power consumption of the system (transmitter, receiver and the I/O) is $19mW$ with an energy efficiency of $4.75pJ/bit$ for the I/O without adaptive tuning, which is further reduced to $12.5mW$ by the adaptive tuning. On average, the power saving of 18.89% and energy efficiency improvement of 15.11% is achieved by adaptive tuning based on the reinforcement Q-learning; and a 12.95% of power saving and 14% of energy-efficiency improvement are achieved when using the basic Q-learning. What is more, on average, accelerated Q-learning

Table 10.4   Power and run-time comparisons for the adaptive tuning with Q-learning under various benchmarks.

| Benchmark | Communication Power ($mW$) | | | Power Saving | | Run Time(s) | | |
|---|---|---|---|---|---|---|---|---|
| | Acc | Conv | Normal | Acc | Conv | Acc | Conv | *Improvement* |
| ammp | 0.231 | 0.250 | 0.296 | 22.11% | 15.54% | 0.0804 | 0.1049 | 23.28% |
| applus | 0.474 | 0.494 | 0.555 | 14.64% | 10.99% | 0.0817 | 0.1051 | 22.25% |
| apsi | 0.652 | 0.674 | 0.744 | 12.34% | 9.41% | 0.0817 | 0.1051 | 22.25% |
| art | 0.199 | 0.216 | 0.255 | 22.01% | 15.29% | 0.0829 | 0.1057 | 21.55% |
| bzip2 | 0.210 | 0.224 | 0.267 | 21.19% | 16.10% | 0.0829 | 0.1056 | 21.50% |
| crafty | 0.381 | 0.419 | 0.461 | 17.39% | 9.11% | 0.0822 | 0.1061 | 22.53% |
| eon | 0.234 | 0.253 | 0.298 | 21.32% | 15.10% | 0.0807 | 0.1051 | 23.24% |
| equake | 0.232 | 0.249 | 0.295 | 21.07% | 15.59% | 0.0821 | 0.1047 | 21.60% |
| facerec | 0.333 | 0.364 | 0.409 | 18.53% | 11.00% | 0.0833 | 0.1060 | 21.42% |
| fft | 0.153 | 0.179 | 0.199 | 23.12% | 10.05% | 0.2002 | 0.2309 | 13.30% |
| file transfer | 0.512 | 0.554 | 0.598 | 14.58% | 7.36% | 0.2057 | 0.2485 | 17.22% |
| fma3d | 0.539 | 0.561 | 0.623 | 13.35% | 9.95% | 0.0850 | 0.1083 | 21.51% |
| galgel | 0.241 | 0.261 | 0.307 | 21.56% | 14.98% | 0.0820 | 0.1069 | 23.29% |
| gap | 0.284 | 0.310 | 0.355 | 19.86% | 12.68% | 0.0808 | 0.1036 | 22.01% |
| gcc | 0.507 | 0.536 | 0.587 | 13.60% | 8.69% | 0.0838 | 0.1064 | 21.24% |
| gzip | 0.238 | 0.257 | 0.302 | 21.08% | 14.90% | 0.0817 | 0.1051 | 22.26% |
| lucas | 0.518 | 0.548 | 0.602 | 13.90% | 8.97% | 0.0818 | 0.1053 | 22.32% |
| mcf | 0.248 | 0.264 | 0.307 | 19.41% | 14.01% | 0.0817 | 0.1056 | 22.63% |
| mesa | 0.228 | 0.243 | 0.287 | 20.45% | 15.33% | 0.0846 | 0.1065 | 20.56% |
| mgrid | 0.274 | 0.297 | 0.346 | 20.83% | 14.16% | 0.0818 | 0.1051 | 22.17% |
| parser | 0.429 | 0.458 | 0.508 | 15.43% | 9.84% | 0.0825 | 0.1084 | 23.89% |
| perlbmk | 0.207 | 0.226 | 0.266 | 22.21% | 15.04% | 0.0822 | 0.1052 | 21.86% |
| sixtrack | 0.341 | 0.357 | 0.412 | 17.39% | 13.35% | 0.0819 | 0.1065 | 23.10% |
| swim | 0.242 | 0.262 | 0.310 | 21.82% | 15.48% | 0.0812 | 0.1052 | 22.81% |
| twolf | 0.216 | 0.235 | 0.279 | 22.38% | 15.77% | 0.0839 | 0.1037 | 19.09% |
| vortex | 0.225 | 0.239 | 0.288 | 21.94% | 17.01% | 0.0832 | 0.1056 | 21.21% |
| vprs | 0.240 | 0.255 | 0.300 | 19.96% | 15.00% | 0.0826 | 0.1059 | 22.00% |
| wupwise | 0.424 | 0.442 | 0.502 | 15.39% | 11.95% | 0.0836 | 0.1058 | 20.98% |
| Average | — | — | — | 18.89 % | 12.95% | 0.0912 | 0.1152 | 20.83% |

takes $0.091s$ for convergence, whereas the conventional Q-learning takes $0.115s$. The reported time includes the training of samples. The run time has also improved with accelerated Q-learning by an average of 20.83%, compared to conventional Q-learning based I/O management as shown in Table 10.4.

## 10.5  Summary

In this chapter, a data-pattern aware reconfigurable memory controller and adaptive voltage-swing tuning for memory-logic communication is demonstrated for a 2.5D microprocessor. With the reconfigurable cross-bar switch-network inside the memory controller, bandwidth demands from many-core to access the shared memory can be managed by reusing the available limited number of TSI I/O channels. A space-time multiplexing based communication between cores and memory is realized by reusing the I/O channels with improved communication efficiency. The adapted architecture is verified by a system-level simulator with benchmarked workloads, which shows up to 58.85% bandwidth balancing and 11.90% QoS improvement. Further, to reduce the I/O communication power, an adaptive tuning of output-voltage swing is adapted at the transmitter. Conventional Q-learning based adaptive voltage swing tuning is employed first, and a accelerated Q-learning based voltage swing tuning is further employed. When compared to the uniform output-voltage swing based I/O management, I/O managements by conventional Q-learning and accelerated Q-learning achieve 12.95% and 18.89% power reduction, and 14% and 15.11% energy efficiency improvement, respectively.

# Chapter 11

# Sensor

## 11.1 Introduction

In this part, we will discuss the design examples of sensors, multi-core processors and memories that are required in the design of a modern system. As discussed in the previous chapters, compared with the two-dimensional (2D) semiconductor integration technology which has been developed over the past three decades [15], three-dimensional (3D) integration technology has been widely recognized as the next generation of manufacturing technology for integrated microsystems with ultra-small form factor [267–269]. This technology can be used to integrate multiple layers of functional electronic blocks into a given chip area. The outstanding advantage of this technology is the possibility of heterogeneous integration [270] which allows for the of building a circuit by stacking active device layers with different fabrication process. Micro-electro-mechanical structures (MEMS) are one of the rapidly growing emerging technologies with several advantages, having feature size at nano-scale level. In this chapter, we present a 3D integration example of heterogeneous devices i.e., CMOS based circuit and MEMS devices.

MEMS-CMOS integration is critical for the future development of multi-sensor data fusion in a low-cost chip size system for internet of things (IoT) application. MEMS-CMOS integration was primarily done using monolithic or hybrid/package approaches until recently [271, 272]. The silicon-on-insulator (SOI) technology has been developed to realize several MEMS devices. It has become a very important fabrication technology with its benefits being a high aspect ratio, precise control of geometry and potential for integration with CMOS circuits via bonding. Bonding technology is a common packaging and integration method for 3D integration. Compared with

Table 11.1   Competitive advantages of 3D integration of MEMS/CMOS.

|  | Monolithic | 3D Integration | Package/Hybrid |
|---|---|---|---|
| Cost | − | + | + |
| Performance | + | + | − |
| Power | + | + | − |
| Functionality | + | + | + |
| Time to market | − | + | + |
| Modularity | − Sequential | + Parallel | + Parallel |
| Hermetic seal | Ex-situ | In-situ | Ex-situ |

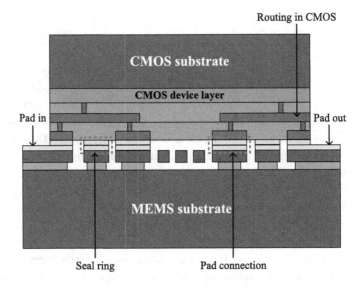

Fig. 11.1   3D stacking of MEMS and CMOS devices.

other bonding technologies, metal thermo-compression bonding (or diffusion bonding) presents competitive advantages in 3D integration as it allows for the formation of hermetic sealing, mechanical, and electrical contact in one step. Furthermore, the selection of metals such as copper (Cu) enables bonding at CMOS–compatible temperature with better electrical conductivity, mechanical strength, electro-migration resistance and low cost [273]. Table 11.1 presents the competitive advantages of 3D MEMS-CMOS integration over package-based or monolithic solutions.

Figure 11.1 shows the cross-sectional view of a TSV-less stacking method for MEMS and CMOS 3D integration. This MEMS-CMOS integration leads to a simultaneous formation of electrical, mechanical, and hermetic bonds [274], eliminates chip-to-chip wire-bonding, and hence presents competitive advantages over hybrid or monolithic solutions. In order to ensure proper operation, the delicate micro-structures (MEMS) should be protected from the ambient, hence, a hermetic seal ring is formed simultaneously during the stacking of CMOS on MEMS. The seal ring is trenched in SOI MEMS wafer during the device layer DRIE etching. The electrode pad of MEMS is bonded to a connection pad on the CMOS die and will be routed to external connection/pads using lower metal layers in the CMOS chip. Metallization process can be realized during the SOI MEMS fabrication. Therefore, 3D stacking can be achieved by the die-to-wafer bonding process which needs alignment, and follows a similar process to that of Cu thermo-compression bonding.

In this chapter, more details about 3D integration technology for MEMS and CMOS process via Cu-Cu bonding are investigated. A SOI-based MEMS accelerometer is designed and fabricated by the deep reactive-ion etch (DRIE) process. A CMOS readout circuit is designed and implemented in the AMS $0.35\mu m$ (2P4M) process. Consequently, simultaneous hermetic sealing and electrical contact by low temperature Cu-Cu thermo-compression bonding has been investigated for the 3D stacking. A TSV-less 3D integration method which uses routing in CMOS metal layers for the I/O pads is designed for process simplicity.

In this chapter, we will first present the design details of the MEMS accelerometer and CMOS readout circuits and evaluate their individual performances. Further, the 3D integrated CMOS-MEMS system is evaluated using the hermeticity test.

## 11.2 3D Sensor Design

We discuss the details on 3D integration of MEMS-CMOS circuits by Cu-Cu bonding in the later parts of this chapter. In this section, we discuss the fabrication of the MEMS accelerometer, CMOS readout circuit and their individual performances, followed by integrated system validation in next section.

### 11.2.1 *SOI MEMS Accelerometer*

As an example we discuss the design of a MEMS–based accelerometer, which can be utilized to sense and measure the motion and vibration under loads. The fabrication process flow by bulk micromachining for MEMS accelerometer is outlined in Figure 11.2. As shown in Figure 11.2(a), it

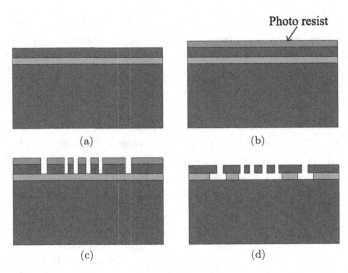

Fig. 11.2   Process flow for fabrication of SOI MEMS: (a) SOI wafer preparation (Si $20\mu m$/SiO$_2$ $1\mu m$; (b) deposition of photo-resist; (c) photo-resist patterning and Si device layer etching; (d) SiO$_2$ etching and photo-resist removal.

starts with a Si/SiO$_2$ ($20\mu m$/$1\mu m$) SOI wafer onto which a photo-resist 1813 is deposited (Figure 11.2(b)). The DRIE etching windows are patterned photo-lithographically and the sacrificial layer releasing channels are etched (Figure 11.2(c)). The $20\mu m$ structural device layer is released by etching away the $1\mu m$ SiO$_2$ buried oxide in HF (Figure 11.2(d)). The released structure results in a comb-drive accelerometer.

The microscopic image of the SOI-based (resistivity $\sim 0.002\Omega/cm$) capacitive MEMS accelerometer (sensitivity $\sim 4.88fF/g$) design, fabricated by DRIE process, is presented in Figure 11.3. A metal layer is patterned on the MEMS prior to etching for electrical contact and hermetic seal. A single layer of patterned metal consists of an electrical contact pad and hermetic seal. Electrical feed-throughs are routed through the on-chip interconnect in the CMOS chip.

The SEM micrograph of the MEMS accelerometer is shown in Figure 11.4. The structure is properly etched and fully released. The top view of the mass in center and the comb-finger are included in the inset at lower left. In order to ameliorate the releasing rate of the sacrificial layer, holes are etched in the center mass. The suspended structure at the edge of the upper mass is included in the inset at lower right. The resonant frequency of SOI MEMS is measured by a laser vibrometer. The

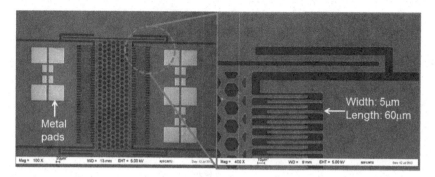

Fig. 11.3   The MEMS capacitive accelerator fabricated on SOI wafer using DRIE and release.

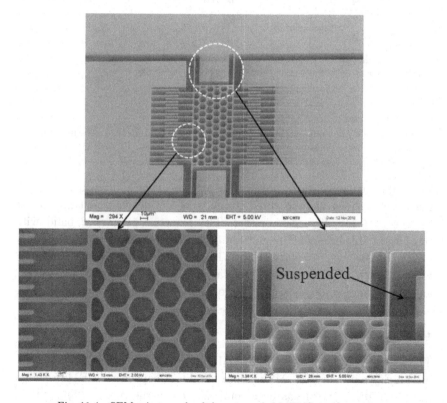

Fig. 11.4   SEM micrograph of the suspended MEMS accelerometer.

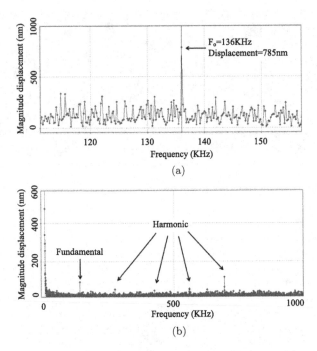

Fig. 11.5   MEMS response during excitation with laser vibrometer: (a) response in narrow band; (b) response in large band.

laser vibrometer measures physical displacement without surface contact. Figure 11.5 shows the measured response of the displacement magnitude for the SOI MEMS accelerometer. The response in narrow band with the largest displacement magnitude of $785nm$ at its resonant frequency (136 KHz) is shown in Figure 11.5(a). The response in large band is shown in Figure 11.5(b). The first mode (fundamental mode) is clearly at 136KHz, and the harmonic modes appear periodically after the fundamental mode.

### 11.2.2   *CMOS Readout Circuit*

The second component in MEMS-CMOS integration is the CMOS circuit, for which we design a MEMS readout circuit. A capacitive MEMS readout circuit ($2mm \times 2mm$) has been implemented in the AMS $0.35\mu m$ 2P4M process. The technique of chopper stabilization [275] has been adopted for low-noise readout. The system block diagram of CMOS readout circuit is shown in Figure 11.6.

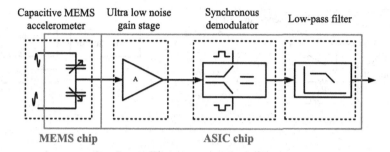

Fig. 11.6   System block diagram of CMOS readout circuit.

The major building blocks of the readout circuit are: (1) low-noise, band-pass gain stage; (2) synchronous demodulator; and (3) low-pass filter. The MEMS is excited using a high frequency sinusoidal signal. The MEMS amplitude-modulates this carrier signal in response to the acceleration experienced. The output of the MEMS goes to a low-noise amplifier, which improves the signal-to-noise ratio. The low-noise, band-pass gain stage is realized using two single-ended output amplifiers, based on the folded-cascode architecture as shown in Figure 11.7, with a single-ended amplifier utilized in the low-noise gain stage. The synchronous demodulator is

These single-ended output amplifiers shown in Figure 11.7 offer much improved common-mode input range and an increased output swing. This allows more flexibility in selecting the gain factor and which can be conveniently set externally via tunable feedback capacitance $C_f$. The flexibility in selecting the gain factor enables the same readout circuit design to work well with a wider range of accelerometer designs, in terms of output capacitance and sensitivity. It is also desirable to have very large feedback resistance $R_f$ since the lower corner of the band-pass gain stage is determined by $1/2\pi R_f C_f$. Using a very large feedback resistor however means an inefficient utilization of on-chip space. This can be taken care of by using pseudo-resistors in place of real resistors which can provide very high feedback resistance, of the order of M$\Omega$, while consuming minimal space. The unity gain bandwidth of the folded-cascode single-ended output amplifiers decides the upper corner frequency of the band-pass gain stage. The lower and the upper corner frequency limits of the gain stage thus determine the working carrier frequency range, which are 10KHz and 18MHz respectively for the reported readout design. The first amplifier of the gain stage boosts up the MEMS sensor's input while the second amplifier works as an

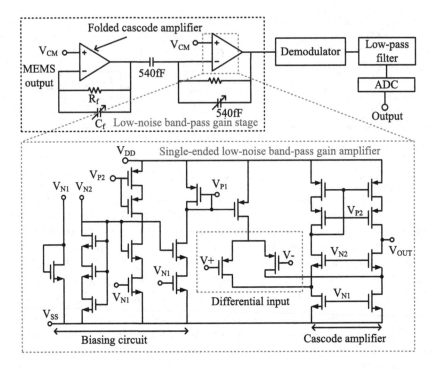

Fig. 11.7   Implementation of one single-ended low noise, band-pass gain amplifier.

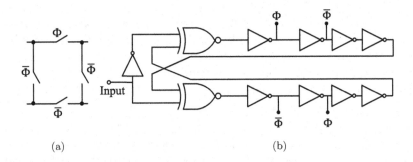

Fig. 11.8   Differential synchronous demodulator.

inverting stage to generate a second differential input for the synchronous demodulator.

A fully differential synchronous demodulator is the second building block of the readout circuit, shown in Figure 11.8. Demodulation is done through envelope detection, by implementing a four-switch full-wave rectifier. The demodulator helps in easing the filtering requirements and reduces the second order harmonic distortion at the same time. The fully differential synchronous demodulator makes it possible to move the low pass filter off-chip which helps greatly in saving on-chip space. Moreover, as a result of the ease in filtering requirements, there is no longer any need of using an active filter and this minimizes the overall circuit's power consumption. The implemented filter is a passive, second order low-pass filter with a cut off frequency of 200Hz. The on-chip low-pass filter is a $2^{nd}$ order Sallen-Key low-pass filter to remove the high frequency carrier and noise components. The readout circuit's chip micrograph with labelled components is shown in Figure 11.9 with (1) indicating low-noise, band-pass gain stage; (2) is synchronous demodulator; and (3) representing a low-pass filter in the chip photograph.

As the readout circuit is designed to read the output from MEMS accelerometer, the readout circuit is excited with a MEMS-like signal (Figure 11.10) using an arbitrary waveform generator. As chopper stabilization has been employed, the input signal will look like an AM-modulated signal with the baseband representing the response to acceleration and the carrier frequency representing the actual chopping carrier at 100 KHz. The readout circuit successfully amplifies, demodulates, low-pass filters and

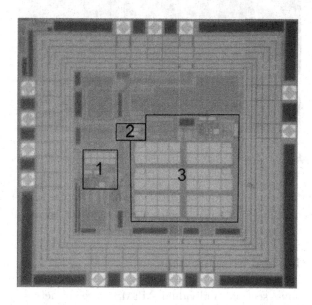

Fig. 11.9   Chip micrograph of the readout circuit.

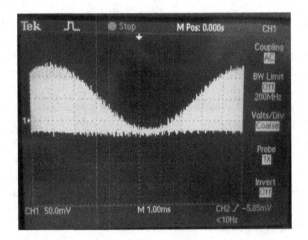

Fig. 11.10   MEMS-like signal input to the readout circuit.

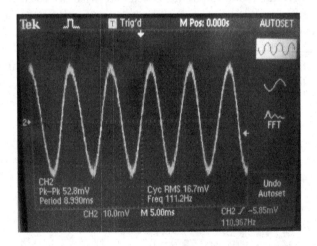

Fig. 11.11   Recovered 111Hz sinusoidal acceleration signal.

recovers the 111 Hz sinusoidal acceleration signal (Figure 11.11). The readout has a sensitivity of 5.1 mV/g with a $\pm 30g$ measurement range.

## 11.3   Testing and Measurement

In the previous section, individual MEMS accelerometers and CMOS readout circuit designs are studied. Here, we study the process of vertical stacking and the performance of 3D integrated 3D MEMS-CMOS structure.

For vertical stacking, both wafer-to-wafer and die-to-wafer are possible. A die-to-wafer approach is chosen as it offers flexibility in the chip size and is accomplished by bonding the CMOS readout circuit die to the MEMS sensor wafer. Although it is commonly known that the quality of thermo-compression bonding can usually be ameliorated when the bonding temperature increases, practical packaging/integration should be achieved at adequately low temperatures (typically $300°C$ or below) for devices that are sensitive to high temperature processing owing to the thermal budget limitation and the post-bonding thermo-mechanical stress control. For the purpose of demonstration, we consider the bonding medium to consist of a Cu ($300nm$) bonding layer and Ti ($50nm$) barrier layer. Cu-Cu can be bonded by the thermo-compression method via parallel application of heat and pressure.

In order to provide a proper operation, MEMS should be protected from oxygen corrosion and water vapor content [276] for the internal micro- and nano-scale devices. Therefore, hermetic packaging is absolutely essential in 3D integration. In this approach, a hermetic seal ring is formed during stacking of CMOS on MEMS and hence eliminate the need for post-processing to obtain hermetic encapsulation. Effectively, the CMOS layer acts as an "active cap". In addition, I/Os to the MEMS chip are routed through the CMOS metal layers to simplify the MEMS process (Figure 11.1). Since the I/O count is low, TSV is not used and electrical feed-through is achieved by peripheral pads. As no solder is applied, the top passivation layer of the CMOS chip is partially recessed to expose the CMOS metal layer for ease of direct bonding with the MEMS metal layer (Figure 11.12). The metal surfaces (MEMS/Au, CMOS/Al) are care-fully treated and bonded (thermo-compression @$300°C/50N/10$ min). The bonded samples were packaged inside a 44-pin J-leaded ceramic package (Figure 11.13) for testing. Table 11.2 summarizes the system specifications. Figure 11.14 shows the cross-sectional view of the bonded layer and elemental mapping.

## 11.3.1 *Hermetic Test*

The cross-sectional SEM image of two bonded non-functional chips is shown in Figure 11.15. A close-up view showing the bonding of the seal ring and the capping wafer is included in the SEM image at lower right. This structure is used to study the properties of the Cu-Cu hermetic seal ring. The bonding

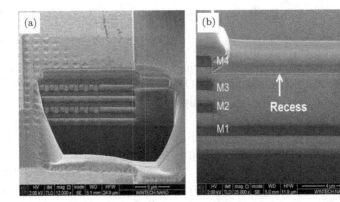

Fig. 11.12   (a) FIB/SEM image showing the on-chip metal layers in the CMOS chip; (b) since no solder bumping is applied, the top passivation layer is intentionally recessed to create a standoff gap for the metal pads to facilitate direct metal bonding (thermo-compression) with the matching pads on the MEMS chip. Bonding is performed at $300^{\circ}C$ for 10 minutes under a force of $50N$.

Fig. 11.13   The CMOS chip is bonded face to face on the MEMS chip. The bonded chip is then wire bonded to the package for electrical testing. The vertically stacked CMOS and MEMS chip has a thickness of 1155 $\mu m$.

interface is clearly marked with an arrow and no void is observed. A high resolution transmission electron microscopy (TEM) image is included in the inset at the lower left. As can be seen from the TEM image, the two Cu layers have fused together during bonding resulting in a homogeneous and seamless interface. The original bonding interface has disappeared and

Table 11.2   Summary of the specifications of the vertically bonded chip.

| Parameters | Measured Values |
| --- | --- |
| Supply voltage | $3.3V$ |
| Power consumption | $1.491mW$ |
| Input referred noise of the readout circuit | $32.663nV/\sqrt{Hz}$ |
| Total harmonic distortion | 0.380% at $50kHz$ carrier frequency at $+1g$ acceleration |
| Minimum detection signal | $\pm0.139g$ |
| Resonant frequency | $136kHz$ |
| Technology | SOI bulk-micromachining for MEMS and AMS $0.35\mu m$ (2P4M) for the readout circuit |

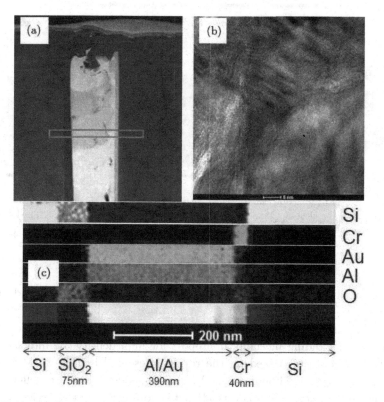

Fig. 11.14   (a) Cross-sectional view of the bonded Al-Au layer from dummy structure; (b) high resolution TEM view; (c) EDX elemental mapping of the bonded layer.

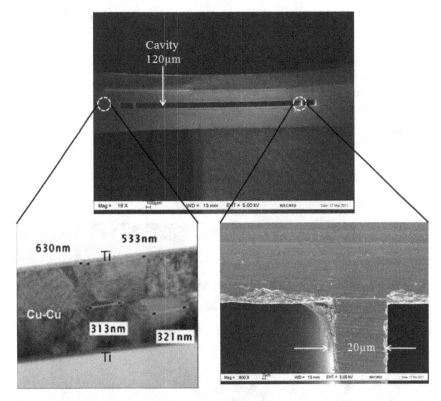

Fig. 11.15   SEM image of cavity sealed with Cu-Cu bonding. (Lower right inset) SEM image shows the close-up view of the bonding interface. (Lower left inset) TEM images of the bonded Cu-Cu layer and the grain structure.

a zig-zag interface is obtained as a result of Cu inter-diffusion and grain growth.

In addition, work in [277] has shown an excellent electrical contact and an outstanding mechanical support for high density 3D interconnect of Cu-Cu contacts. Based on the results presented in this work, Cu-Cu thermo-compression bonding can also simultaneously provide excellent hermetic encapsulation. This is very attractive for packaging the devices which need high hermeticity and for the heterogeneous integration of MEMS and CMOS. For example, one can use the CMOS layer as an 'active cap' to protect the sensitive MEMS structure [278] using Cu-Cu seal bonding at low temperature.

### 11.3.2  *Bonding Reliability*

The metal bonding quality is investigated using shear and helium leak tests using the requirements in MIL-STD-883E, 1014.9 ($> 5kgf$ shear force and $< 5 \times 10^{-8} atm.cm^3/s$ Helium leak rate). The mechanical strength evolves with bonding temperature as shown in Figure 11.16. The hermeticity is tested using sealed cavities by bubble test in the first instant followed by helium leak test. The hermeticity is maintained at $\sim 2.8 \times 10^{-8} atm.cm^3/s$.

### 11.3.3  *Measurement Results*

The frequency range over which the readout circuit works is decided by the corner frequencies of the low noise, band-pass gain stage (Figure 11.7). These corners for the reported readout circuit are 10KHz and 18MHz, respectively. The frequency of excitation carriers was kept within 50KHz-1MHz during testing to ensure that the results were obtained in the flatband region. The bonded chip is subjected to a shock test, prescribed in JESD22-B104C ($500g$, $1ms$, 10 cycles) to ensure its reliability. The measurement results for the bonded MEMS-CMOS chip are shown in Figure 11.17, when the chip was excited by $1V_{pp}$, 50KHz differential sinusoids. The variation in the peak-to-peak amplitude of the gain stage output with respect to an excitation carrier is observed as the chip is flipped between $-1g/+1g$

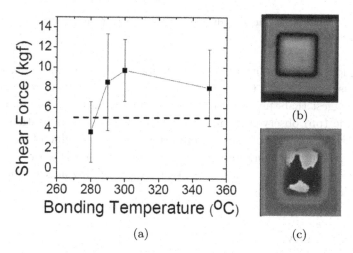

(a)　　　　　　　　　　　(c)

Fig. 11.16　(a) The bonding quality is evaluated for its mechanical strength shear test on dummy structures. The evolution of shear strength with bonding temperature is shown; C-SAM images of: (b) successfully bonded chip; (c) and failed chip.

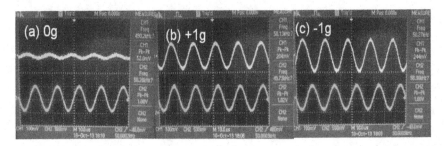

Fig. 11.17   The gain stage output (yellow, upper trace) when excitation carriers are at $1V_{pp}/50\,\text{kHz}$ (blue, lower trace) under: 0g, +1g and −1g orientations.

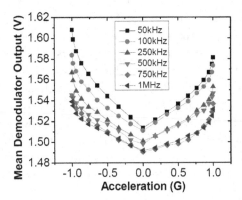

Fig. 11.18   Variation in the demodulator output with $g$ at various carrier frequencies for the bonded chip.

orientations. The amplitude of the gain stage output grows in-phase and anti-phase with respect to carrier in $+g$ and $-g$ flip directions, respectively, which suggest that the bonded chip is working as desired. The maximum amplitude (pp) observed in the two flip directions is an indication of the symmetrical behavior of the accelerometer. The mean value of one of the differential rectified sinusoids at the demodulator output is recorded within the measurement range $-1g$ to $+1g$ at various carrier frequencies, and plotted in Figure 11.18.

## 11.4   Summary

In this chapter, the design of a silicon-on-insulator (SOI) micro-electro-mechanical system (MEMS) accelerometer, complementary metal oxide semiconductor (CMOS) readout circuit and simultaneous hermetic encapsulation using low temperature Cu-Cu bonding are investigated for 3D heterogeneous integration of MEMS and CMOS without TSV. Bulk micromachining technology is used for SOI MEMS accelerometer fabrication. The resonant frequency of this accelerometer is measured by a Laser vibrometer. The response in large band indicates the resonant frequency at 136 KHz. A CMOS readout circuit is designed in a $0.35\mu m$ CMOS process employing chopper stabilization. It has a sensitivity of 5.1 mV/g. A TSV-less I/O routing 3D integration method is proposed for future integration of CMOS and MEMS. Since Cu-Cu thermo-compression bonding can provide electrical, mechanical and hermetic bonds in one step, this TSV-less 3D integration method will simplify the fabrication process and improve yield. A helium leak test demonstrates that the samples encapsulated (bonding between non-functional cavity wafer and cap wafer) achieve a superior helium leak rate below $4.9 \times 10^{-9} atm.cm^3/sec$, which is at least one order of magnitude smaller than the reject limit of $5 \times 10^{-8} atm.cm^3/sec$, defined by the MILSTD-883E standard.

# Chapter 12

# I/O

## 12.1 Introduction

The three-dimensional (3D) memory-logic-integration by through-silicon-via (TSV) is one promising solution for many-core memory-logic integration, but as discussed in previous chapters, it has limited thermal and mechanical reliability when many layers are stacked [17, 21, 25, 50]. Through-silicon interposer (TSI), or the so called 2.5D integration [24], emerged as an alternative solution for memory-logic integration with significantly improved thermal and mechanical reliability compared to the 3D TSVs [12, 22]. However, the design of TSI I/O links is one of the challenges to be addressed. I/Os contribute significantly to the overall power consumption [58]. Hence, a low power I/O design has to be implemented. In addition, the data recovery at the receiver side has to be implemented at high frequencies to overcome the data loss due to channel noise.

However, different from the RC-interconnect based I/O scheme by 3D TSV, the challenge of a 2.5D TSI is to design a high-speed and low-power I/O, i.e., a serial data link with high data-rate per Joule. The recent microprocessor I/O in [279] has achieved high-speed performance with less than $1pJ/bit$ in 2D integration. But there are no works to address the 2.5D I/O. Hence, in this chapter we discuss the low–power I/O buffer designs for a 2.5D integrated memory-logic system, shown in Figure 12.1 with TSIs deployed underneath the common substrate.

In addition to the normal TSI, adaptive I/O design needs to be performed. As discussed in the previous chapters, use of the adaptive I/O can help in reducing the communication power. However, the adaptive I/O needs to have an driver with tunable tail current. In this chapter, a design of an adaptive I/O with tunable voltage-swing is discussed.

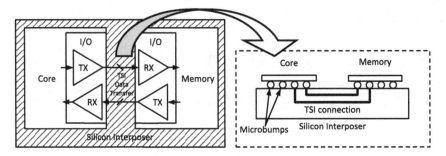

Fig. 12.1    2.5D TSI I/O interconnection between core and memories its cross section view.

What is more, source synchronous links with forwarded-clock (FC) architecture [280–286] are widely deployed in parallel I/O interfaces due to their low power consumption, inherent correction of clock and data jitter, and appropriate jitter tracking bandwidth (JTB). In the FC receiver, the static phase offset (SPO) between input data and sampling clock is corrected at start-up; while the dynamic phase error (DPO)/jitter is tracked by the forwarded clock with jitter correction.

The model of a FC receiver is shown in Figure 12.2. The data and the clock are sent to the receiver simultaneously. However, due to PCB traces mismatch and frequency dependent delay from the channels, the data and the clock have a time misalignment at the receiver especially for high data rates. As a result, the SPO has to be corrected before sampling, which is realized by the phase interpolator (PI) here. Due to the appropriate JTB introduced by the FC receiver structure, the DPO can be restrained as well. The PI presented here can generate a wide-range (0°-360°) of clock deskew which can cover the phase misalignment and make sure the sampling at the center of the data, as shown in Figure 12.2. As a result, the low bit error rate can be achieved and make the FC receiver insensitive to the jitter.

Moreover, continuous-time linear equalizer (CTLE) is widely utilized in FC receivers [286–288] due to its compact structure and better high frequency performance for middle-distance interconnects (such as interposer based memory-logic integration) without decision feedback equalizer (DFE) taps. The CTLE equalizer is usually followed by a sampler in traditional data recovery circuits. But the sampler always has limited bandwidth and speed due to a voltage sampling structure that seriously degrades the speed even though the equalizer provides a gain-boost at high frequency to compensate channel loss [288–290]. The design we discuss in this chapter makes

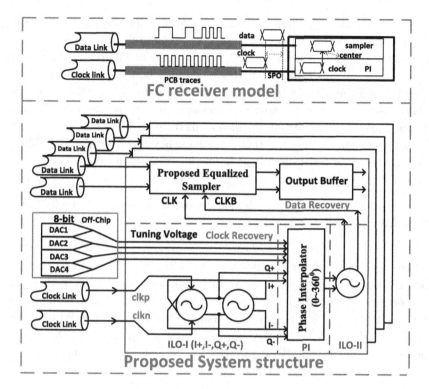

Fig. 12.2   FC receiver model and the proposed full-rate and energy-efficient FC receiver architecture.

use of a current sampling structure sampler merged with equalized function to realize high speed sampling.

In this chapter, we first present the design of CML and LVDS I/O buffer designs to over come the channel loss by implementing the pre-emphasis stage. In addition to the normal CML buffer, we present the design of an adaptive I/O, which can help in reducing communication power. Further, the design of clock data recovery circuits with CTLE at the receiver end is discussed with measurement results.

## 12.2   2.5D I/O Buffer Design

The 2.5D integration can significantly improve the bandwidth and reduce power consumption compared to the conventional multi-chip modules

(MCM), in which wire-bonding and PCB-level T-line introduces a significant latency and power in-efficiency [291]. The architecture of the high-speed low-power I/O for 2.5D TSI is illustrated in Figure 12.1. One can observe that signals from the digital logic of the core is transferred to I/O circuits, and then transmitted or received as differential signals at the two chips. All the I/Os in the core and memory are attached to microbumps, which makes TSI connection in silicon substrate.

At very high frequencies, there is a frequency-loss from the TSI T-line. For example, the resistance of the T-line increases with frequency as,

$$G_{skin}(f) = e^{-(1+j)l\sqrt{\pi\mu\sigma f}} \tag{12.1}$$

where $l$ is T-line length, $\mu$ is the permeability, and $\sigma$ is the conductivity. Similarly, the dielectric loss is

$$G_{di}(f) = e^{-l\sqrt{\varepsilon_r}tan\delta f/c} \tag{12.2}$$

where $\varepsilon_r$ is the dielectric constant; $tan\delta$ is the loss tangent of material, and $c$ is the speed of light [292]. As such, one needs to develop methods to compensate for the frequency-dependent loss for the 2.5D TSI T-line based I/O designs when operating in lossy mode. The drivers that will be mostly used for T-line can be a low-voltage differential signalling (LVDS) buffer and current-mode logic (CML) buffers. In the following, the design of low power I/O with LVDS and CML buffers are discussed.

### 12.2.1  *LVDS I/O Buffer*

A typical LVDS interface as shown in Figure 12.3 consists of a current source with a nominal current of $3.5mA$ to drive the differential TSI T-lines terminated with $100\Omega$ load. The LVDS receivers have high input impedance, and the driving current mainly flows through the $100\Omega$ termination resistor, thereby creating a valid logic "one" or "zero" [293]. The resistor at the receiver end can not only provide the optimum T-line impedance matching, but also act as the load resistor for the current source to generate a low-voltage swing of $250mV$-$400mV$.

Considering the loss of the TSI T-line discussed previously, to improve the margins of the data eye-diagram, an amplitude pre-emphasis is used in the LVDS design. The schematic of a LVDS transmitter with pre-emphasis is shown in Figure 12.4. On the left, transistors $M_5$-$M_{12}$ contribute to a pre-emphasized signal generation. The signals IP and IN are differential data signals coming from a digital logic input, and IPD and IND are the reverse input data delayed by a time constant $\tau$.

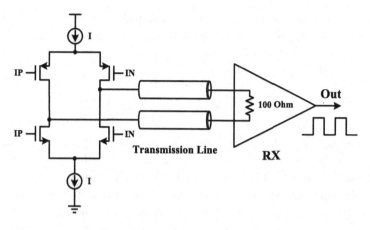

Fig. 12.3   Typical LVDS I/O architecture.

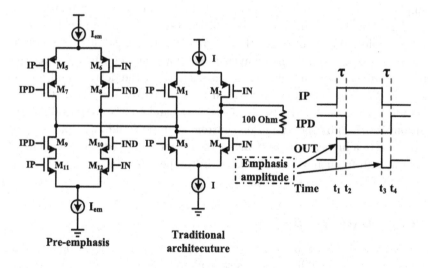

Fig. 12.4   A LVDS transmitter with pre-emphasis technique.

Assume that the amplitude of pre-emphasis gain is $G$, i.e., the ratio of current in the pre-emphasis driver $I_{em}$ and the normal driver current $I$. The transfer function for the pre-emphasized stage can be expressed as

$$H_{pre-em} = \frac{\tau A(s)}{A(s)} = I \cdot (1 - G e^{-s\tau}) \qquad (12.3)$$

where $A(s)$ is the transfer function of normal LVDS transmitter. When a first-order Pade approximation is used in this function, one can have

$$
\begin{aligned}
H_{pre-em} &= I \cdot (1 - G\frac{1 - s(\tau/2)}{1 + s(\tau/2)}) \\
&= I \cdot (1 - G)\frac{1 + s(\tau/2)(1 + G)/(1 - G)}{1 + s(\tau/2)}.
\end{aligned}
\tag{12.4}
$$

Equation (12.4) reveals that the amplitude pre-emphasis can reduce the DC gain, but enhance high frequency response. The transfer function has an additional pole at $1/\tau$. The high frequency compensation, which is $\pm(1+G)$ in (12.4), is performed when there is a data transition. In this design, use of the digital inverter chain delays the input digital signals with a time constant $\tau = 45ps$. In order to achieve low-power consumption, the current source of the LVDS buffer can be controlled to drive about only $350mV$ peak-to-peak voltage swing at the termination resistor. The ratio of pre-emphasis current gain is set as $G = 0.28$ for the purpose of evaluation.

The timing graph in Figure 12.4 illustrates the operation of the overall LVDS buffer. When the input IP is high at the time instant $t_1$, transistors $M_{2,3,6,11}$ are turned on. The main current $I$ flows into a termination resistor to generate the output voltage. At the same time, the delayed and reverse signal IPD is high enough to turn on transistors $M_8$ and $M_9$ at the time instant $t_2$. So, during the period between $t_1$ and $t_2$, the amplitude of the output is driven by the sum of two current sources, $I$ and $I_{em}$. As such, one can achieve a higher voltage swing at the output rising-edge to compensate for the high frequency loss. Similarly, in the period between $t_3$ and $t_4$, the output falling-edge is emphasized.

### 12.2.2   CML I/O Buffer

In addition to LVDS buffer, the design of the current-mode-logic (CML) buffer for the 2.5D TSI T-line is also explored. CML is a differential digital logic circuit that can transmit data at high frequency.

A two-stage CML I/O buffer as transmitter with a pre-amp stage is shown in Figure 12.5. Two CML buffers are cascaded such that the output of the pre-amp stage is connected to the input of driving stage. When the digital logic input is IN, the differential branches of CML buffers are ON and OFF that are controlled by two input transistors $M_1$ and $M_2$. Note that the CML buffer requires a full current switching and that the current flows through the ON branch only. In this design, there is no tail current used to

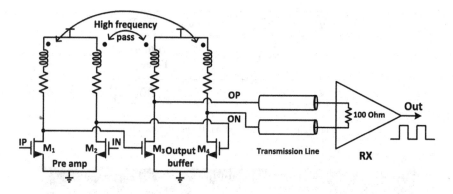

Fig. 12.5   A schematic of the proposed CML I/O interface.

achieve a full current swing. The currents through the resistor and inductor loads are controlled by the input voltage swing of two transistors ($V_{GS}$) and their size. Thus, the size of two pairs of transistors are designed carefully in order to provide enough driving ability with low current consumption at the same time.

Moreover, the CML buffer is required to drive load through the microbump pad with a $3mm$ TSI T-line, which has a smaller load than the bonding wire with package trace. This means that NMOS transistors of the second stage CML buffer in Figure 12.5 can have smaller sizes and smaller gate-to-channel capacitance. Thus, voltage level of power supply can be lowered to $0.6V$ to decrease the power consumption.

Additionally, to compensate for the high frequency loss of a 2.5D TSI T-line, inductor loads are designed as cross–coupled. As such, there is a fast data-pass through inductors when the CML buffer transmits high speed signals. The first-stage of the CML transmitter works as a pre-amplifier and provides a common mode voltage input to the next stage to guarantee the two input pairs of NMOS in the second stage working in saturation region all the time.

### 12.2.3   Simulation Results

For the evaluation of the 2.5D TSI based integration by CML and LVDS buffers, all chips are assembled face down and are attached to microbumps. They are connected by a $\sim 20\mu m$ width TSI T-line through the common substrate. The I/O circuits are the communication units that deliver data signals between cores and memories. The TSI T-line is made of aluminum,

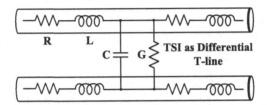

Fig. 12.6   TSI realized as Transmission line.

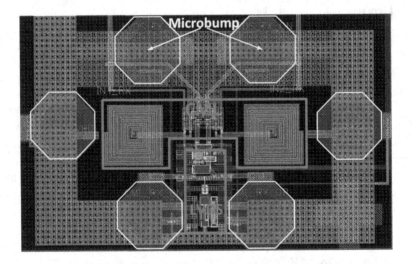

Fig. 12.7   Layout of two I/O Interface.

and the dielectric is silicon dioxide. The T-line length, width and character-istic impedance are $3mm$, $20\mu m$ and $50\Omega$, respectively. The TSI modeled as a transmission line (T-line) is shown in Figure 12.6.

CML and LVDS I/O Interface circuits are designed in UMC $65nm$ CMOS process using 1P6M layers. Figure 12.7 illustrates the layout of LVDS and CML transceiver circuits. The layout consists of six microbumps distributed in the surrounding (the yellow octagons), which are used for microbumps fabricated and TSI connection between chips. The whole chip area is $550\mu m \times 330\mu m$, with about $60\mu m \times 120\mu m$ area for the LVDS transceiver. In CML circuits, the inductors cost most of the chip area, which is $90\mu m \times 90\mu m$ for each. The layout of a $3mm$ TSI interconnect is shown in Figure 12.12.

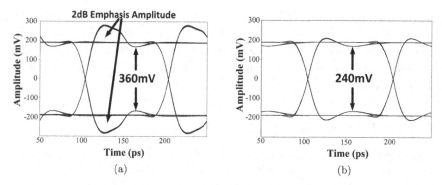

Fig. 12.8   (a) Simulation waveform of LVDS output eyediagram; (b) waveform of LVDS passed through TSI connection.

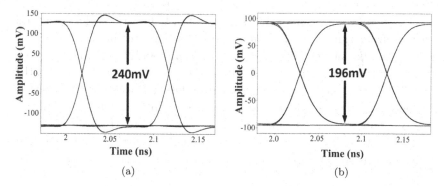

Fig. 12.9   (a) Simulation waveform of CML output eyediagram; (b) waveform of CML passed through TSI connection.

Figures 12.8 and 12.9 shows the post layout simulation results of the LVDS and CML I/O buffers respectively. From Figure 12.8, it can be observed that the LVDS I/O interface can achieve $360mV$ peak-to-peak output swing and $563fs$ cycle-to-cycle jitter at 10GHz bandwidth. This waveform consists of about $2dB$ amplitude pre-emphasis, with 0.28 ratio of current between emphasis and mean circuit. The power consumption is only $4.8mW$ under $1.2V$ supply, which is about $0.48pJ/b$ in energy efficiency. In Figure 12.8(b), when the signals are transmitted through the TSI transmission lines between chips, the signal has $30mV$ loss in simulation. But the $2dB$ amplitude pre-emphasis part, which compensate for the loss exactly, can protect the performance of high frequency edges. For the CML circuits, it can work under only $0.6V$ power supply, with $1.6mA$ total current consumption due to amplitude boosted inductors. As shown in Figure 12.9, it can provide $240mV$ peak-to-peak differential signal swing and $453fs$ jitter with $12.8Gb/s$ bandwidth, and $196mV$ peak-to-peak voltage after TSI T-line loss.

### 12.2.3.1 *EM Simulation of TSV and TSI I/O*

EM simulations are carried out for 3D TSV and 2.5D TSI I/Os in Ansoft HFSS. Frequencies up to 20GHz are swept to measure the insertion loss for TSV and TSI I/Os. EM simulation is performed for a pair of TSVs and TSIs. The substrate is a lightly doped silicon with bulk conductivity of $5Seimens/m$, and the bulk conductivity of copper TSV is set as $5.7 \times 10^7 Seimens/m$.

The simulation setup for TSVs is shown in Figure 12.10. TSVs of radii starting from $1\mu$m to $1\mu$m is simulated to evaluate the insertion loss performance. The height of the TSV is set as $100\mu m$ and the distance between two TSVs is set as $50\mu m$. From Figure 12.11, one can observe that with the increase in frequency of signal the insertion loss increases also. This is due to the increase in Ohmic loss with frequency. An insertion loss of nearly 0.5dB is observed for a $1mu$m TSV at a frequency of 8GHz.

Similar to TSVs, EM simulations are carried out for TSI I/Os. The TSI I/O length is set as $3mm$, shown in Figure 12.12. The dielectric layer is

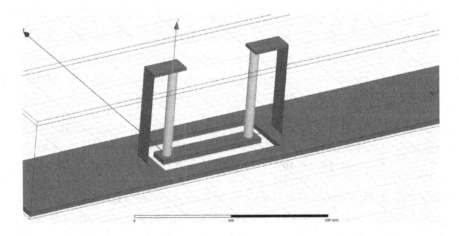

Fig. 12.10   EM simulation setup for TSV I/O.

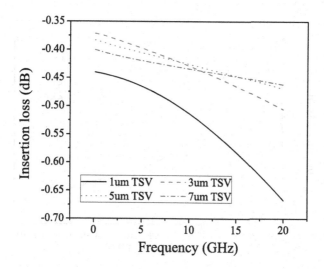

Fig. 12.11 Insertion loss for TSVs with various diameters.

Fig. 12.12 Layout of TSI I/O.

polymer with an effective permittivity of $3.5F/m$ and 0.02 loss tangents, and the bulk conductivity of transmission line is set as $5.7 \times 10^7 Seimens/m$. Similar to TSVs, with the increase in frequency, the insertion loss increases, due to the increase in Ohmic loss of the metal material as well as the radiation loss into the substrate. For example, at 8GHz, the insertion loss of TSI is found to be nearly as shown in Fig. 12.13. TSI has larger loss compared to TSV due to its longer length and the conductive substrate.

## 12.3   2.5D Adaptive I/O

To achieve a low power communication, an adaptive I/O interconnect can be implemented. In this section, we will discuss the design of an adaptive I/O link which we have discussed in the previous chapters. The overall 2.5D I/O link with transmitter and receiver in the adaptive I/O design

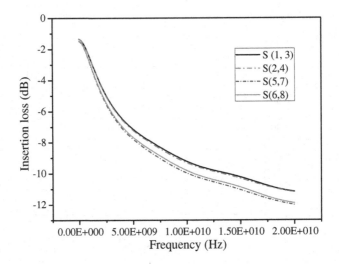

Fig. 12.13   Insertion loss for $3mm$ TSI I/O.

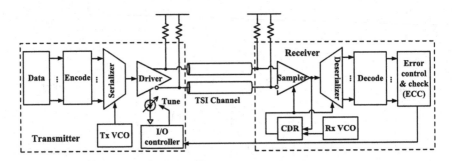

Fig. 12.14   Adaptive tuning I/O with error checking and correction.

is shown in Figure 12.14. Both the core and memory blocks are equipped with transmitter and receiver blocks to support duplex communication. However, the major difference lies in the design of transmitter driver and the additional error correction and check (ECC) block at the receiver. two blocks.

To operate with high bandwidth using a single channel of 2.5D TSI I/O, a 8:1 serializer in the transmitter (Tx) and 1:8 deserializer at the receiver (Rx) are employed. Each of the Tx and Rx has a voltage-controlled-oscillator (VCO) to generate the required clock signal (2GHz). Both the Tx and Rx are terminated for the 2.5D TSI based T-line with matched $50\Omega$ resistance. At the Rx, the serial bit stream is sampled and deserialized;

and is re-synchronized by the recovery clock from the clock data recovery (CDR) block.

### 12.3.1 *Transmitter and Receiver*

The transmitter (Tx) employs a 8:1 serializer to convert 8-bit parallel data into serial data as shown in Figure 12.15(a). Four digital D flip-flops are implemented as a shift-register chain for each of the odd ($D_1$, $D_3$, $D_5$, $D_7$) and even ($D_0$, $D_2$, $D_4$, $D_6$) bits of data. This is followed by a 2:1 MUX to combine them altogether. A current-mode logic (CML) output driver is used to drive the TSI T-line from the Tx to the Rx on the common substrate. The CML output stage is powered by the fixed supply (1.2V). The I/O communication power $Pw$ depends on the output-voltage swing and the tail current of the driver. One can generate control bits to tune the tail current of the CML driver and alter the output-voltage swing, as shown in Figure 12.15(b).

Compared to the traditional serial I/Os based on the backplane PCB trace [102, 103], the 2.5D TSI I/Os do not need the complex equalizer circuits at the receiver due to the small signal loss in the TSI T-line channel.

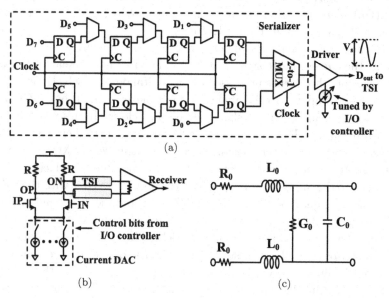

Fig. 12.15   (a) Transmitter with 8:1 serializer; (b) adaptive tuning of driver tail current; (c) TSI realized by a T-line.

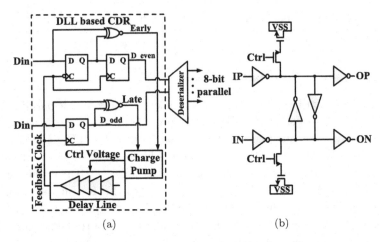

Fig. 12.16   (a) The architecture of DLL based CDR; (b) voltage controlled delay cell in the DLL.

A sampler at the front-end receiver is employed to sample the received signals and convert them into digital signals, as shown in Figure 12.14.

After the data decision, this data is processed in the digital domain, which saves more power compared to analog de-multiplexer. A delay-locked loop (DLL) based CDR at the receiver is implemented to de-skew the sampling clocks, as shown in Figure 12.16(a).

At the receiver end, a sampler connected to the CDR is employed to convert the current-mode signals into digital CMOS level signals. After the data decision, this data is processed in the digital domain, which could save more power compared to the analog de-multiplexer. A DLL based CDR at receiver is implemented to de-skew the sampling clocks. In this CDR design, a half-rate clock architecture is employed to decrease digital circuit working frequency and save power consumption. Two exclusive-or (XOR) gates in Figure 12.16(a) form a phase detector to judge the sampling clock position compared to the input data. It compares the input data edge with the rising edge sampled signal to obtain the *"early"* pulse and the *"late"* pulse. Then a charge-pump block converts these pulses into variable voltage to control the DLL delay line, which can tune the delay phase of the clocks and also provide feedback to the sampler. The schematic of the voltage-controlled delay cell is illustrated in Figure 12.16(b), which is based on the inverter chain for reducing the constant current consumption. This implementation of DLL in the CDR circuit makes it inherently stable and avoids jitter accumulation.

### 12.3.1.1 *Error Correcting Code*

To recover the received data and perform the recovery, data is encoded using the hamming code [294] and transmitted along with the parity check bits. As shown in Figure 12.17, 32-bit parallel data $(D-32)$ is initially stored in the output FIFO of the transmitter. For the 32-data bits, 7 parity bits are generated by the parity generator and an additional MSB of parity check vector is set as 0. The parity generator uses the code generator matrix $C$ to generate the parity bit vector $P$ as shown in Figure 12.18(a), where the parity generator consists of a set of $AND$ and $XOR$ gates. Thus, the total encoded data to be transmitted will be 40-bit for every 32-bit of data. One multiplexer (MUX) is implemented for serial transmission. The data format of the transmitted data is presented in Figure 12.18(b).

At the receiver end, the 32-bit data is stored in the input FIFO $(D-32$ bits) and the last 7-bit of the 8-bit (parity) is utilized for error checking and correction. The checking result vector $(R)$ is generated from the parity code $P$. By summing the result vector, one can detect if any bit is wrong and a left-shifter is used to correct bit error. The current implementation of the error-correcting code (ECC) has a capability to correct the 1-bit error but detect multiple bit errors.

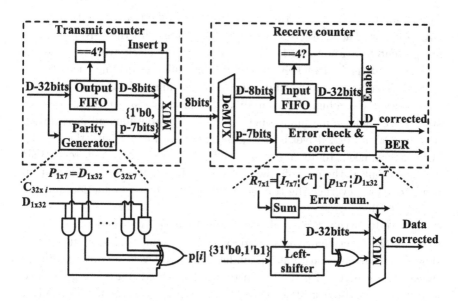

Fig. 12.17   Encoding and decoding at transmitter and receiver.

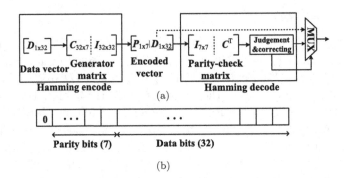

Fig. 12.18   (a) Data encoding and decoding; (b) transmitted data format.

As shown in Figure 12.15(b), the CML driver is set by the DAC current and load resistor. The DAC tail current source is comprised of a group of current sources connected in parallel, with switches. When the driver tail current is varied, the output-voltage swing will change. Generally, the load resistor is set at $50\Omega$ for the TSI T-line impedance matching. Here the tail current source is varied from $2mA$ to $5mA$.

### 12.3.2   *Simulation Results*

The simulation results for the above discussed adaptive I/O link are discussed below. The characteristics of eye-diagrams with the driver currents are presented in Figure 12.19 by introducing 10% clock cycle-to-cycle jitter (noise) at the TSI I/O channel. Note that different driving currents can make different eye openings under the noise in channel. A larger eye opening is associated with a higher current driving ability (or a larger output-voltage swing), which has a minimum opening of $76mV$ amplitude and $77ps$ timing margin with a $2mA$ driver current, and increases further with the driving current. One needs to observe that with the increase in driver current, BER decreases, but at the cost of power. As BER does not need to be low at all times, one can leverage the trade-off between the power reduction and the necessary BER.

We study further the eye-diagram under the control of adaptive tuning to verify the functionality of the adaptive I/O circuit tuning. Figure 12.20 shows the current consumption under different levels of output-voltage swing. The sources of error are introduced in three stages: stage 1 is to introduce 20% of clock jitter; stage 2 is to add an additional 10% receiver offset; and stage 3 is to further add an additional 10% power supply noise.

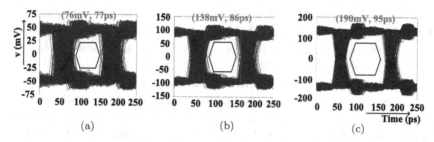

Fig. 12.19 Eye diagram of output data with different driver current (or output-voltage swing) levels: (a) $2mA$; (b) $3mA$; (c) $4mA$.

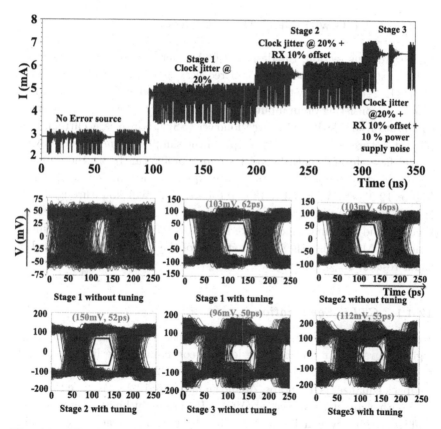

Fig. 12.20 The eye-diagrams under adaptive current (or power) adjustment by output-voltage swing tuning.

As discussed previously, with the increase in noise, the tail current at the CML buffer is varied to tune the output-voltage swing. For example, for stage 1, which only has clock jitter, the current is increased to $5mA$ to increase the eye-opening ($103mV$, $62ps$). With the increase in noise to, i.e., for stage 3, the current is increased adaptively. The difference in eye-diagrams with tuning the output-voltage swing and without tuning the output-voltage swing is shown in Figure 12.20. One can observe that for stage 3, without tuning the tail current, the eye opening is $96mV$, but the eye opening increases to $112mV$ by adaptively changing the current. Similar improvement in eye openings is shown for other stages as well.

## 12.4   2.5D I/O with Clock Data Recovery

In the previous section, the design of I/O buffers is discussed. In this section, we discuss the clock and data recovery at the receiver with CTLE. Figure 12.21(a)-(b) show the current-sampler merged with the active CTLE equalizer, consisting of an input buffer with inductive loading $L_1$ for active equalization and switched source follower (SSF), which is a current sampling structure. The merging principle of the sampler is that when CLK=1, $I_1$ will flow through path-I, and the input buffer can boost the high frequency part of the data to realize the equalization function as shown in Figure 12.21(a). When CLKB=1, the current $I_1$ will flow through path-II, and $M_2$ will be turned off to hold the data. As such, the equalization function and sampling function are realized by the proposed circuit simultaneously. Meanwhile, the input matching of the FC receiver is realized by shunt resistor $R_{match}$.

### 12.4.1   *CTLE Equalization*

For middle-distance interconnects ($<10\,cm$) such as interposers for memory-logic integration at the inter-die level, a continuous-time linear equalizer (CTLE) is sufficient enough for data recovery [286–288] without decision feedback equalizer (DFE) taps. As shown in Figure 12.21(b), when the input data with channel loss arrives at (VIN, VIP) the input buffer, compensation at high frequency can be achieved by the inductive load $L_1$ with gain-boosting. The gain of the input buffer is targeted to have peaked at 10GHz for the compensation. As such, the value of its inductor load $L_1$ must be optimized. As shown in Figure 12.22(a), $L_1$ is $1.2nH$ obtained by sweeping

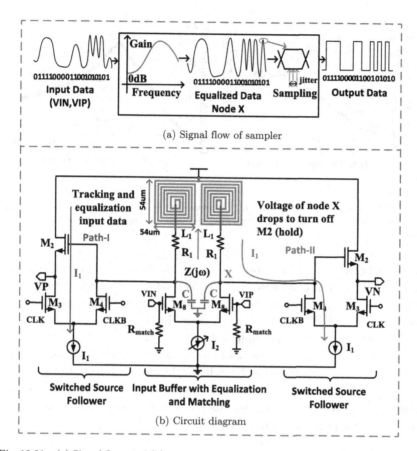

(a) Signal flow of sampler

(b) Circuit diagram

Fig. 12.21 (a) Signal flow; and (b) circuit diagram of data recovery by equalized sampler with inductor load ($1.2nH$).

from $0.3nH$ to $2.7nH$, and is realized within a compact area of $50\mu m \times 50\mu m$. Moreover, the current source $I_2$ can be tuned from $0.6mA$ to $2.4mA$ for an adaptive equalization as shown in Figure 12.22(b).

### 12.4.2 *Current Sampling*

Compared to the voltage sampling, the current sampling can achieve superior sampling speed [290, 295]. To implement current sampling, the SSF structure is commonly utilized.

As shown in Figure 12.21(a), the equalized data can be recognized as "0" or "1" at point X and will be further sampled by SSF. Note that the

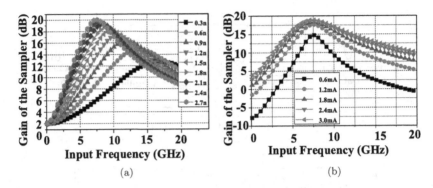

Fig. 12.22 Simulation results of CTLE equalization of sampler: (a) gain-peaking is above $18dB$ at 10GHz; (b) tunable gain-boost at high-frequency from $0.6mA$ to $3mA$.

input buffer transfers the input data from the voltage domain to current domain by the transconductance of $M_8$. When CLK=1, the current $I_1$ flows through $M_2$ by path-I, and the sampler tracks the equalized data at track-mode; when CLKB=1, the current $I_1$ flows through $M_4$ and $R_1$ by path-II, and the sampler holds the input data due to the low voltage of node X that turns off transistor $M_2$ at hold mode. Moreover, the bandwidth of the SSF is also improved because the inductor $L_1$ can absorb part of parasitic capacitor $C$ at the node X [295]. As a result, the equalized current sampler can realize both of the sampling and equalization functions at the same time with low power and high energy efficiency.

### 12.4.3 *Phase Interpolator*

In the conventional clock recovery design, the clock deskew is realized by a single ILO and the deskew is highly dependent on the offset frequency between the injected frequency and the Injection locked oscillator (ILO) free running frequency. What is worse, it can only provide a 90° phase deskew. In order to achieve a larger phase deskew to cover the phase misalignment between data and clock, a phase interpolator (PI) is applied here to generate clock deskew, instead of utilizing the single ILO.

Figure 12.23 shows the circuit diagram of clock recovery. Dual Injection-locked oscillators (ILOs) with phase interpolator (PI) are developed for clock de-skew. It reduces the JTB variation introduced by the offset frequency between ILO's free-running frequency and injected frequency. As shown in Figure 12.23(a), the ILO-I is locked to an injection clock, and it can provide four phases to realize phase interpolation with low power

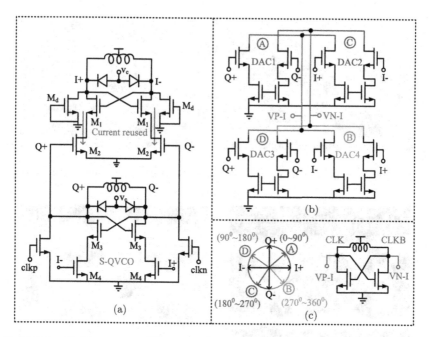

Fig. 12.23  Clock recovery: (a) ILO with four phases and locked to input clock; (b) Phase interpolator; (c) ILO locked to PI clock.

due to the current-reused S-QVCO structure. As shown in Figure 12.23(b), by adjusting the relative strength of the injection phase (I+,I-,Q+,Q-) with off-chip DAC (1,2,3,4) control, a complete 0-360° phase interpolation is realized to maintain JTB as constant, as shown in Figure 12.23(c). The ILO-II is locked to interpolated phases (VP-I, VN-I), and its output directly drives the sampler to further reduce power consumption.

### 12.4.4  *Measurement Results*

The prototype of the proposed FC receiver was fabricated in the UMC 65$nm$ CMOS process. The channel length is 4-5$cm$ on a FR-4 substrate of a 2-layer PCB. The test setup is shown in Figure 12.24. The random data is generated and transmitted by Agilent J-BERT N4903A. The chip area is 0.16$mm^2$ with die photo as shown in Figure 12.24.

#### 12.4.4.1  *Data Recovery*

Firstly, the data recovery is measured. The eye diagrams of the recovered 5$Gbps$ and 9.8$Gbps$ data with $2^{15} - 1$ random data patterns are measured

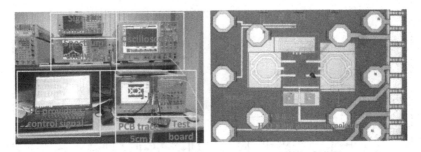

Fig. 12.24   Measurement setup; die photo of the FC receiver in 65 nm CMOS process.

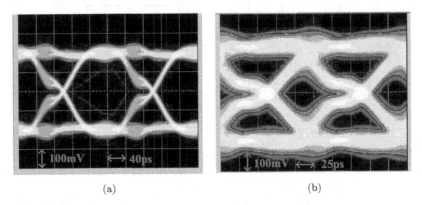

(a)                                                    (b)

Fig. 12.25   Measured eye diagrams: (a) recovered data at 5Gbps; (b) recovered data at 9.8Gbps.

by Agilent J-BERT N4903A as shown in Figure 12.25(a)-(b). The eye is well open with $200mV$, and the BER is below $10^{-12}$ at $5Gbps$ and below $10^{-10}$ at $9.8Gbps$.

### 12.4.4.2   *Clock Recovery*

Secondly, the clock recovery is measured. The transient I/Q signals of ILO-I in the FC receiver are measured to check the jitter performance by Agilent Infiniium 90008 with $40GSps$ sampling rate and 13GHz bandwidth. The measured result of the 8GHz I/Q signals is shown in Figure 12.26(a), and its peak-to-peak jitter is around $20ps$ as shown in Figure 12.26(b). The measured BER with phase deskew is shown in Figure 12.27.

Lastly, Table 12.1 shows the comparison of recently published FC receivers. The proposed FC receiver achieves the data rate of $9.8Gbps$

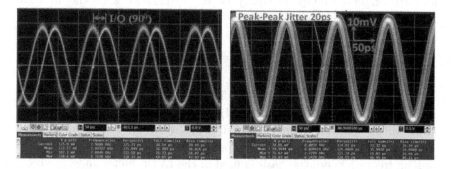

Fig. 12.26   Measured ILO-I: (a) IQ signals; (b) peak-to-peak jitter measurement.

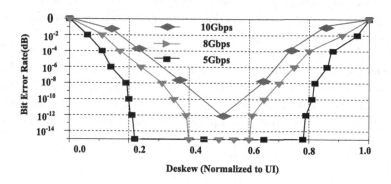

Fig. 12.27   The measured BER vs. clock deskew at 5*Gbps*, 8*Gbps* and 10*Gbps*.

Table 12.1   Comparison with state-of-art forward-clock I/O Receivers.

|  | [280] | [281] | [282] | [283] | Discussed work |
|---|---|---|---|---|---|
| Technology | 40nm CMOS | 65nm CMOS | 65nm CMOS | 65nm CMOS | 65nm CMOS |
| Supply ($V$) | 1 | 1 | 1 | 1 | 1/0.8* |
| Architecture | MSSC | ILO+DJM | ILO | DCA+ILO | ILO+PI |
| Data rate (*Gbps*) | 5.6 | 9.6 | 7.4 | 12 | 9.8 |
| Clocking | 1/2 rate | 1/2 rate | 1/2 rate | 1/4 rate | Full rate |
| Power (*mW*) | 13.5 | 11.8 | 6.8 | 11 | 5-6.5 |
| FoM (*mW/Gbps*) | 2.4 | 1.22 | 0.92 | 0.917 | 0.65 |

*Supply voltage: 1*V* for sampler and buffer, and 0.8*V* for ILOs.

and the highest energy efficiency of $0.65mW/Gbps$ with the full-rate architecture.

## 12.5  Summary

In this chapter, we investigated the 2.5D I/O circuit designs realized as TSI T-lines. Two high-speed low-power I/O buffer designs: LVDS buffer and CML buffer, both for a $3mm$ TSI T-line towards multi-core memory-logic integration, were discussed. To compensate for the loss of the TSI T-line in high frequency, pre-emphasis is used in the LVDS buffer and wide-band inductor matching is used in the CML buffer. In addition to the normal buffer, we also presented the design of adaptive I/O which can be utilized to reduce the communication power. Further, a forwarded-clock receiver by equalized current sampler for data recovery and phase-interpolation for clock recovery is implemented in the UMC $65nm$ CMOS process. Post layout simulation results show that the proposed LVDS buffer and CML buffer are capable of providing speeds of $10Gb/s$ and $12.8Gb/s$ with $0.48pJ/bit$ and $0.075pJ/bit$ efficiency. The current sampler has merged CTLE function with $18dB$ gain at $10GHz$ and $10GSps$ sampling speed with 20GHz bandwidth. Moreover, the PI can provide $0 - 360°$ clock deskew. The measurement results show that a data rate of up to $9.8Gbps$ with the energy efficiency of $0.65mW/Gbps$ is achieved.

# Chapter 13

# Microprocessor

## 13.1 Introduction

To meet the computing demands from data-oriented applications, multiple cores with memory have to be integrated on a single chip. Multi-core processors outperform single-core processors in terms of energy efficiency, speed, area and flexibility with low implementation cost [296, 297]. There are multi-core microprocessors from the industry as well as academia such as Intel SCC [298] with 48 cores, Tilera Tile64 [299] having 64 cores, Intel Polaris chip accommodating 80 cores [31], a 24-core microprocessor from Furan [297] and a 3D multi-core microprocessor from Georgia Tech university [16]. This high core counts is an indicator of better performance, but comes with additional complexity of design and management. With the increase in number of integrated core and memory blocks, communication between cores and cores/memory is the bottleneck of the system.

A network-on-chip (NoC) based communication scheme helps in improving the performance of the system [296, 297, 300]. A network-on-chip paradigm uses on-chip routers for communication between cores, memory, and other components. For communication, every core first contacts the router attached to it and further, based on the routing protocol embedded in the router, the routing is done and communication is carried out. Different kinds of routing such as circuit-based, packet switching or wormhole routing can be implemented in the router. Router based designs outperform bus-based systems mainly in terms of communication, because more messages will be in flight for NoC based communication compared to that of bus-based communication [300]. What is more, router based designs are scalable and have distributed communication. Hence, we present a router-based

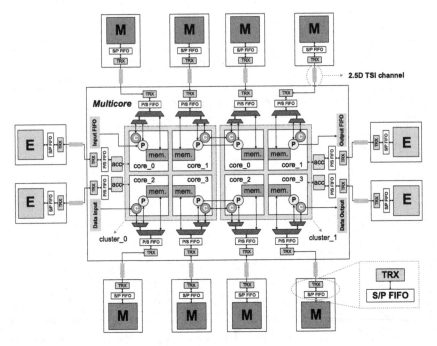

Fig. 13.1    Overall system architecture for multi-core memory 2.5D integrated microprocessor.

design to implement a multi-core architecture integrated with memory by 2.5D integration.

In addition to the I/O link design, we demonstrate the design of the 2.5D many-core integrated system with TSI I/O links in this chapter. Integration of a 8-core MIPS microprocessor with off-chip memory and hardware accelerators is shown in Figure 13.1. Each MIPS core denoted by P is designed with a 0.5KB word bank for both instruction and data memory, denoted by B. The core-to-core and core-to-memory communication is carried out through on-chip routers. Further, to reduce the complexity of system, the 8 cores are divided into 2 clusters (enclosed in dotted line), with each cluster having 4 cores. In order to expedite the processing, we employ 4 special hardware processing units i.e., accelerators, denoted by E, in this design. The system is integrated with 8 off-chip memory blocks, denoted by M, each having a capacity of 16KB and designed for on-chip data memory expansion, which can be accessed by pipeline or direct-memory access (DMA).

As discussed in Section 1.3, the loss due to backplane PCB wire trace is large, hence, we demonstrate the multi-core memory-logic integrated design by 2.5D TSI interconnects. The above discussed many-core memory-logic integrated system architecture is presented in Figure 13.1. The TSIs are shown by the green colored rectangular boxes.

As shown in Figure 13.1, routers are employed for communication between MIPS cores and memory units. The hardware accelerators are designed specially for multi-media and communication applications, which can help to improve the performance of some frequently used operations. The logic module is composed of an entropy decoder in H.264, FFT and complex multiplier. We discuss each of the components and evaluate their performance in next sections.

## 13.2 Building Blocks

The architecture of individual components presented in Figure 13.1 is discussed in this section. Firstly, we start with the MIPS core architecture, followed by router and I/O links.

### 13.2.1 *Core*

Core is one of the main blocks in the multi-core microprocessor design. We design a microprocessor without a interlocked pipeline stages (MIPS) core, as shown in Figure 13.2. The design is a typical 6-stage pipeline structure of MIPS core and has an enriched instruction set to support SIMD and direct-memory access (DMA) operations. The six-stage pipeline structure is presented in Figure 13.2. Data-level parallelism (DLP), configurability

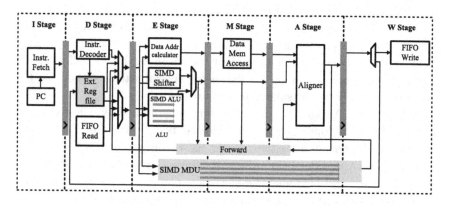

Fig. 13.2 Pipeline implementation inside MIPS core.

and simplicity are the major principles kept in mind while designing. The single instruction multiple data (SIMD) instruction set architecture (ISA) is utilized for DLP. The used SIMD supports data of different widths. Computing modes in SIMD supports scalar-scalar, scalar-vector and vector-vector operations as in [297]. The operation of the pipeline stages can be explained as follows.

Instruction fetch stage (Istage): The first stage of the pipeline stage is the instruction fetch, which fetches the instructions. The instructions are fetched according to the program counter (PC). The main component of this stage of the pipeline is the instruction cache, which stores instructions.

Decode stage (Dstage): The second stage of the pipeline is the instruction decode. The decode stage generates the control signals and fetches operands from the register file or the input FIFO. The decode stage is implemented by decoders, ROMs (FIFO, registers) and ad-hoc circuitry, whose main objective is to identify the main attributes of instructions such as type (eg., control flow) and resources that are required (eg., register ports, functional units).

Execution stage (Estage): After being issued, instructions are acted upon in the execution phase. Multiple kinds of instructions with different operands such as floating-point, integer and logical kinds of operations are carried out. The execution stage consists of functional blocks including ALU, shifter and MDU (multiplication division unit).

Memory stage (Mstage): The execution stage is followed by the memory stage. It is mostly associated with the data memory access. It consumes one clock cycle to finish a private data memory access, while consuming two control cycles for shared memory access, when no contention occurs. Another component of equal importance is the bypass logic in the memory-access phase (indicated as forward function in Figure 13.2). It is basically composed of wires that can transfer results from one unit to the input of other units, and an associated logic is utilized to determine whether the result from bypass or the data coming from the incoming FIFO buffer should be used. The bypass structure design needs special attention, because the wires do not scale at the same pace as gate delays.

Align stage (Astage): As the processing is performed by different units, the output result/data needs to be aligned before providing the output or storing the output into the register file or output in the FIFO. This alignment is carried out before the write stage, during the align stage.

Write back and commit: The last stage of the pipeline stage is write back and commit stage. The main purpose of this stage is to finish the execution and write the result into the register file.

To improve data locality and reduce power consumption, the register file of MIPS core is set as 64 words. This extended register file results in more available registers, indicating higher capacity to allocate data used by SIMD instructions and improve performance. In addition, the data locality of the processor is enhanced with more registers, resulting in less memory access and power consumption. Lastly, the extended register file serves as FIFO mapping ports.

### 13.2.2 *Router*

In a multi-core microprocessor, the communication between cores or core and memory is important and is often the bottleneck in a many-core system. In our design, we make use of a NoC topology. The communication is carried out with the help of a router. The data from the core is sent to the router for the purpose of routing the data to the destination in out architecture. A five port router is designed for the purpose of communication in mesh topology. The central port of the router is connected to the core and four other ports connecting to an adjacent router, as shown in Figure 13.3. Each router consists of three components: input buffer to store the data with simple FIFO logic, an arbiter to determine the data transferring sequence and direction according to the routing algorithm, and a crossbar switch for the purpose of making switches based on the result from the arbiter. The message-passing mechanism is widely utilized due to its better scalability

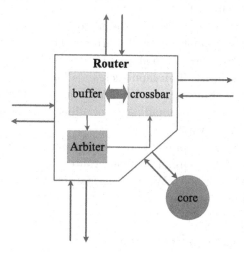

Fig. 13.3   An inner view of implemented 5-port router.

[32]. However, we discuss a hybrid communication mechanism similar to that used in [296] to support both message passing and shared memory to achieve higher performance and energy efficiency.

Let us first discuss the general working principle of a router. A message is initially divided into packets of varying sizes, and a packet is further divided into flits. A flit can be termed as smallest unit of data transmitted over the network. The size of a flit indicates the width of the link connecting the routers. Further, the router directs the flit to the destination based on the information encoded in the packet, which is usually the first flit. In our implementation, a flit-based switching is implemented, which requires a small amount of channel buffer, and wormhole routing flow control is implemented for efficient buffer utilization.

The operation of the router can be described as follows. When a head flit (first flit of packet) arrives at the router's input channel, the router stores the flit in the FIFO buffer and determines the next hop for the packet (router computation phase). An XY-ordered dimension order routing (DOR) algorithm is implemented by fixed logic, to generate a deadlock-free route in implemented mesh topology [32, 301].

Once the next hop is determined, the router allocates a virtual channel (virtual-channel allocation phase). Two virtual channels are implemented at each port with time division multiplex access (TDMA) implementation to enhance the channel utilization. When the flit competes for switch (switch allocation phase) and the next hop is ready to accept the flit, the packet is moved to the output port.

Switch allocation (SA) phase is the major and critical stage in router design, due to the complexity of the arbiter. During the SA phase, the arbiter issues switch traversal to all input ports requesting output ports for which they have priority. If an input demands for an output port, and the priority holder of that output port is idle or requesting for another port, the priority is allocated to that input. Arbiter adjusts the priority to promote fairness and avoid starvation.

We have two kinds of routers in our design, one being a packet-switched router, and the other a circuit-switched router. The NoC organization is inherited from work in [297]. We next discuss the core-to-core and core-to-memory communication protocols.

### 13.2.2.1   *Core-to-core Communication*

Core-to-core communication is different from core-to-memory communication. The communication between core-to-core happens through the on-chip

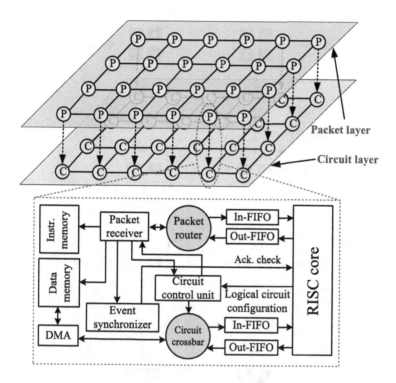

Fig. 13.4   Packet-controlled circuit-switched double-layer NoC.

routers. The router is provided with input in form of a message from the core or other adjacent components. The router is partitioned into two layers i.e., packet layer and circuit layer. Depending on the kind and size of data to be transferred through the on-chip routers, the router's layer is utilized. Packet switched NoC is utilized to transfer small amounts of data and in a synchronized communication. However to transfer continuous and large data transfers like multi-media and big data applications, a circuit-switched router (NoC) is utilized, which is flexible and configures to all combinational wire connections, resulting in less latency and better energy efficiency. This communication is shown in Figure 13.4.

### 13.2.2.2   *Core-to-memory Communication*

Each core has a 4kB on-chip data memory and 16kB more for expansion in off-chip SRAM. All memory is shared/can be accessed by pipeline of all cores within a cluster. Note that inter-chip access has a grand delay in I/O bandwidth bottleneck leading to significant performance degradation.

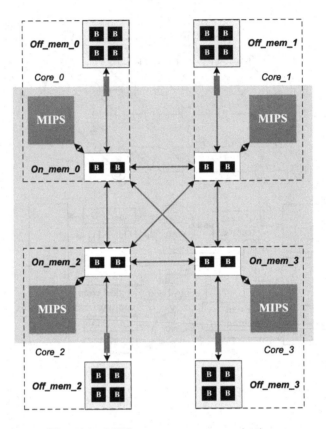

Fig. 13.5   MIPS core-memory communication.

Hardware architecture provides abundant types of direct memory access (DMA) operation to hide off-chip access latency, such as: on-off, off-off data DMA between memories; DMA between different multi-core chips when at core-core expansion mode; and DMA between local memory and off-chip accelerator. Core-to-memory communication happens through the dedicated pipeline memory access ports by hard wired connection, as shown in Figure 13.5. The arbiter helps to resolve the conflicts when simultaneous communication from multiple cores to same port happens.

### 13.2.3   *Accelerator*

Accelerators help in speeding up the processing, but is limited for one kind of application. A framework of array of accelerators, called execution array,

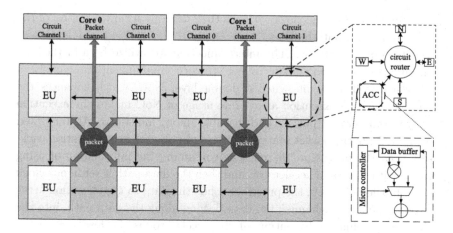

Fig. 13.6    Framework of the execution array.

is shown in Figure 13.6. The execution array (EA) is a mixed-granular spatio-temporal hardware reconfigurable framework. Similar to cores, it consists of an array of execution units (EUs) composed of processing element (PE), which is an ASIC and a router. The accelerator communicates with the external environment using a configurable hierarchical interconnect architecture. The target applications to be mapped to the accelerator are mostly compute-intensive tasks, such as multimedia, communication, and scientific computations, like H.264, LTE etc. As the hardware supports multi-input multi-output tasks, more tasks can be executed on the accelerator in parallel.

Processing element: Processing element, also accelerator, is one of the key component parts in execution array, and carries out critical computing tasks. It can be divided into two classes: one is the basic accelerator and large granularity accelerator. Basic accelerators such as adder, multiplier and shifter have small-scope granularity. To finish a complete task, they must be connected with other accelerators through interconnects. As such, these accelerators have high flexibility but complex configurations. Large granularity accelerators are designed for some special applications, like FFT, matrix multiplication, DCT and so on. They can be invoked simply but have lower flexibility. Whatever the kind of accelerator is, they have similar architectures. A local data buffer is responsible for storing the temporary operands and results. The sequence of operations is controlled by a $\mu$-controller referred to as a schedule table. Most of the tasks cannot be accomplished in one clock cycle, as they need an accurate control of the $\mu$-controller and data buffer.

Interconnect network: The interconnection we utilize here in the execution array is very similar to core-to-core communication, namely, a packet-controlled circuit-switched double-layer NoC, which enhances the scalability of the execution array. The usage of NoC makes the execution array reconfigurable, parallel and sharable among multi-core. As illustrated in Figure 13.6, the packet router receives and parses the configuration packages from the processor; the package containing the information of using the circuit router and accelerator, and then the data path is set in one clock cycle. In the next step, the processor sends the data to the execution array through the circuit channel; the data flowing along the data path, into the accelerator. Finally, the output of the accelerator is the feedback to the processor. The communication between core and accelerator can be DMA or normal register type instructions, the DMA having better transmission efficiency.

### 13.2.4   *I/O Link*

Another important component in the multi-core memory-logic integrated system is the interconnections. We make use of TSI I/Os to achieve higher speed with low power and better energy efficiency. Here we will discuss the I/O utilized in our microprocessor design. An overview of the I/O design is presented in Figure 13.7.

The I/O design consists of a transmitter (TX), channel and receiver (Rx) parts. The input data is first serialized with the help of the serializer, and

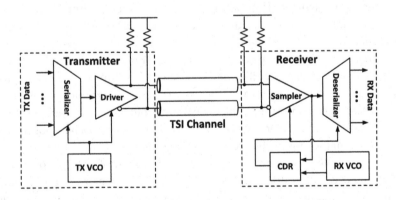

Fig. 13.7   I/O link architecture based on TSI interconnection.

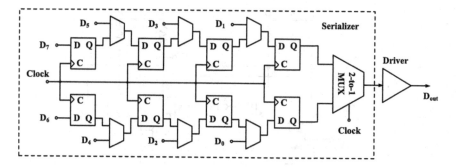

Fig. 13.8 Transmitter with 8:1 serializer.

the further voltage-controlled oscillator (VCO) is utilized to maintain the clock frequency. A 8:1 serializer is employed at the transmitter to serialize the data. Further, to drive the TSI channel, a driver is implemented after the serializer.

### 13.2.4.1 *Transmitter*

The transmitter TX a 8:1 serializer, converts 8-bit parallel data into serial data, as shown in Figure 13.8. Four digital D flip-flops are implemented as a shift-register chain for each of the odd ($D_1$, $D_3$, $D_5$, $D_7$) and even ($D_0$, $D_2$, $D_4$, $D_6$) bits of data. This is followed by a 2:1 MUX to combine them altogether. A current-mode logic (CML) output driver is used to drive the TSI T-line from the Tx to the Rx on the common substrate.

The driver at the transmitter utilized for the 2.5D I/O in Figure 13.8 is presented in Figure 13.9. The driver circuit implemented at the transmitter of the multi-core I/O consists of two stage cascaded CML buffers. To alleviate the mismatching of impedance, $50\Omega$ resistors are utilized for the impedance matching of the transmission line.

### 13.2.4.2 *Receiver*

Compared to the traditional serial I/Os based on backplane PCB trace [102, 103], the 2.5D TSI I/Os do not need the complex equalizer circuits at the receiver due to the small signal loss in the TSI T-line channel. A sampler at the front-end receiver is employed to sample the received signals and convert them into digital signals, as shown in Figure 13.7. The design of the sampler is shown in Figure 13.10. At the beginning of the sampler, an

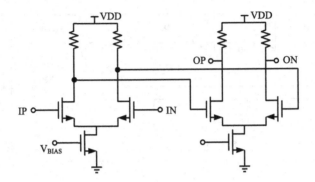

Fig. 13.9　Driver implemented at the transmitter side.

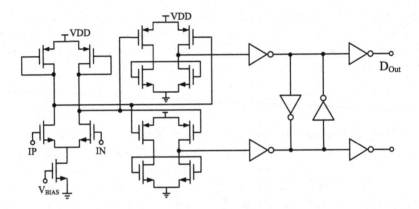

Fig. 13.10　Sampler implemented at the receiver side.

amplifier is used to amplify the input signals, followed by two current mode comparators to convert the signals into differential logic digital signals.

After the data decision, this data is processed in the digital domain, which saves more power compared to the analog de-multiplexer. A DLL based CDR at the receiver is implemented to de-skew the sampling clocks, as shown in Figure 12.6(a).

At the receiver end, a sampler connected to the clock-data recovery (CDR) is employed to convert the current-mode signals into digital CMOS level signals. After the data decision, this data is processed in the digital domain, which could save more power compared to the analog de-multiplexer. A delay-locked loop (DLL) based CDR at receiver is implemented to de-skew the sampling clocks. In this CDR design, a half-rate clock architecture is employed to decrease the digital circuit working frequency and save power consumption. Two exclusive-or (XOR) gates in Figure 12.6(a) form a phase detector to judge the sampling clock position compared to the input data. It compares the input data edge with the rising edge sampled signal to obtain the *"early"* pulse and the *"late"* pulse. Then a charge-pump block converts these pulses into variable voltages to control the DLL delay line, which can tune the delay phase of the clocks and also provide feedback to the sampler. The schematic of the voltage-controlled delay cell is illustrated in Figure 12.6(b), which is based on an inverter chain for reducing the constant current consumption. This implementation of DLL in the CDR circuit makes it inherently stable and avoids jitter accumulation.

### 13.2.4.3 *Voltage-controlled Oscillator*

For the voltage-controlled oscillator (VCO) design, a ring VCO is employed to get eight different phases for the whole system, as shown in Figure 13.11. Compared to the LC tank oscillator, the ring VCO could get worse phase noise and low frequency, but saves more power consumption and chip area, which can be considered as a trade-off. Hence, we choose the ring VCO for this design. The VCO architecture and delay cell unit are shown in Figure 13.11.

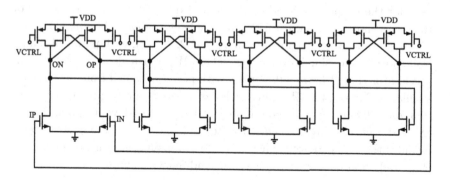

Fig. 13.11   Voltage-controlled oscillator design.

## 13.3    2.5D Multi-core Microprocessor

### 13.3.1    *System Design*

The whole system simulation is performed with a test-bench, multi-core, 4 accelerators and memory blocks. A test bench is initially loaded to the multi-core block, and the communication and other characteristics are correspondingly evaluated. Simulations are carried out using the TSI I/Os.

The co-simulation platform is composed of a multi-core die, 8 off-chip memory and 4 off-chip accelerator units with corresponding TSI I/O pairs. The whole system is designed in a Global Foundries (GF) $65nm$ low power process. The I/O circuit and TSI model are designed with Cadence IC design tools. The other design parts are RTL design for functional simulation and netlist for post synthesis simulation with delay information from a standard delay format (SDF) file. The digital part is simulated on a Synopsys Verilog compiler simulator (VCS) for fast verification. Co-simulation is performed for the whole system at 500MHz frequency. 2.5D TSI integrated system verification is performed in the Ultrasim Verilog simulator with analog design environment (ADE) to introduce I/O and TSI I/O features. Post layout simulation is set in Spectre simulator for analog circuit simulation.

### 13.3.2    *Fabricated Chip*

For the purpose of fabrication, a chip with one set of multi-core (8-cores), one accelerator and one memory block of SRAM is designed. Each of the blocks are allocated with the base to attach the gold microbumps, which is used for the purpose of bonding to the TSI die. The layout and picture of the die is shown in Figure 13.12. The whole chip area is $3.3mm \times 3.3mm$, which contains a multi-core chip ($3.3mm \times 2.35mm$), with power obtained from post-layout simulation being $456.4mW$, and a delay of $2.05ns$; an accelerator chip ($1.3mm \times 0.83mm$), with power obtained from post-layout simulation being $24.3mW$, and a delay of $1.34ns$; an off-chip memory chip ($1.3mm \times 0.83mm$) with power obtained from post-layout simulation of $18.0mW$, and a delay of $1.39ns$. Please note that the area indicated above includes the I/O pad area. There are 246 bumps in total on the multi-core chip, and 28 bumps on the accelerator and memory chips.

### 13.3.3    *Simulation Results*

The system comprises of five pairs of TSI I/Os to connect off-chip memory/accelerator and core blocks. The TSI I/Os consist of three pairs of

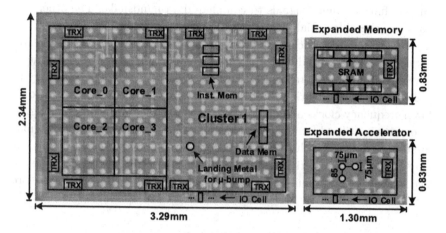

Fig. 13.12 Fabricated multi-core processor with memory and accelerator blocks.

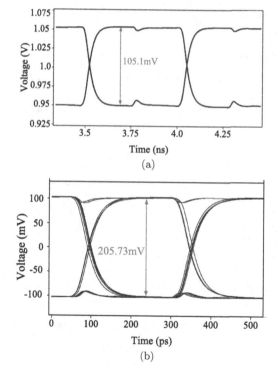

Fig. 13.13 The differential signals of transmitter: (a) output signal of transmitter; (b) corresponding eye diagram.

enable, full and index signals to complete data handshake. The other two pairs are the differential high-speed data signals.

The system global clock is provided by one VCO in I/O, and the frequency can be adjusted by control voltage. The multi-core, off-chip memory and off-chip accelerator use multi-clock and gated clock technology to reduce power consumption. Multiple clock domains are implemented, some lower frequency clocks are employed to initialize the multi-core, and higher frequency is utilized for normal operation. Many asynchronous reset signals are designed for each cluster (a set of four cores), off-chip memory and off-chip accelerator respectively. For the convenience of the test chip, there are two sequential data input and output signals implemented in the multi-core

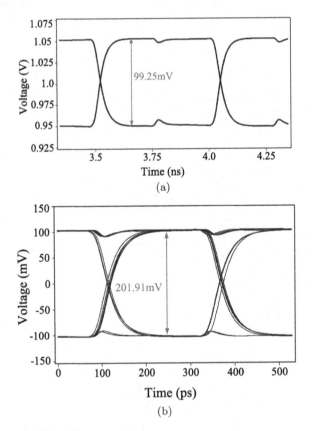

Fig. 13.14 The differential signals of receiver: (a) input signal at receiver; (b) corresponding eye diagram.

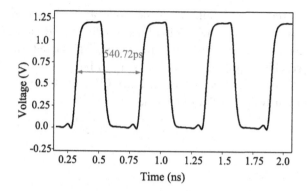

Fig. 13.15    The simulation results of VCO.

and bonded to the test board. The function of sequential data is similar with scan chains in chip. One can set a sequential to input data and collect the result from the output, and check the status of the inner key register, initialization of instruction cache and so on.

With the main focus on I/O and VCO design, we study the signals at the receiver and transmitter. With power supply $V_{DD}$ of 1.2V for I/O, the output signal at the transmitter is studied. Figure 13.13 shows the time domain output differential signals and the eye diagram of the transmitter. The amplitude of a single-ended signal is about $105mV$.

When the signal is transmitted from the transmitter through the designed normal TSI I/O, a minimal loss is expected at the receiver. Figure 13.14 shows the differential signals received after TSI transmission lines, and its eye diagram. An amplitude of about $99mV$ for a single-ended signal is observed with a loss of $6mV$ over a $3mm$ T-line.

Figure 13.15 shows the VCO post layout simulation. When the VCTRL signal of the VCO (shown in Figure 13.11) is set at $830mV$, the frequency of the VCO is nearly 1.9GHz.

## 13.4    Summary

In this chapter, a multi-core microprocessor design integrated with memory and specially designed hardware accelerator blocks using specially designed TSI interconnects is discussed. To alleviate the communication bottleneck in a large system, a mesh topology with router based communication is employed in the design. The architecture of the router, MIPS core and the I/O is presented in detail. A loss of less than $6mV$ is observed when transmitted over a $3mm$ transmission line. Further, the design of the VCO with a frequency of nearly 1.9GHz is presented. The layout and picture of the fabricated chip is presented, with testing to be carried out. The whole chip area is $3.3mm \times 3.3mm$, which includes the I/O pad area, multi-core, memory and hardware accelerator blocks.

# Chapter 14

# Non-volatile Memory

## 14.1 Introduction

In the previous chapters, we have discussed I/O design and multi-core microprocessor design. Memory is another block of equal importance in multi-core design. Memory technologies have undergone rapid advances from traditional CMOS SRAM/DRAM to nano-scale non-volatile memory (NVM) [302–307], which can provide low power and high performance memory solutions for future computing systems. Characterized by different features, the emerging NVMs may fit in different applications. For example, spin-torque-transfer based memory (STT-RAM) exhibits performance close to DRAM and virtually unlimited endurance, but requires a large footprint and high driving current [302, 308], and thus is considered to be cache-class memory. Phase-change-memory (PCRAM) has high integration density, 10~1000ns scale performance but limited endurance, and is thus categorized as storage-class memory. The recent designs are also for a hybrid memory structure [309] towards high integration density, fast accessing time, long endurance time, and low leakage power by leveraging strengths of different memory technologies. This chapter shows one possible hybrid memory design by 3D integrated SRAM/DRAM with CBRAM-crossbar.

From the perspective of device technology, the recent conductive-bridge random-access-memory (CBRAM) has been introduced as one emerging resistive random-access-memory (ReRAM) device [304, 310–314]. Based on the fast ion-migration process, the conductive filament of CBRAM is able to grow and shrink extremely quickly under low supply voltage. As a result, compared to other types of NVMs, CBRAM has advantages in higher speed and lower power. Moreover, the CBRAM fabrication process can easily result in an extremely high density crossbar structure. The CBRAM is categorized as storage-class memory due to high density and limited endurance ($\sim 10^6$) [315].

From the perspective of system architecture, 3D die-stacking [316–320] promises to integrate hybrid memory components with high density and low latency. One can design a hybrid memory system with each tier implemented by different memory technologies and stacked by through-silicon-vias (TSVs) [316]. However, as leakage power is the primary concern of a memory system, one needs to have a well-designed data-retention scheme such that the power gating can be effectively deployed to reduce leakage power yet without degrading performance. Traditionally, the common approach for data retention of SRAM/DRAM [321] is to deploy a small retention voltage for all memory cells in sleep-mode, which still has non-negligible leakage power. In this chapter, we introduce a 3D hybrid memory architecture, in which the CBRAM-crossbar is stacked as a tier for data retention of the other SRAM/DRAM tiers under power gating.

## 14.2   3D Hybrid Memory with ReRAM

In this section, we introduce the proposed architecture for the 3D hybrid memory system based on the ReRAM CBRAM-crossbar with block-level data retention ability.

Figure 14.1 illustrates the overall system architecture of the proposed 3D hybrid memory, which is composed of embedded DRAM, SRAM and CBRAM-crossbar located at three layers (or tiers) connected by TSVs. Similar to the common memory organization [191], the entire CBRAM-crossbar memory is broken into *banks*, where each bank can be accessed independently with dedicated data and address buses. Each bank is further broken into a $M \times N$ *mat* array, where each mat consists of one $m \times n$ CBRAM-crossbar. In addition, the terms *sleep mode* and *active mode* are used to denote whether the system is power-gated or not. The active to sleep mode transition is called the *hibernating transition*, and the term *wakeup transition* is used vice versa.

To reduce the sneak-path power in crossbar structure [322], we propose to distribute the multi-bit data into different mats of the same row concurrently, where each mat (i.e., one CBRAM-crossbar) accepts one bit of the written data at each cycle. To achieve this, the data address is decoded to find the exact same crossing point of each CBRAM-crossbar

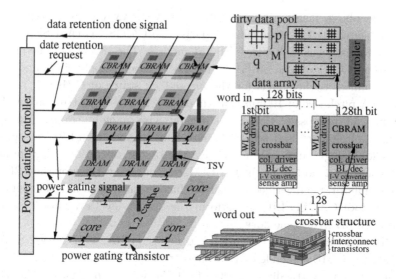

Fig. 14.1 3D hybrid memory system with CBRAM-crossbar based data retention.

mat where one bit of the arriving data is to be kept. For read operation, the same address decoding is carried out and the bits from each mat are combined together as output. As such, during one hibernating transition, each SRAM/DRAM bank is associated with one dedicated CBRAM-crossbar bank of the same capacity for data retention. Therefore, power gating is employed at the bank-level (i.e., block-level). Once hibernating transition begins, all the data in the specified SRAM/DRAM bank must be copied or migrated to the corresponding CBRAM-crossbar memory bank through dedicated data and address buses implemented by TSVs in the vertical direction. On the other hand, all data must be migrated from CBRAM-crossbar back to the SRAM/DRAM during the wakeup transition.

However, the primary design challenges to develop the above-mentioned 3D hybrid memory with CBRAM-crossbar are addressed below. Firstly, there is no design platform for CBRAM device and CBRAM-crossbar circuit such that the delay, area and power can be estimated and optimized. Secondly, there is no memory controller developed and verified for the CBRAM-crossbar based data retention, which can perform efficient data migration for SRAM/DRAM. In the following sections, we show the development of one design platform for the CBRAM device and CBRAM-crossbar circuit. Moreover, we show one memory controller design for the CBRAM-crossbar using an incremental block-level data retention.

The above mentioned architecture of CBRAM-crossbar platform is implemented by C++ on Linux. Firstly, one CMOS-CBRAM SPICE-like simulator is developed with CBRAM models. The model parameters $k_1$, $k_2$, $k_3$ and $k_4$ are based on [310] with 1.8V voltage supply. 2M$\Omega$ and 1G$\Omega$ are assumed as thresholds for the *on* and *off* states of CBRAM, respectively. A pitch size of $100nm$ is assumed for the crossbar structure, with a $50nm \times 50nm$ cross-section a area of nano-wire made by copper. Multiple supply voltages are used, where 65nm technology node with 1.2V is assumed for CMOS logic and 1.8V for CBRAM-crossbar operations. The power, delay and area models for the CBRAM-crossbar are verified by the SPICE-like simulator. Such verified models are further integrated into the memory design tool CACTI [191] to perform the system-level evaluation for the proposed 3D hybrid memory system.

The system under investigation is composed of one 2MB level-2 SRAM cache, one 64MB embedded DRAM and one 66MB CBRAM for data retention, which are vertically connected by TSVs with $50\mu m$ length. The TSV power and delay models in [22] are integrated into the extended CACTI. Here we adjust the design parameters $C_b$, $W_d$ and $f_b$ under constraints of a maximal 10W transition power, maximal $60/mm^2$ TSV density and maximal $3ms$ transition speed.

## 14.3   ReRAM-crossbar Memory Design

As CBRAM is a kind of ReRAM, we explore the design of a CBRAM-crossbar based memory in this section. Figure 14.4 illustrates the design of a CBRAM-crossbar structure. Compared to the 1T-1R structure [323], the crossbar structure has two major advantages. First, the extremely high integration density can be realized due to the small pitch-size of nano-bars [322]. Second, the crossbar structure can be stacked on top of the active transistor layer in a 3D fashion [324–326], which further reduces the area overhead and improves communication bandwidth. Figure 14.2(a) shows the approaches to fabricate the CBRAM-crossbar structure within the interconnect layers, where the CBRAM devices are deposited at the bottom of the copper vias [324]. Another approach in CBRAM-crossbar fabrication, shown in Figure 14.2(b), is to stack the crossbar structure on top of the interconnect layer, which incurs the least modifications on the conventional CMOS process [325].

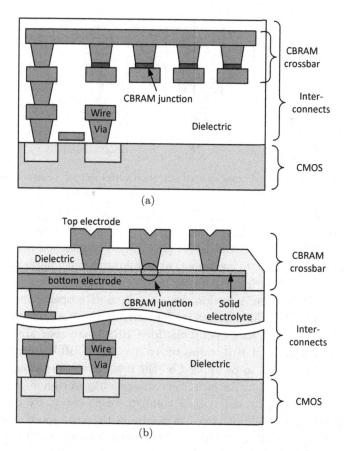

Fig. 14.2 Two approaches for 3D stacked CBRAM-crossbar (ReRAM) fabrication: (a) crossbar structure integrated within the interconnect layers, and CBRAM devices fabricated at the bottom of vias (b) crossbar structure fabricated on the top of interconnect layers.

The conventional voltage-divider based readout design is shown in Figure 14.3, where $R_l$ follows $\frac{R_{off}}{R_l} = \frac{R_l}{R_{on}}$. As such, $V_o$ will be logic-1 when target-cell is in low-resistance state (LRS) and logic-0 when in high-resistance state (HRS). Note that all the unselected word-lines and bit-lines need to be floated to avoid sensing current branches before reaching $R_l$. This will incur a sneak-path issue. When the target-cell is in HRS and is surrounded by other cells in LRS, there will be a significant leakage current flowing through the neighboring cells, which may lead to misinterpretation

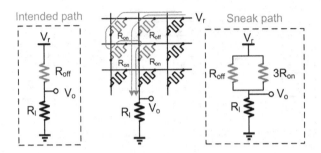

Fig. 14.3 Conventional crossbar read operation with incurred sneak-path issue.

of the stored bit. The sneak-path issue can be addressed by adding a selection device for each CBRAM device [327, 328]. The selection device works like bidirectional-diode to ensure no current flows through paths with more than one CBRAM device.

Instead of applying selection devices, alternative operations are proposed for the CBRAM-crossbar readout circuit to avoid the sneak-path issue. The *write* and *read* operations for CBRAM-crossbar are shown in Figures 14.4(a) and 14.4(b), respectively. To write CBRAM-crossbar, the bias-voltage $v_w$ needs to be applied on the designated cell through a multiplexed voltage selector. As such, the corresponding word-line and bit-line are applied with $v_w/2$ and $-v_w/2$ voltages, respectively. By controlling the voltage level and polarity, one can determine the write-speed and change the on/off states of the cell. The cells that are half-selected will not change their states. To read the CBRAM-crossbar, the bias-voltage $v_r$ (normally $v_r < v_w$) is applied on the corresponding word-line. By measuring the current of the designated column, the on/off state of the target-cell can be detected. Note that because all unselected word-lines and bit-lines are grounded rather than floated, the voltage-divider based readout scheme is incapable of function. As such, a more sophisticated readout circuit design is required and is discussed later in this section.

Figure 14.4(c) further shows the peripheral circuit design for one CBRAM-crossbar. During write operation, the corresponding word-line and bit-line are applied with writing voltage through address decoding. The voltage selection is done by the row drivers and column drivers, which switch among different voltage supplies according to the command issued. During read operation, word-lines still need voltage selection while bit-lines are switched to the current sensing circuit.

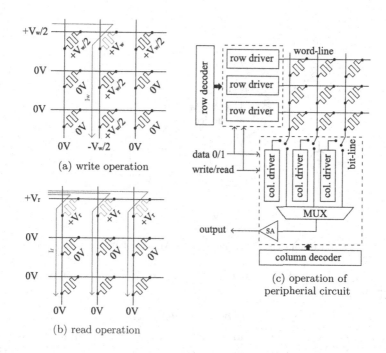

+Vw/2

0V

0V

0V   -Vw/2   0V

(a) write operation

+Vr

0V

0V

0V   0V   0V

(b) read operation

row decoder

row driver
row driver
row driver

word-line

data 0/1
write/read

col. driver
col. driver
col. driver

bit-line

MUX

output   SA

column decoder

(c) operation of
peripherial circuit

Fig. 14.4   Crossbar operations and peripheral circuits.

To sense the current of the designated column in the crossbar structure, a readout circuit illustrated in Figure 14.5 is proposed. The sensing procedure is done in two steps here. Firstly, a current mirror is deployed to amplify the current determined by the state of the target-cell. The bias current $I_{bias}$ is applied to ensure that the transistor M1 works in the saturation region. The bias voltage $V_{bias}$ has to be deliberately chosen according to $I_{bias}$ to achieve a virtually grounded node $A$, i.e., a 0V column voltage required in Figure 14.4(b). Otherwise, the current to detect will have branched at node $A$ to other cells in the same column, which will weaken the current signal and possibly degrade sensing accuracy. Secondly, the amplified current signal is converted into a voltage signal, and compared with the reference voltage to decide an output on/off state denoted by different logic levels.

Another issue in readout operation is the severe device resistance variations. It is reported that $5\times - 100\times$ $R_{off}$ and $2\times - 10\times$ $R_{on}$ variations can be expected based on a variety of measurement data [329]. For one crossbar array, there exists a valid reference voltage in the readout circuit only when

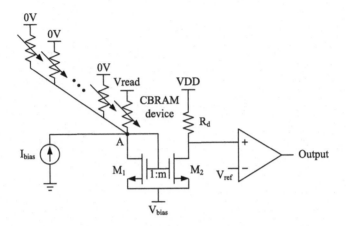

Fig. 14.5   Readout circuit design for CBRAM crossbar structured memory.

the minimum $R_{off}$ of all CBRAM devices is greater than the maximum $R_{on}$ of all devices, i.e.,

$$min(R_{off}) > max(R_{on}) \qquad (14.1)$$

which, however, may not hold true if the variation is significant. A crossbar with device resistance variations that cannot meet the above condition is called a failure. As such, the sense amplifier is not able to guarantee successful readout for every operation. The transient analysis for CBRAM switching is conducted in a SPICE-like simulator [330]. The CBRAM is initialized in the off-state, and $1.8V$ voltage is applied to set the device (off to on).

The Monte Carlo simulation with 5000 times is conducted in order to investigate the impact of device resistance variations on readout operation, which is shown in Figure 14.6. In each experiment, a $100 \times 100$ crossbar-array is generated, where $R_{off}$ and $R_{on}$ of all CBRAM devices have normal distribution with $\mu$ of $1G\Omega$ and $2M\Omega$, and deviation $\delta$ that follows the variation in [329], i.e., $10\times$ for $R_{on}$ and $100\times$ for $R_{off}$. The $max(R_{on})$ and $min(R_{off})$ are calculated and then used as the x-axis and y-axis values for each point. The whole domain is divided into two regions, pass region and fail region, separated by the dashed line $max(R_{on}) = min(R_{off})$. It can be observed that there are 9 failed cases, which leads to a 0.18% failure rate for the $100 \times 100$ crossbar size. In other words, 9 out of 5000 crossbars in this size do not have the ideal reference voltage value.

The Monte Carlo simulations for crossbar sizes starting at $50 \times 50$ to $600 \times 600$ with a step of 50 are conducted with results shown in Table 14.1.

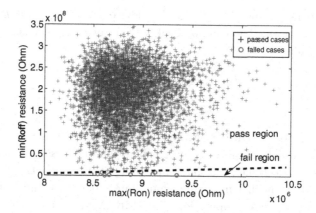

Fig. 14.6    Monte Carlo simulation of 100 × 100 crossbar-array with device resistance variations.

Table 14.1    Device resistance variation caused sense amplifier failure rate for different crossbar-array sizes.

| Array size | $50^2$ | $100^2$ | $150^2$ | $200^2$ | $250^2$ | $300^2$ |
|---|---|---|---|---|---|---|
| Failure rate | 0.05% | 0.18% | 0.85% | 0.95% | 2.65% | 3.20% |
| Array size | $350^2$ | $400^2$ | $450^2$ | $500^2$ | $550^2$ | $600^2$ |
| Failure rate | 4.30% | 5.50% | 5.55% | 7.65% | 10.85% | 13.40% |

It can be observed that the failure rate increases as the crossbar size enlarges. For large crossbar arrays where the failure rate is not negligible, robust readout schemes, namely the self-reference [331] and runtime ECC/ECP [332], can be applied to increase the readout reliability. Thus it is favorable to limit the crossbar size to within a few hundreds by a few hundreds.

### 14.3.1    *Performance Modeling*

Based on the circuit design for the CBRAM-crossbar, we discuss its system-level delay, power and area models like the CMOS memory design tool CACTI [191]. We further verify such models using the transistor-level SPICE-level simulator developed in [330].

#### 14.3.1.1    *Delay Model*

Generally, the time needed for write operation in CBRAM-crossbar is the sum of the signal propagation delay on wires (i.e., word-line and bit-line), denoted by $D_{wire}^w$, and the device switching time $D_{switching}$; for read operation, it is composed of wire delay $D_{wire}^r$ and sensing delay $D_{sensing}$. Thus

$$D_{write} = D_{wire}^w + D_{switching} \qquad (14.2)$$

$$D_{read} = D_{wire}^r + D_{sensing}. \qquad (14.3)$$

Different from DRAM/SRAM cell, the CBRAM device has asymmetrical write/read-delays. Since the write operation requires a physical change of the CBRAM cell, i.e., the shape-morphing of the conductive filament, it usually takes a much longer time than simply detecting the resistance of CBRAM cell in the read operation. The CBRAM switching time $D_{switching}$ can be obtained from the CMOS-CBRAM simulator with high accuracy, with the use of the physical model in [330].

Due to the existence of leakage current in the crossbar structure, CBRAM cells at different positions suffer from different amounts of reduced applied voltages introduced by IR-drop along the word-line and bit-line. Since the switching time is very sensitive to even the slightest of voltage deviation from the expected value, CBRAM cells at different positions of the crossbar array exhibit different values of switching time. Therefore, the exponential relation between $D_{switching}$ and the applied voltage $v_w$, represented as a lookup table, is built into CACTI.

Different from SRAM/DRAM sensing schemes where the subtle voltage/current swing or capacitor driving signal needs to be detected, the current signal in Figure 14.5 is amplified and driven by a power source, thus, the sensing can be performed really fast. Note that the sensing delay can also be obtained by performing SPICE-level simulation. In the following, we focus on the modeling and calculation of wire delay $D_{wire}$ for read/write operations.

For conventional 1T-1R structures, the word-line delay can be calculated by distributed RC-line delay, and the bit-line delay can be estimated with the Seevinck model [333]. However, although the same approach has been applied to estimate the wire delay of crossbar structure in [334], it lacks accuracy due to the following reasons. Firstly, since there is no transistor in a crossbar, the word-line and bit-line delays are symmetric. In other words, the word-line and bit-line delays have to be modeled in the same manner. Moreover, the leakage current of cells along the word-line and bit-line is a phenomenon specific to crossbar structures, which is not considered in the conventional RC-line delay model. The leakage current will weaken the

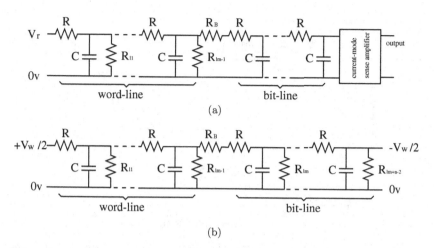

Fig. 14.7 RC-delay model of one CBRAM-crossbar for: (a) read operation; (b) write operation.

driving ability of row and column drivers, and hence, a longer delay can be predicted compared to the conventional RC-line delay.

Figure 14.7(a) and Figure 14.7(b) illustrate the crossbar delay model for read/write operation, with the leakage path of cell $i$ along the word-line and bit-line modeled as parallel $R_{li}$, whose value is its corresponding CBRAM on/off resistance. For the read operation, the bit-line is virtually grounded as illustrated in Figure 14.4, thus, only the word-line delay for the propagation of $v_r$ contributes to the wire delay. For the write operation, both word-line driving voltage $v_w/2$ and bit-line $-v_w/2$ will propagate to target cell, and the total wire delay is determined by the slower one. Therefore, for a CBRAM-crossbar with $m$ rows and $n$ columns, the worst case wire delay for read and write operation can be calculated respectively by

$$D^r_{wire} = \alpha RCn^2 \tag{14.4}$$

$$D^w_{wire} = max(\alpha RCn^2, \alpha RCm^2) \tag{14.5}$$

where $R$ and $C$ are parasitic resistor and capacitor in unit length similar to the distributed RC-line model. Note that when the CBRAM device is scaled down, $R$ and $C$ will be reduced accordingly, which produces a much smaller wire delay. In addition, the location-dependent switching speed issue incurred by the IR-drop along the word-line and bit-line will be relieved as well.

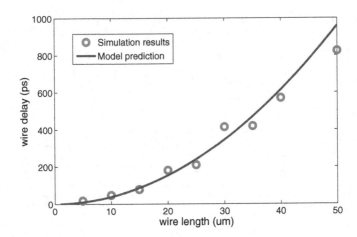

Fig. 14.8    Verification of proposed CBRAM-crossbar specific wire delay model against simulation results, with fitting parameter $\alpha = 1.2$.

Compared to conventional RC-line delay expression, a fitting parameter $\alpha$ has been added to approximate the expected longer delay due to the CBRAM-crossbar structure, such as the effect introduced by $R_{li}$ in Figure 14.7. Practically, $\alpha$ can be obtained by fitting with a few samples obtained by simulating the entire CBRAM-crossbar in different sizes using the SPICE-like simulator.

Figure 14.8 shows the verification of the crossbar specific delay model against accurate simulation results obtained through the SPICE-like simulator [330]. It can be observed that the proposed model with fitting parameter $\alpha = 1.2$ is able to capture the delay behavior of the crossbar. As expected, the crossbar wire delay, calculated by $1.2RCn^2$, therefore, is more than twice as long compared to the conventional distributed RC-line delay calculated by $0.5RCn^2$. In other words, an error of more than 50% will be incurred if the conventional distributed RC-line delay model is used instead.

### 14.3.1.2    *Power Model*

The energy per write-access for the CBRAM-crossbar is composed of: the energy consumed to switch the target cell state, and the energy dissipated along the word-line and bit-line. Consider a CBRAM-crossbar with $m$ rows and $n$ columns, where the write-access energy $E_{write}$ can be calculated as:

$$E_{write} = E_{switching} + E_{static}^w + E_{dynamic}^w \tag{14.6}$$

where $E_{switching}$ is the energy for changing the target CBRAM cell state, and needs to be obtained through the developed SPICE-like simulator since its dynamics are difficult to be approximated using the resistor; $E_{dynamic}^w$ is the energy for charging parasitic capacitors along the word-line and bit-line; and $E_{static}^w$ is the Joule heat dissipated on the half-selected CBRAM cells along the word-line and bit-line, and can be modeled as resistors since their resistance values remain constant during the write operation.

The $E_{static}^w$ and $E_{dynamic}^w$ can be given by

$$E_{static}^w = \left( k \cdot \frac{v_w^2}{4 \cdot R_{on}} + l \cdot \frac{v_w^2}{4 \cdot R_{off}} \right) \cdot D_{write} \qquad (14.7)$$

$$E_{dynamic}^w = \frac{1}{8} C \cdot v_w^2 \cdot (m + n - 2) \qquad (14.8)$$

where $v_w$ is the write-voltage, $R_{on}$ and $R_{off}$ are the *on/off* state resistances of CBRAM, $D_{write}$ is the crossbar write-delay, $C$ the distributed unit capacitance of crossbar wire, $k$ and $l$ the numbers of CBRAM cells in on-state and off-state respectively along the path, following $k + l = m + n - 2$.

Similarly, the read-access energy $E_{read}^r$ for the crossbar can be calculated by

$$E_{read} = E_{static}^r + E_{dynamic}^r \qquad (14.9)$$

Note that for read operation, all the bit-lines are virtually grounded and only cells in the target row consume power as in Figure 14.4, thus

$$E_{static}^r = \left( k \cdot \frac{v_r^2}{R_{on}} + l \cdot \frac{v_r^2}{R_{off}} \right) \cdot D_{read} \qquad (14.10)$$

$$E_{dynamic}^r = \frac{1}{2} C \cdot v_r^2 \cdot n \qquad (14.11)$$

where $v_r$ is the read-voltage, and $D_{read}$ is the crossbar read-delay. Moreover, $k$ and $l$ are the numbers of CBRAM cells in an on-state and off-state respectively in the target row, where $k + l = n$. Note that the scalability of the CBRAM device is beneficial to dynamic power reduction. When scaled down, the smaller pitch size will reduce the wire capacitance with the power reduction at the word-line and bit-line.

The verification of the power model against simulation results is shown in Figure 14.9. The simulation results are obtained by simulating $n \times n$ square CBRAM-crossbars with different $n$ values, where $n$ is used to denote crossbar size. It can be observed that the proposed model for read/write

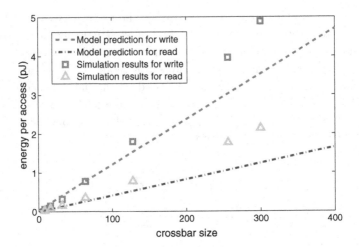

Fig. 14.9   Verification of proposed CBRAM-crossbar specific power model for read/write operations against simulation results.

operations is able to capture the trend with minor errors when crossbar size increases.

### 14.3.1.3   *Area Model*

The area model is comprised of two parts: $A_c$ for the pure area crossbar structure and $A_p$ for its corresponding CMOS peripheral circuits. Utilizing the 3D integration technique shown in Figure 14.2, the crossbar is stacked over the active layer where its peripheral circuits are located. As such, the total area becomes

$$A = max(A_p, A_c). \tag{14.12}$$

For a CBRAM-crossbar with $M$ rows and $N$ columns, its area can be calculated by

$$A_c = M \cdot N \cdot L_{pitch}^2 \tag{14.13}$$

where $L_{pitch}$ is the nano-bar pitch size, determined by the technology node of the CBRAM device. Therefore, at the advanced technology node, an extremely small CBRAM-crossbar area can be achieved due to the scalability of the CBRAM device. Besides the reduced wire delay inside the CBRAM-crossbar, the addressing delay outside the CBRAM-crossbar can be greatly reduced as well, which together will lead to significant memory access latency reduction. Note that for the peripheral circuits the similar area model can be developed based on [191].

Table 14.2 Performance comparison of 16MB SRAM, DRAM, PCRAM, and CBRAM memories.

| Feature | SRAM | DRAM | PCRAM | CBRAM |
|---|---|---|---|---|
| area (mm$^2$) | 27.4 | 4.19 | 3.77 | 2.33 |
| read latency (ns) | 3.43 | 2.25 | 2.54 | 3.90 |
| write latency (ns) | 3.43 | 2.55 | RESET 42.54 SET 102.54 | 8.01 |
| read energy (nJ) | 0.83 | 0.61 | 0.73 | 1.8 |
| write energy (nJ) | 0.75 | 0.61 | RESET 6.66 SET 2.28 | 2.0 |

#### 14.3.1.4 *Stacked CBRAM-crossbar Memory*

Table 14.2 evaluates and compares the performance of the stacked CBRAM-crossbar memory with other memory technologies for the same capacity of 16MB. The data for SRAM and DRAM are generated by CACTI with default settings, and PCRAM data is extracted from PCRAM-sim [323]. As shown in Table 14.2, mainly due to the fast device-level accessing speed by the CBRAM device as well as the high density of the crossbar structure, the CBRAM-crossbar performance, especially the accessing latency for read/write operations, is already close to DRAM, which shows its potential for future application as the main memory. Moreover, compared to PCRAM, the CBRAM-crossbar shows 9× faster write-latency, 1.6× smaller area, 4.5× less write-energy per access, and 1.5× slower read-latency on average. Only a slightly slower read-latency is observed for the CBRAM-crossbar.

### 14.3.2 *Design Space Exploration*

In this part, based on developed CBRAM-crossbar models, we further show how to optimize the 3D hybrid memory system by performing design space exploration under different optimization objectives.

For a given design freedom of parameters, the total capacity of SRAM, DRAM or CBRAM is calculated by

$$C_M = N_b \cdot C_b = N_b \cdot N_m \cdot C_m = N_b \cdot M \cdot N \cdot m \cdot n \qquad (14.14)$$

where $M \cdot N$ is the dimension of the mat array for each bank, and $m \cdot n$ is the dimension of the crossbar array for each mat. Different combinations of bank number, mat array dimension and crossbar array dimension need to be explored for optimal performance. There are several design constraints to be considered as illustrated below one by one:

*Data bandwidth constraint:* as aforementioned, since each CBRAM-crossbar supports only one-bit-write for each cycle, we need to design the multi-bit data to be distributed into different crossbars in the same row. Consequently, the data-bus width $W_d$ must equal to $N$, which is the number of crossbars (i.e., mats) that each row contains.

*Transition power constraint:* during the hibernating and wakeup transition, concurrent data migration will induce high transition power

$$\sum P_{bi} = \sum f_b W_{di} E_{mi} \leq P \qquad (14.15)$$

where $W_{di}$ and $E_{mi}$ is the bus-width and energy consumed of bank $i$. The transition power needs to be limited by reliability concerns for hot-spots of thermal and supply-current.

*TSV density constraint:* since the TSVs for data transmission occupy certain area, a diameter of $5\mu m$ for example [4], they lead to larger footprint as well as difficulties in placement and routing. As a result, the average TSV density must be limited, which leads to an upper-bound on data bandwidth

$$\frac{W_D + W_A}{A} \leq D \qquad (14.16)$$

where A is the chip area; $W_A$ and $W_A$ represents the total address and data bus width respectively.

For our 3D hybrid memory system, one can perform design space exploration based on design parameters such as bank capacity $C_b$, data-migration bus bandwidth $W_d$ and CBRAM-crossbar write-frequency $f_b$. Figure 14.10 shows the performance-objective with different tradeoffs bounded by design constraints, which can be summarized as follows:

- For speed optimization i.e., minimizing hibernating/wakeup transition time, large bandwidth is desirable. As such, a small $C_b$, large $W_d$ and high $f_b$ are preferred. However, it is limited by TSV density and power constraints.
- For power optimization i.e., minimizing transition power, a small $C_b$ and $W_d$ architecture working in a low $f_b$ is favorable, which is mainly limited by TSV density and speed constraints.
- For memory performance optimization i.e., minimizing memory SRAM/DRAM performance degradation due to high TSV density, a small $W_d$ and large $C_b$ help reduce TSV density, which in turn alleviates memory performance degradation, mainly limited by speed constraint.

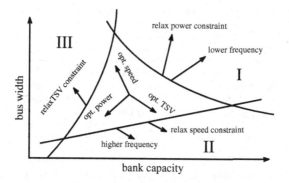

Fig. 14.10 Design space exploration of 3D hybrid memory with CBRAM-crossbar.

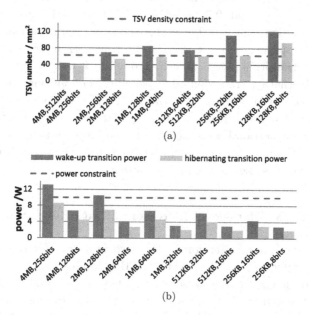

Fig. 14.11 (a) TSV density and (b) mode transition power under different architecture-level parameters.

For example, regions I, II, III in Figure 14.10 correspond to the limiting design constraints of TSV density, power and speed, respectively.

As an example, we show the procedure of transition time (i.e.,speed) optimization. Figure 14.11 shows the trend of TSV density and $P_H/P_W$ with $C_b/W_d$. Among the available combinations, the one with a $C_b$ of 256KB and $W_d$ of 16-bits is chosen since it results in the largest $W_D$ and power margin when increasing frequency and when satisfying the previously defined

Table 14.3 Optimized performance for different design objectives.

| Objective | Speed | Power | MP |
|---|---|---|---|
| $C_b$ | 256KB | 256KB | 4MB |
| $W_d$ (bits) | 16 | 8 | 128 |
| $P_H/P_W$ (W) | 2.3/9.0 | **1.8/1.8** | 4.3/7.1 |
| $T_H/T_W$ (ms) | **0.5/1.5** | 3/3 | 1.9/3 |
| TSV density ($/mm^2$) | 60 | 48 | **26** |

constraints. As such, when hybrid memory works at its maximal allowed frequency, $0.5ms$ $T_W$ and $1.5ms$ $T_h$ can be achieved.

Table 14.3 demonstrates the optimal results for different performance objectives. An optimal $0.5ms$ hibernating transition time and $1.5ms$ wakeup transition time is achieved for the speed optimization. For transition power optimization, $5\times$ less wakeup transition power and $1.6\times$ less hibernating transition power are achieved with $6\times$ wakeup transition time and $2\times$ hibernating transition time penalties. For memory performance (MP in Table 14.3) optimization, the lowest TSV density is obtained as the minimized memory performance degradation in normal active mode. The design exploration leaves the designer the freedom to choose different design parameters based on system requirements. The configuration with optimized speed is used in the data retention.

## 14.4  ReRAM Block-level Incremental Data Retention

Here, *block-level* data retention is discussed in details for the 3D hybrid memory architecture with ReRAM i.e., CBRAM-crossbar.

The data is migrated between memory blocks (i.e., banks) sequentially through dedicated 3D TSV buses. Compared to the *bit-level* data retention scheme [335], where each SRAM memory cell is associated with one neighboring FeRAM cell with cell-wise controllers, our block-level approach can achieve much smaller area overhead since the data migration controller is shared by all memory cells of the same block. In addition, our block-level approach will not degrade the SRAM performance since no change is made inside the SRAM memory cell. Furthermore, [336] has a system-level data retention by updating check-points. Because the use of PCRAM limits the frequency and amount of data that are retained, the system in [336] can only keep system check-points between relatively longer time intervals, which is insufficient when more fine-grained data retention is required. Based on the CBRAM-crossbar structure, this section introduces one block-level data retention, which includes the block-level memory controller with two operations: dirty-bit set-up and incremental write-back.

### 14.4.1 *Dirty Bit Set-up*

The target for data retention is to synchronize the data of the CBRAM-crossbar with its corresponding SRAM/DRAM contents at block-level. For any SRAM/DRAM bank $i$, the time needed to copy the data to CBRAM-crossbar is decided by

$$T_{Hi} = \frac{M_b}{f_b \cdot W_d} \qquad (14.17)$$

where $f_b$ is the write-frequency of the CBRAM-crossbar memory limited by its latency, $W_d$ is the bank-level data bandwidth, and $M_b$ is the amount of data to be migrated, which equals to the bank capacity $C_b$ in the brute-force approach.

Clearly, reducing $M_b$ directly reduces the $T_{Hi}$. Considering the common law of locality [337], the system tends to access relatively local-memory regions during a given period of time. As such, between two successive power gating stages, only part of the content in CBRAM-crossbar and SRAM/DRAM becomes unsynchronized, which is denoted as *dirty data*. By incrementally writing dirty data to the CBRAM-crossbar, a significant amount of migration data and power can be saved.

As shown in Figure 14.1, to keep the dirty-data status information, an extra CBRAM-crossbar called *dirty-data pool* is embedded to each CBRAM-crossbar bank, where each bit in the pool, referred to as *dirty bit*, indicates the dirty status of a few continuous bytes of data in SRAM/DRAM, referred to as the *dirty-data group*. Empirically, we design the group granularity $G_d$ as the cache line size $B_C$ for SRAM or the page size $B_D$ for DRAM.

The dirty-bit set-up occurs simultaneously each time the content of SRAM/DRAM is changed during active mode. As shown in Figure 14.12, each CBRAM-crossbar bank *listens* to all memory write operations issued to its corresponding SRAM/DRAM bank to update the dirty pool during active mode. Once the SRAM/DRAM write-action is detected, the corresponding data group becomes dirty.

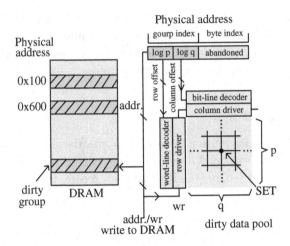

Fig. 14.12    Circuit diagram for dirty bit set-up at active mode.

As such, the corresponding bit in the dirty pool needs to be SET. The dirty pool size $C_p$ is decided by

$$C_p = \frac{C_b}{G_d} = p \cdot q \qquad (14.18)$$

where $p$ and $q$ are the CBRAM-crossbar dimensions of the dirty pool. Therefore, the designated dirty bit position can be located by decoding the first $log(p)$ and the following $log(q)$ bits of the physical memory write-address, respectively.

### 14.4.2   *Incremental Write-back*

The flow to write back dirty SRAM/DRAM data to the CBRAM-crossbar during hibernating transition is illustrated in Figure 14.13. Specifically, one address counter is used to check the status of all dirty bits in the dirty-data pool. Once the dirty bit in SET state is detected, the corresponding data group needs to be copied to the CBRAM-crossbar. Due to the limited data-bus bandwidth $W_d$, the group data of size $G_d$ needs to be written back to the CBRAM-crossbar in several cycles. As such, the address counter generates the memory address of the next piece of data to be copied from the SRAM/DRAM by adding $W_d$-offset each cycle, and the read-signal to DRAM and write-signal to CBRAM-crossbar are issued for data migration. Once finished, the corresponding dirty bit is RESET. During wakeup

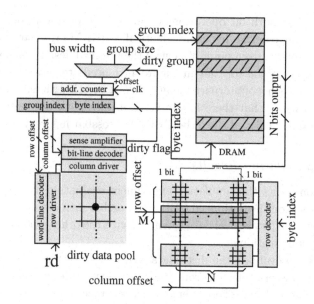

Fig. 14.13  Circuit diagram for dirty-data write-back at hibernating transition.

transition, all data will be copied from the CBRAM-crossbar back to the SRAM/DRAM with similar hardware supports.

Here we will discuss how the use of the CBRAM-crossbar is able to instinctively facilitate the dirty-data write-back without additional design efforts. As discussed above, we can see that intensive bit operations are required for dirty flags update and read out. For the conventional 1T-1R structured memory, accesses are done in the unit of byte or word. Therefore, to read a bit, the byte or word which contains the target bit is first read out, then the OR, AND or SHIFT instruction is further applied to obtain one bit; similarly, to write a bit, byte or word is first read out, then merged with the bit by an OR or AND operation; and finally the bit-modified byte or word can be written back.

As such, the bit operations completed in multiple cycles may not be able to meet the real-time dirty flag update requirement. Also, significant power overhead may be incurred. Therefore, without additional design efforts, the byte-addressable or word-addressable conventional memory is not suitable for dirty-data write-back where intensive bit operations are required. On the other hand, as previously discussed, bit operation is the instinctive way in which the CBRAM-crossbar is operated. Byte or word operations are achieved by multiple identical CBRAM-crossbar units working in parallel.

In other words, both bit operations required by dirty-data pool and word operations can co-exist by using identical CBRAM-crossbar units naturally. Additionally, from point of view of the physical design, each CBRAM-crossbar block at the top tier in Figure 14.1 needs to have a smaller size compared to their counterpart memory block in other tiers. This requirement is to ensure that there is one vertical data path between the pairs, and this can be achieved since the CBRAM-crossbar has very high density.

### 14.4.3 *Simulation Results*

This section shows the simulation experiment evaluation of the 3D hybrid memory based on the optimized CBRAM-crossbar memory design, the comparison between block-level data retention with the other schemes discussed in the previous section.

In order to evaluate the dirty-data write-back strategy, a set of benchmark programs are selected from the SPEC2006 suite and are run in the gem5 simulator [189], where memory-access traces are generated. As the advantage of the dirty-data write-back strategy may depend on the memory access patterns of executed programs, the benchmarks with different memory access characteristics are picked in general. For examples, *mcf* and *lbm* have high cache-miss rates while *h264* and *namd* have low cache-miss rates; *perlbench* and *gcc* have intensive store-instruction intensive while *astar* and *namd* have low store-instruction [338]. For each benchmark, its dirty-data flags are updated according to the memory-access trace. Then, the dirty-data write-back strategy is deployed to evaluate the power saving and speedup during data migration. Figure 14.14 compares the hibernating transition power and time between systems that use and do not use the

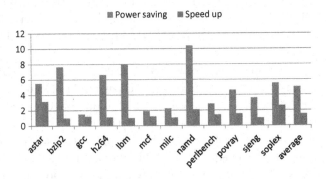

Fig. 14.14    Hibernating power and time reduction by incremental dirty-data write-back.

dirty-data write-back strategy. On the system with the dirty-data write-back strategy achieves 5× reduction in power and 1.5× reduction in time during the hibernating transition.

Additionally, we compare the performance of data retention and leakage power in sleep mode for the following data retention schemes:

- *STD:* standard SRAM/DRAM without power gating.
- *DPG:* data-retentive power gating (DPG) of both SRAM [339] and DRAM [340] with reduced supply voltage.
- *PCRAM:* system-level data retention by PCRAM [336] which is used to keep the entire SRAM/DRAM contents.
- *FeRAM:* bit-level data retention by FeRAM [335].
- *CBRAM:* our proposed block-level data retention with incremental dirty-data write-back by CBRAM.

Table 14.4 shows the full comparison results in terms of the sleep-mode leakage power for SRAM and DRAM $(P_s^L/P_d^L)$, hibernating and wakeup transition time $(T_H/T_W)$, and hibernating and wakeup transition power $(P_H/P_W)$. The leakage power for SRAM/DRAM under the standard scheme is generated by CACTI at the $65nm$ technology node. The leakage power under the DPG scheme is calculated by the leakage reduction factors reported in [339, 340] for both SRAM and DRAM. Block-level CBRAM based data-retention performance is derived from our platform, combining the architecture optimization results and power and time reduction results. PCRAM based scheme performance estimation is based on [336], with PCRAM memory performance obtained from PCRAMsim [323]. For the FeRAM based scheme, the bit-to-bit data retention performance is extracted from [335]. Because all data is migrated concurrently in this scheme, data retention power performance can be estimated by multiplication, and the speed performance is the same with bit-to-bit performance.

Table 14.4   Data-retention performance comparison for different leakage reduction schemes.

| Scheme | $P_s^L/P_d^L$ (mW) | $T_H/T_W$ (ms) | $P_H/P_W$ (W/MB) |
|--------|--------------------|----------------|-------------------|
| STD    | 209/220            | NA             | NA                |
| DPG    | 21/22              | 1e-4           | 0                 |
| PCRAM  | 0/0                | 3.7/1.3        | 0.072/0.1         |
| FeRAM  | 0/22               | 1              | 0.75              |
| CBRAM  | 0/0                | 0.33/1.5       | 0.007/0.14        |

Due to the use of NVM for data retention, PCRAM, FeRAM and CBRAM based memory systems all outperform the STD and DPG schemes in terms of leakage power reduction. Moreover, our proposed 3D hybrid CBRAM-crossbar memory system allows a shut-down of both SRAM and DRAM during the power gating. As a result, compared to the block-level PCRAM based data retention, we achieve a 11× faster hibernating transition time and 10× smaller hibernating transition power with the same number of TSVs. As shown in Table 14.2, the hibernating transition performance improvement comes from the advantageous CBRAM-crossbar memory performance and also the block-level incremental dirty-data write-back strategy. However, its the wakeup transition time and power is slightly inferior to the PCRAM based scheme. This is mainly because the incremental dirty-data write-back strategy does not apply to the wakeup transition. In addition, the proposed CBRAM system also outperforms the FeRAM by 107×/5.4× hibernating/wakeup transition power saving, and around 3 × faster hibernating transition time.

Further illustrated in Table 14.5, when compared with the block-level CBRAM based data retention, the bit-level FeRAM based data retention induces the overwhelming cache performance degradation in normal active mode due to the bit-wise controllers embedded in the SRAM. For the bit-level FeRAM based data retention scheme, the area overhead for retaining one bit data is $6.1\mu m^2$, which is only $0.36\mu m^2$ in our CBRAM system, estimated by our developed CBRAM-crossbar based CACTI. Due to such a large area overhead, the FeRAM based system shows an 84% SRAM access time degradation and 74% SRAM access energy overhead on average. The

Table 14.5  Cache performance comparison between block-level data retention (CBRAM) and bit-level data retention (FeRAM) in active mode.

| Cache capacity | Feature | Block-level CBRAM | Bit-level FeRAM |
|---|---|---|---|
| 128KB | access time (ns) | 1 | 1.8 (−80%) |
| | access energy (nJ) | 0.13 | 0.21 (−62%) |
| | area (mm$^2$) | 0.82 | 7.6 (−827%) |
| 512KB | access time (ns) | 1.5 | 2.8 (−87%) |
| | access energy (nJ) | 0.33 | 0.6 (−82%) |
| | area (mm$^2$) | 3.8 | 28.6 (−642%) |
| 2MB | access time (ns) | 2.7 | 4.3 (−59%) |
| | access energy (nJ) | 0.6 | 1.3 (−117%) |
| | area (mm$^2$) | 14.2 | 113 (−696%) |

performance of the bit-level FeRAM based cache is derived from CACTI by replacing the SRAM cell with a FeRAM-SRAM cell pair.

## 14.5 Summary

In this chapter, non-volatile CBRAM memory is introduced and area, access time and power models are developed for the optimization of CBRAM-crossbar based memory design. As such, the design of 3D hybrid memory is explored based on the CBRAM-crossbar memory for the block-level data retention of SRAM/DRAM with incremental dirty-data write-back. Experimental results show that the proposed hybrid memory achieves 11× faster migration time and 10× less migration power for hibernating transition, when compared to the system-level 3D PCRAM-based data retention in [336] with the same amount of TSVs used. Moreover, the proposed hybrid memory also achieves 17× area reduction along with 2× SRAM performance improvement, averaging 56× transition power reduction respectively, with similar transition time, when compared to the bit-level FeRAM-based data retention in [335].

# Bibliography

[1] K. Uchiyama. Power-efficient heteregoneous parallelism for digital convergence. In *IEEE Symp. on VLSI Circuits* (2008).

[2] H. H. Goldstine and A. Goldstine, The electronic numerical integrator and computer (eniac), *IEEE Annals of the History of Computing*. **18**(1), 10–16 (March, 1996).

[3] J. Wang, S. Ma, M. Sai, M. Yu, R. Weerasekera, and H. Yu. High-speed and low-power 2.5D I/O circuits for memory-logic-integration by through-silicon interposer. In *IEEE Int. 3D Systems Integration Conf. (3DIC)* (2013).

[4] B. Davidsen. ITRS Roadmap (2013). http://www.itrs.net/Links/2013IT RS/2013Chapters/2013ExecutiveSummary.pdf.

[5] K. Banerjee, S. Souri, P. Kapur, and K. Saraswat, 3-D ICs: a novel chip design for improving deep-submicrometer interconnect performance and systems-on-chip integration, *Proceedings of the IEEE*. **89**(5), 602–633 (May, 2001).

[6] C. S. Tan, R. J. Gutmann, and L. R. Reif, *Overview of Wafer-Level 3D ICs*. Springer US (2008).

[7] J. Cong, J. Wei, and Y. Zhang, A thermal-driven floorplanning algorithm for 3D ICs. In *IEEE/ACM Int. Conf. on Computer Aided Design (ICCAD)* (2004).

[8] B. Goplen and S. Sapatnekar, Thermal via placement in 3D ICs. In *IEEE/ACM Int. Symp. on Physical Design (ISPD)* (2005).

[9] H. Yu, Y. Shi, L. He, and T. Karnik, Thermal via allocation for 3D ICs considering temporally and spatially variant thermal power. In *Int. Symp. on Low Power Electronics and Design (ISLPED)* (2006).

[10] H. Yu, J. Ho, and L. He, Simultaneous power and thermal integrity driven via stapling in 3D ICs. In *ACM/IEEE Int. Conf. on Compter-Aided Design (ICCAD)* (2006).

[11] H. Yu, Y. Shi, L. He, and T. Karnik, Thermal via allocation for 3-D ICs considering temporally and spatially variant thermal power, *IEEE Trans. on Very Large Scale Integration (TVLSI) Systems*. **16**(12), 1609–1619 (Dec, 2008).

[12]  H. Yu, J. Ho, and L. He, Allocating power ground vias in 3D ICs for simultaneous power and thermal integrity, *ACM Trans. on Design Automation of Electronic Systems (TODAES).* **14**(3), 41:1–41:31 (Jun, 2009).

[13]  Y. Xie, G. H. Loh, B. Black, and K. Bernstein, Design space exploration for 3D architectures, *ACM J. on Emerging Technologies in Computing Systems (JETCAS).* **2**(2), 65–103 (Apr, 2006).

[14]  T. Kgil, S. D'Souza, A. Saidi, N. Binkert, R. Dreslinski, T. Mudge, S. Reinhardt, and K. Flautner. Picoserver: Using 3D stacking technology to enable a compact energy efficient chip multiprocessor. In *Int. Conf. on Architectural Support for Programming Languages and Operating Systems (ASPLOS)*, ACM (2006).

[15]  M. Motoyoshi, Through-silicon via (TSV), *IEEE Proceedings.* **97**(1), 43–48 (2009).

[16]  M. Healy, K. Athikulwongse, R. Goel, M. Hossain, D. H. Kim, Y.-J. Lee, D. Lewis, T.-W. Lin, C. Liu, M. Jung, B. Ouellette, M. Pathak, H. Sane, G. Shen, D. H. Woo, X. Zhao, G. Loh, H.-H. Lee, and S. K. Lim. Design and analysis of 3D-MAPS: a many-core 3D processor with stacked memory. In *IEEE Custom Integrated Circuits Conf. (CICC)* (2010).

[17]  H. Qian, X. Huang, H. Yu, and C. H. Chang. Real-time thermal management of 3D multi-core system with fine-grained cooling control. In *IEEE Int. 3D Systems Integration Conf. (3DIC)* (2010).

[18]  X. Huang, H. Yu, and W. Zhang. NEMS based thermal management for 3D many-core system. In *ACM/IEEE Int. Symp. on Nanoscale Architectures*, pp. 218–223 (June, 2011).

[19]  H. Qian, X. Huang, H. Yu, and C. Chang, Cyber-physical thermal management of 3D multi-core cache-processor system with microfluidic cooling, *ASP J. of Low Power Electronics.* **7**(1), 110–121 (Feb, 2011).

[20]  H. Qian, C.-H. Chang, and H. Yu, An efficient channel clustering and flow rate allocation algorithm for non-uniform microfluidic cooling of 3D integrated circuits, *Integration VLSI Journal.* **46**(1), 57–68 (Jan, 2013).

[21]  Y. Shang, C. Zhang, H. Yu, C. S. Tan, X. Zhao, and S. K. Lim. Thermal-reliable 3D clock-tree synthesis considering nonlinear electrical-thermal-coupled TSV. In *Asia and South Pacific Design Automation Conf. (ASP-DAC)*, ACM/IEEE (2013).

[22]  M. Sai, H. Yu, Y. Shang, C. S. Tan, and S. K. Lim, Reliable 3-D clock-tree synthesis considering nonlinear capacitive TSV model with electrical-thermal-mechanical coupling, *IEEE Trans. on Computer-Aided Design of Integrated Circuits and Systems.* **32**(11), 1734–1747 (Nov, 2013).

[23]  S. M. P. D., H. Yu, and K. Wang, 3d many-core microprocessor power management by space-time multiplexing based demand-supply matching, *IEEE Trans. on Computers.* **PP**(99), 1–1 (2015). doi: 10.1109/TC.2015.2389827.

[24]  J. Cubillo, R. Weerasekera, Z. Z. Oo, E.-X. Liu, B. Conn, S. Bhattacharya, and R. Patti. Interconnect design and analysis for through silicon interposers (TSIs). In *IEEE Int. 3D Systems Integration Conf. (3DIC)* (2012).

[25]  S.-S. Wu, K. Wang, M. Sai, T.-Y. Ho, M. Yu, and H. Yu. A thermal resilient integration of many-core microprocessors and main memory by

2.5D TSI I/Os. In *Design, Automation and Test in Europe Conf. (DATE)*, ACM/IEEE (2014).

[26] D. Xu, M. Sai, H. Huang, N. Yu, and H. Yu. An energy-efficient 2.5D through-silicon interposer I/O with self-adaptive adjustment of output-voltage swing. In *ACM/IEEE Int. Symp. on Low Power Electronics and Design (ISLPED)* (2014).

[27] H. Hantao, P. Sai Manoj, D. Xu, H. Yu, and Z. Hao. Reinforcement learning based self-adaptive voltage-swing adjustment of 2.5D I/Os for many-core microprocessor and memory communication. In *ACM/IEEE Int. Conf. on Computer-Aided Design* (2014).

[28] A. Agarwal and M. Levy, Going multicore presents challenges and opportunities, *Embedded Systems Design*. **20**(4) (Apr, 2007).

[29] M. B. Taylor et al., *Tiled Multicore Processors*, In *Multicore Processors and Systems*, chapter 1, pp. 1–34. Integrated Circuits and Systems. Springer (2009).

[30] M. Taylor, J. Psota, A. Saraf, N. Shnidman, V. Strumpen, M. Frank, S. Amarasinghe, A. Agarwal, W. Lee, J. Miller, D. Wentzlaff, I. Bratt, B. Greenwald, H. Hoffmann, P. Johnson, and J. Kim. Evaluation of the raw microprocessor: an exposed-wire-delay architecture for ILP and streams. In *Int. Symp. on Computer Architecture (ISCA)* (2004).

[31] S. Vangal, J. Howard, G. Ruhl, S. Dighe, H. Wilson, J. Tschanz, D. Finan, P. Iyer, A. Singh, T. Jacob, S. Jain, S. Venkataraman, Y. Hoskote, and N. Borkar. An 80-Tile 1.28TFLOPS network-on-chip in 65nm CMOS. In *Int. Solid-State Circuits Conf. (ISSCC)*, IEEE (2007).

[32] J. Howard, S. Dighe, S. Vangal, G. Ruhl, N. Borkar, S. Jain, V. Erraguntla, M. Konow, M. Riepen, M. Gries, G. Droege, T. Lund-Larsen, S. Steibl, S. Borkar, V. De, and R. Van Der Wijngaart, A 48-core IA-32 processor in 45nm CMOS using on-die message-passing and DVFS for performance and power scaling, *IEEE J. of Solid-State Circuits*. **46**(1), 173–183 (Jan, 2011).

[33] K. Sankaralingam, R. Nagarajan, R. McDonald, R. Desikan, S. Drolia, M. S. Govindan, P. Gratzf, D. Gulati, H. Hanson, C. Kim, H. Liu, N. Ranganathan, S. Sethumadhavan, S. Shariff, P. Shivakumar, S. Keckler, and D. Burger. Distributed microarchitectural protocols in the TRIPS prototype processor. In *IEEE/ACM Int. Symp. on Microarchitecture* (2006).

[34] D. Wentzlaff, P. Griffin, H. Hoffmann, L. Bao, B. Edwards, C. Ramey, M. Mattina, C.-C. Miao, J. F. Brown III, and A. Agarwal, On-chip interconnection architecture of the tile processor, *IEEE Micro*. **27**(5), 15–31 (Sep, 2007).

[35] M. Baron. Tilera's core communicate better. In *Microprocessor Report* (Nov, 2007).

[36] M. Baron. Low-key intel 80-core intro: The tip of the iceberg. In *Microprocessor Report* (Apr, 2007).

[37] J. Shin, K. Tam, D. Huang, B. Petrick, H. Pham, C. Hwang, H. Li, A. Smith, T. Johnson, F. Schumacher, D. Greenhill, A. Leon, and A. Strong. A 40nm 16-core 128-thread CMT SPARC SoC processor. In *IEEE Int. Solid-State Circuits Conf. (ISSCC)* (2010).

[38] N. Sturcken, E. O'Sullivan, N. Wang, P. Herget, B. Webb, L. Romankiw, M. Petracca, R. Davies, R. Fontana, G. Decad, I. Kymissis, A. Peterchev, L. Carloni, W. Gallagher, and K. Shepard. A 2.5D integrated voltage regulator using coupled-magnetic-core inductors on silicon interposer delivering 10.8A/mm². In *IEEE Int. Solid-State Circuits Conf. (ISSCC)* (2012).

[39] Nvidia Tesla GPU http://www.nvidia.com/object/tesla-servers.html.

[40] R. C. Whaley, A. Petitet, and J. J. Dongarra, Automated empirical optimization of software and the ATLAS project, *Parallel Computing.* **27**, 3–35 (2000).

[41] A. Jantsch, M. Grange, and D. Pamunuwa, *The Promises and Limitations of 3-D Integration*, In *3D Integration for NoC-based SoC Architectures*, pp. 27–44. Integrated Circuits and Systems. Springer (2011).

[42] J. Bulzacchelli, M. Meghelli, S. Rylov, W. Rhee, A. Rylyakov, H. Ainspan, B. Parker, M. Beakes, A. Chung, T. Beukema, P. Pepeljugoski, L. Shan, Y. Kwark, S. Gowda, and D. Friedman, A 10-Gb/s 5-tap DFE/4-tap FFE transceiver in 90-nm CMOS technology, *IEEE J. of Solid-State Circuits (JSSC).* **41**(12), 2885–2900 (2006).

[43] R. Palmer, J. Poulton, W. Dally, J. Eyles, A. Fuller, T. Greer, M. Horowitz, M. Kellam, F. Quan, and F. Zarkeshvari. A 14mW 6.25Gb/s transceiver in 90nm CMOS for serial chip-to-chip communications. In *IEEE Int. Solid-State Circuits Conf.* (2007).

[44] W. J. Dally and J. Poulton, *Digital Systems Engineering.* Cambridge Univ. Press (1998).

[45] M.-J. Lee, W. Dally, and P. Chiang, Low-power area-efficient high-speed I/O circuit techniques, *IEEE J. of Solid-State Circuits.* **35** (11), 1591–1599 (Nov, 2000).

[46] H. Hatamkhani, F. Lambrecht, V. Stojanovic, and C.-K. K. Yang. Power-centric design of high-speed I/Os. In *ACM/IEEE Design Automation Conf. (DAC)* (2006).

[47] G. Balamurugan, J. Kennedy, G. Banerjee, J. Jaussi, M. Mansuri, F. O'Mahony, B. Casper, and R. Mooney, A scalable 5-15 Gbps, 14–75 mW low-power I/O transceiver in 65 nm CMOS, *IEEE J. of Solid-State Circuits.* **43**(4), 1010–1019 (April, 2008).

[48] A. Papanikolaou, D. Soudris, and R. Radojcic, *Introduction to Three-Dimensional Integration*, In *Three Dimensional System Integration*, pp. 1–13. IC Stacking Process and Design. Springer (2011).

[49] J. H. Choi, A. Bansal, M. Meterelliyoz, J. Murthy, and K. Roy. Leakage power dependent temperature estimation to predict thermal runaway in fin-fet circuits. In *IEEE/ACM Int. Conf. on Computer-Aided Design (ICCAD)* (2006).

[50] A. Vassighi and M. Sachdev, Thermal runaway in integrated circuits, *IEEE Trans. on Device and Materials Reliability.* **6**(2), 300–305 (2006).

[51] T. Burd *et al.*, A dynamic voltage scaled microprocessor system, *IEEE J. of Solid-State Circuits.* **35**(11), 1571–1580 (Nov, 2000).

[52] P. Choudhary and D. Marculescu, Power management of voltage/frequency island-based systems using hardware-based methods, *IEEE Trans. on VLSI Systems.* **17**(3), 427–438 (Feb, 2009).

[53]  J. Zhao, X. Dong, and Y. Xie. An energy-efficient 3D CMP design with fine-grained voltage scaling. In *ACM/IEEE DATE Conf.* (2011).

[54]  R. David, P. Bogdan, R. Marculescu, and U. Ogras. Dynamic power management of voltage-frequency island partitioned network-on-chip using intel's single-chip cloud computer. In *ACM/IEEE Int. Symp. on Network on Chip (NoCs)* (2011).

[55]  K. Wang, H. Yu, B. Wang, and C. Zhang. 3D reconfigurable power switch network for demand-supply matching between multi-output power converters and many-core microprocessors. In *ACM/IEEE Design, Automation Test in Europe Conf. (DATE)* (2013).

[56]  P. Sai Manoj, K. Wang, and H. Yu. Peak power reduction and workload balancing by space-time multiplexing based demand-supply matching for 3D thousand-core microprocessor. In *ACM/IEEE Design Automation Conf.* (2013).

[57]  A. Gulati, A. Merchant, M. Uysal, P. Padala, and P. Varman. Workload dependent IO scheduling for fairness and efficiency in shared storage systems. In *IEEE Int. Conf. on High Performance Computing (HiPC)* (2012).

[58]  N. Magen, A. Kolodny, U. Weiser, and N. Shamir. Interconnect-power dissipation in a microprocessor. In *Int. Workshop on System Level Interconnect Prediction* (2004).

[59]  M. Mondal, A. Ricketts, S. Kirolos, T. Ragheb, G. Link, N. Vijaykrishnan, and Y. Massoud. Thermally robust clocking schemes for 3D integrated circuits. In *Design, Automation Test in Europe Conf. (DATE)*, ACM/IEEE (2007).

[60]  I. Ndip, B. Curran, K. Lobbicke, S. Guttowski, H. Reichl, K. Lang, and H. Henke, High-frequency modeling of TSVs for 3-D chip integration and silicon interposers considering skin-effect, dielectric quasi-TEM and slow-wave modes, *IEEE Trans. on Components, Packaging and Manufacturing Technology*. 1(10), 1627–1641 (2011).

[61]  W. Sandra, *Advanced IC Packaging*. Newyork: Springer (2007).

[62]  A. Rahman and R. Reif. Thermal analysis of three-dimensional (3D) integrated circuits (ICs). In *IEEE Int. Interconnect Technology Conf.* (2001).

[63]  S. Ramaswami, J. Dukovic, B. Eaton, S. Pamarthy, A. Bhatnagar, Z. Cao, K. Sapre, Y. Wang, and A. Kumar, Process integration considerations for 300 mm TSV manufacturing, *IEEE Trans. on Device and Materials Reliability*. 9(4), 524–528 (Dec, 2009).

[64]  A. Yu, J. Lau, S. W. Ho, A. Kumar, H. W. Yin, J. M. Ching, V. Kripesh, D. Pinjala, S. Chen, C.-F. Chan, C.-C. Chao, S. Chiu, C.-M. Huang, and C. Chen. Three dimensional interconnects with high aspect ratio TSVs and fine pitch solder microbumps. In *Electronic Components and Technology Conf. (ECTC)*, pp. 350–354 (May, 2009).

[65]  L. Zhang, H. Li, S. Gao, and C. Tan. Design, fabrication and electrical characterization of TSV. In *Electronics Packaging Technology Conf. (EPTC)*, pp. 765–768 (Dec, 2010).

[66]  A. Mercha, G. Van der Plas, V. Moroz, I. de Wolf, P. Asimakopoulos, N. Minas, S. Domae, D. Perry, M. Choi, A. Redolfi, C. Okoro, Y. Yang, J. Van Olmen, S. Thangaraju, D. Sabuncuoglu Tezcan, P. Soussan,

J. Cho, A. Yakovlev, P. Marchal, Y. Travaly, E. Beyne, S. Biesemans, and B. Swinnen. Comprehensive analysis of the impact of single and arrays of through silicon vias induced stress on high-k/metal gate CMOS performance. In *IEEE Int. Electron Devices Meeting (IEDM)* (2010).

[67] A. D. Trigg, L. H. Yu, C. K. Cheng, R. Kumar, D. L. Kwong, T. Ueda, T. Ishigaki, K. Kang, and W. S. Yoo, Three dimensional stress mapping of silicon surrounded by copper filled through silicon vias using polychromator-based multi-wavelength micro Raman spectroscopy, *Applied Physics Express*. 3: 086601 (2010).

[68] C. Okoro, K. Vanstreels, R. Labie, O. Luhn, B. Vandevelde, B. Verlinden, and D. Vandepitte, Influence of annealing conditions on the mechanical and microstructural behavior of electroplated Cu-TSV, *Journal of Micromechnics and Microengineering*. 20(4): 054032 (2010).

[69] J. Lin, W. Chiou, K. Yang, H. Chang, Y. Lin, E. Liao, J. Hung, Y. Lin, P. Tsai, Y. Shih, T. Wu, W. Wu, F. Tsai, Y. Huang, T. Wang, C. Yu, C. Chang, M. Chen, S. Hou, C. Tung, S. Jeng, and D. Yu. High density 3D integration using CMOS foundry technologies for 28 nm node and beyond. In *IEEE Int. Electron Devices Meeting (IEDM)* (2010).

[70] S. Thompson, G. Sun, Y. S. Choi, and T. Nishida, Uniaxial-process-induced strained-Si: extending the CMOS roadmap, *IEEE Trans. on Electron Devices*. 53(5), 1010–1020 (May, 2006).

[71] K. Ghosh, J. Zhang, L. Zhang, Y. Dong, H. Li, C. Tan, G. Xia, and C. Tan. Strategy for TSV scaling with consideration on thermo-mechanical stress and acceptable delay. In *Microsystems, Packaging, Assembly and Circuits Tech. Conf. (IMPACT)*, IEEE (2012).

[72] R. Hoofman, J. Michelon, P. Bancken, R. Daamen, G. Verheijden, V. Arnal, O. Hinsinger, L. Gosset, A. Humbert, W. Besling, C. Goldberg, R. Fox, L. Michaelson, C. Guedj, J. Guillaumond, V. Jousseaume, L. Arnaud, D. Gravesteijn, J. Torres, and G. Passemard. Reliability challenges accompanied with interconnect downscaling and ultra low-k dielectrics. In *IEEE Int. Interconnect Technology Conf. (IITC)* (2005).

[73] P. Neuzil, Y. Liu, H.-H. Feng, and W. Zeng, Micromachined bolometer with single-crystal silicon diode as temperature sensor, *IEEE Electron Device Letters*. 26(5), 320–322 (May, 2005).

[74] L. Su, J. Chung, D. Antoniadis, K. E. Goodson, and M. Flik, Measurement and modeling of self-heating in SOI nMOSFET's, *IEEE Trans. on Electron Devices*. 41(1), 69–75 (Jan, 1994).

[75] M. Puech, J. Thevenoud, J. Gruffat, N. Launay, N. Arnal, and P. Godinat. Fabrication of 3D packaging TSV using DRIE. In *IEEE Symp. on Design, Test, Integration and Packaging of MEMS/MOEMS* (2008).

[76] G. Katti, M. Stucchi, K. de Meyer, and W. Dehaene, Electrical modeling and characterization of through silicon via for three-dimensional ICs, *IEEE Trans. on Electron Devices*. 57(1), 256–262 (Jan, 2010).

[77] A. Boogaard, A. Kovalgin, and R. Wolters, Net negative charge in low-temperature sio$_2$ gate dielectric layers, *Microelectronic Engineering*. 86(7–9), 1707–1710 (2009).

[78] W.-J. Cho and Y.-C. Kim, Characterization of annealing effects of low temperature chemical vapor deposition oxide films as application of 4H-SiC metal-oxide-semiconductor devices, *Journal of Vacuum Science & Technology.* **20**(1), 14–18 (2002).

[79] D. Malta, C. Gregory, D. Temple, T. Knutson, C. Wang, T. Richardson, Y. Zhang, and R. Rhoades. Integrated process for defect-free copper plating and chemical-mechanical polishing of through-silicon vias for 3D interconnects. In *IEEE Electronic Components and Technology Conf. (ECTC)* (2010).

[80] D. K. Schroder, *Semiconductor Material and Device Characterization.* Newyork:Wiley (1988).

[81] K. Nomura, K. Abe, S. Fujita, Y. Kurosawa, and A. Kageshima. Hierarchical cache system for 3D-multi-core processors in sub 90 nm CMOS. In *ACM/IEEE DATE* (2009).

[82] D. Archard, K. Giles, A. Price, S. Burgess, and K. Buchanan. Low temperature PECVD of dielectric films for TSV applications. In *IEEE Electronic Components and Technology Conf. (ECTC)* (2010).

[83] T. Bandyopadhyay, R. Chatterjee, D. Chung, M. Swaminathan, and R. Tummala. Electrical modeling of through silicon and package vias. In *IEEE Int. Conf. on 3D System Integration (3DIC)* (2009).

[84] J. Benick, B. Hoex, M. C. M. van de Sanden, W. M. M. Kessels, O. Schultz, and S. W. Glunz, High efficiency n-type Si solar cells on Al2O3-passivated boron emitters, *Applied Physics Letters.* **92**(25), 253504–253504 (2008).

[85] P. Saint-Cast, J. Benick, D. Kania, L. Weiss, M. Hofmann, J. Rentsch, R. Preu, and S. Glunz, High-efficiency c-Si solar cells passivated with ALD and PECVD aluminum oxide, *IEEE Electron Device Letters.* **31**(7), 695–697 (July, 2010).

[86] T. Seidel. Atomic layer deposition: a film technology for the nano device era. In *IEEE Int. Conf. on Solid-State and Integrated Circuits Tech.* (2004).

[87] F. Liu, R. Yu, A. Young, J. Doyle, X. Wang, L. Shi, K. N. Chen, X. Li, D. Dipaola, D. Brown, C. Ryan, J. Hagan, K. Wong, M. Lu, X. Gu, N. Klymko, E. Perfecto, A. Merryman, K. Kelly, S. Purushothaman, S. Koester, R. Wisnieff, and W. Haensch. A 300-mm wafer-level three-dimensional integration scheme using tungsten through-silicon via and hybrid Cu-adhesive bonding. In *IEEE Int. Electron Devices Meeting* (Dec, 2008).

[88] B. Swinnen, W. Ruythooren, P. De Moor, L. Bogaerts, L. Carbonell, K. De Munck, B. Eyckens, S. Stoukatch, D. Sabuncuoglu Tezcan, Z. Tokei, J. Vaes, J. Van Aelst, and E. Beyne. 3D integration by Cu-Cu thermocompression bonding of extremely thinned bulk-si die containing 10 $\mu m$ pitch through-Si vias. In *IEEE Int. Electron Devices Meeting* (Dec, 2006).

[89] E. Ege and Y.-L. Shen, Thermomechanical response and stress analysis of copper interconnects, *J. of Electronic Materials.* **32**(10), 1000–1011 (2003).

[90] Y.-L. Shen and E. S. Ege. Thermomechanical stresses in copper interconnect/low-k dielectric systems. In *MRS Proceedings* (2004).

[91] W. Kang, M. Zhang, Y. Zhu, S. Ma, M. Miao, and Y. Jin. A stress relief method for copper filled through silicon via with parylene on sidewall. In *IEEE Int. Conf. on Electronic Packaging Technology High Density Packaging (ICEPT-HDP)* (2010).

[92] Y.-L. Shen, Thermo-mechanical stresses in copper interconnects-a modeling analysis, *Microelectronic Engineering.* **83**(3), 446–459 (2006).

[93] S. S. Hwang, H. C. Lee, H. W. Ro, D. Y. Yoon, and Y. C. Joe. Porosity content dependence of TDDB lifetime and flat band voltage shift by Cu diffusion in porous spin-on low-k. In *IEEE Int. Reliability Physics Symposium (IRPS)* (2005).

[94] H.-J. Lee, C. L. Soles, D.-W. Liu, B. J. Bauer, E. K. Lin, and W.-l. Wu. Structural characterization of methylsilsesquioxane-based porous low-k thin films using X-ray porosimetry. In *IEEE Int. Interconnect Technology Conf.* (2003).

[95] C. Wilson, C. Zhao, L. Zhao, Z. Tokei, K. Croes, M. Pantouvaki, G. Beyer, A. Horsfall, and A. O'Neill. Synchrotron measurement of the effect of dielectric porosity and air gaps on the stress in advanced Cu/Low-k interconnects. In *IEEE Int. Interconnect Technology Conf.* (2009).

[96] S.-K. Ryu, T. Jiang, K. H. Lu, J. Im, H.-Y. Son, K.-Y. Byun, R. Huang, and P. S. Ho, Characterization of thermal stresses in through-silicon vias for three-dimensional interconnects by bending beam technique, *Applied Physics Letters.* **100**(4): 041901 (2012).

[97] K. Kimoto, K. Usami, H. Sakata, and M. Tanaka, Measurement of strain in locally oxidized silicon using convergent-beam electron diffraction, *Japanese J. of Applied Physics.* **32**, L211–L214 (1993).

[98] W. S. Kwon, D. T. Alastair, K. H. Teo, S. Gao, T. Ueda, T. Ishigaki, K. T. Kang, and W. S. Yoo, Stress evolution in surrounding silicon of cu-filled through-silicon via undergoing thermal annealing by multiwavelength micro-raman spectroscopy, *Applied Physics Letters.* **98**(23): 232106 (2011).

[99] B. P. Asthana and W. Kiefer, Deconvolution of the lorentzian linewidth and determination of fraction Lorentzian character from the observed profile of a Raman line by a comparison technique, *Applied Spectroscopy.* **36**(3), 250–257 (Jun, 1982).

[100] K. Zoschke, J. Wolf, C. Lopper, I. Kuna, N. Jurgensen, V. Glaw, K. Samulewicz, J. Roder, M. Wilke, O. Wunsch, M. Klein, M. Suchodoletz, H. Oppermann, T. Braun, R. Wieland, and O. Ehrmann. TSV based silicon interposer technology for wafer level fabrication of 3D SiP modules. In *IEEE Electronic Components and Technology* (2011).

[101] H. Y. Li, L. Ding, G. Katti, J. R. Cubillo, S. Bhattacharya, and G. Q. Lo, Through-silicon interposer technology for heterogeneous integration, *Future Fab International.* **45** (Apr, 2013).

[102] S. Gondi and B. Razavi, Equalization and clock and data recovery techniques for 10-Gb/s CMOS serial-link receivers, *IEEE J. of Solid-State Circuits.* **42**(9), 1999–2011 (2007).

[103] M. Pozzoni *et al.*, A multi-standard 1.5 to 10Gb/s latch-based 3-tap DFE receiver with a SSC tolerant CDR for serial backplane communication, *IEEE J. of Solid-State Circuits.* **44**(4), 1306–1315 (2009).

[104]  J. Cong and Y. Zhang. Thermal-driven multilevel routing for 3D-ICs. In *ACM/IEEE ASP-DAC* (2005).

[105]  G. Katti, A. Mercha, M. Stucchi, Z. Tokei, D. Velenis, J. Van Olmen, C. Huyghebaert, A. Jourdain, M. Rakowski, I. Debusschere, P. Soussan, H. Oprins, W. Dehaene, K. De Meyer, Y. Travaly, E. Beyne, S. Biesemans, and B. Swinnen. Temperature dependent electrical characteristics of through-siicon-via (TSV) interconnections. In *IEEE Int. Interconnect Technology Conf. (IITC)* (2010).

[106]  T. Ishii, H. Ito, M. Kimura, K. Okada, and K. Masu. A 6.5-mW 5-Gbps on-chip differential transmission line interconnect with a low-latency asymmetric tx in a 180nm CMOS technology. In *IEEE Asian Solid-State Circuits Conf. (ASSCC)* (2006).

[107]  A. Carpenter, J. Hu, O. Kocabas, M. Huang, and H. Wu. Enhancing effective throughput for transmission line-based bus. In *Int. Symp. on Computer Architecture (ISCA)*, ACM/IEEE (2012).

[108]  J. seok Yang, K. Athikulwongse, Y.-J. Lee, S. K. Lim, and D. Pan. TSV stress aware timing analysis with applications to 3D-IC layout optimization. In *ACM/IEEE Design Automation Conf. (DAC)* (2010).

[109]  D. F. Lim, K. C. Leong, and C. S. Tan. Selection of underfill material in Cu hybrid bonding and its effect on the transistor keep-out-zone. In *3D-IC*, IEEE (2011).

[110]  J. Cong and Y. Zhang. Thermal via planning for 3-D ICs. In *IEEE/ACM Int. Conf on Computer-Aided Design (ICCAD)* (2005).

[111]  NVM SPICE http://www.nvmspice.org.

[112]  S. Zhao, K. Roy, and C. K. Koh, Decoupling capacitance allocation and its application to power supply noise aware floorplanning, *IEEE Trans. on Computer Aided Design (TCAD) of Circuits and Systems.* pp. 81–92 (2002).

[113]  H. Su, K. Gala, and S. Sapatnekar, Analysis and optimization of structured power/ground networks, *IEEE Trans. on Computer-Aided Design of Integrated Circuits and Systems (TCAD).* **22**(11), 1533–1544 (Nov, 2003).

[114]  H. Zheng, B. Krauter, and L. Pileggi. On-package decoupling optimization with package macromodels. In *IEEE Custom Integrated Circuits Conf. (CICC)* (2003).

[115]  C.-C. Teng, Y.-K. Cheng, E. Rosenbaum, and S.-M. Kang, iTEM: a temperature-dependent electromigration reliability diagnosis tool, *IEEE Trans. on Computer-Aided Design of Integrated Circuits and Systems (TCAD).* **16**(8), 882–893 (Aug, 1997).

[116]  T.-Y. Chiang, K. Banerjee, and K. Saraswat. Compact modeling and SPICE-based simulation for electrothermal analysis of multilevel ULSI interconnects. In *IEEE/ACM Int. Conf. on Computer Aided Design (ICCAD)* (2001).

[117]  V. Tiwari, D. Singh, S. Rajgopal, G. Mehta, R. Patel, and F. Baez. Reducing power in high-performance microprocessors. In *ACM/IEEE Design Automation Conf. (DAC)* (1998).

[118]  W. Liao, L. He, and K. Lepak, Temperature and supply voltage aware performance and power modeling at microarchitecture level, *IEEE Trans.*

on *Computer-Aided Design of Integrated Circuits and Systems (TCAD)*. **24**(7), 1042–1053 (Ju, 2005).

[119] H. Yu, Y. Hu, C. Liu, and L. He. Minimal skew clock embedding considering time variant temperature gradient. In *ACM Int. Symp. on Physical Design (ISPD)* (2007).

[120] H. Yu, Y. Shi, and L. He. Fast analysis of structured power grid by triangularization based structure preserving model order reduction. In *ACM/IEEE Design Automation Conf. (DAC)* (2006).

[121] H. Yu, C. T. Chu, and L. He. Off-chip decoupling capacitor allocation for chip package co-design. In *ACM/IEEE Design Automation Conf. (DAC)* (2007).

[122] E. J. Grimme, *Krylov projection methods for model reduction (Ph. D Thesis)*. Univ. of Illinois at Urbana-Champaign (1997).

[123] A. Odabasioglu, M. Celik, and L. Pileggi, PRIMA: passive reduced-order interconnect macromodeling algorithm, *IEEE Trans. on Computer-Aided Design of Integrated Circuits and Systems (TCAD)*. **17**(8), 645–654 (Aug, 1998).

[124] P. Astrid, S. Weiland, K. Willcox, and T. Backx, Missing point estimation in models described by proper orthogonal decomposition, *IEEE Trans. on Automatic Control*. **53**(10), 2237–2251 (Nov, 2008).

[125] T.-H. Chao, Y.-C. Hsu, J.-M. Ho, and A. Kahng, Zero skew clock routing with minimum wirelength, *IEEE Trans. on Circuits and Systems II*. **39**(11), 799–814 (Nov, 1992).

[126] R. S. Tsay, An exact zero-skew clock routing algorithm, *IEEE Trans. on Computer-Aided Design of Integrated Circuits and Systems*. **12**(2), 242–249 (Feb, 1993).

[127] J. Cong, A. B. Kahng, C.-K. Koh, and C.-W. A. Tsao, Bounded-skew clock and steiner routing, *ACM Trans. on Design Automation of Electronic Systems*. **3**(3), 341–388 (Jul, 1998).

[128] M. Cho, S. Ahmed, and D. Pan. TACO: Temperature aware clock-tree optimization. In *ACM/IEEE Int. Conf. on Computer-Aided Design (ICCAD)* (2005).

[129] J. Long, J. C. Ku, S. Memik, and Y. Ismail, SACTA: A self-adjusting clock tree architecture for adapting to thermal-induced delay variation, *IEEE Trans. on VLSI Systems*. **18**(9), 1323–1336 (Sep, 2010).

[130] X. Zhao and S. K. Lim. Power and slew-aware clock network design for through-silicon-via (TSV) based 3D ICs. In *IEEE/ACM Asia and South Pacific Design Automation Conf. (ASP-DAC)* (2010).

[131] S. Basir-Kazeruni, H. Yu, F. Gong, Y. Hu, C. Liu, and L. He, SPECO: Stochastic perturbation based clock tree optimization considering temperature uncertainity, *Elsevier Integration, the VLSI Journal*. **46**(1), 57–68 (Jan, 2013).

[132] X. Zhao, J. Minz, and S. K. Lim, Low-power and reliable clock network design for through-silicon via (TSV) based 3D ICs, *IEEE Trans. on Components, Packaging, and Manufacturing Technology*. **1**(2), 247–259 (Feb, 2011).

[133] D. Luenberger and Y. Ye, *Linear and nonlinear programming.* vol. 116, Springer Verlag (2008).

[134] 3D-ACME, 3D-IC steady state temperature simulator http://www. 3dacme.allalla.com/.

[135] Hotspot http://lava.cs.virginia.edu/hotspot/.

[136] SPEC 2000 CPU benchmark suits http://www.spec.org/cpu/.

[137] J. Minz, X. Zhao, and S. K. Lim. Buffered clock tree synthesis for 3D ICs under thermal variations. In *IEEE/ACM Asia and South Pacific Design Automation Conf. (ASP-DAC)* (2008).

[138] IBM clock tree benchmarks http://vlsicad.ucsd.edu/GSRC/bookshelf/ Slots/BST/.

[139] C. Zhang, M. Jung, S. K. Lim, and Y. Shi. Novel crack sensor for TSV-based 3D integrated circuits: design and deployment perspectives. In *IEEE/ACM Int. Conf. on Computer-Aided Design (ICCAD)* (2013).

[140] E. J. Marinissen. Challenges and emerging solutions in testing TSV-based 1/2D-and 3D-stacked ICs. In *ACM/IEEE Design Automation and Test in Europe Conf. (DATE)* (2012).

[141] B. Noia and K. Chakrabarty. Pre-bond probing of TSVs in 3D stacked ICs. In *IEEE Int. Test Conf. (ITC)* (2011).

[142] E. J. Marinissen and Y. Zorian. Testing 3D chips containing through-silicon vias. In *IEEE Int. Test Conf. (ITC)* (2009).

[143] J. Xie, Y. Wang, and Y. Xie. Yield-aware time-efficient testing and self-fixing design for TSV-based 3D ICs. In *IEEE Asia and South Pacific Design Automation Conf. (ASP-DAC)* (2012).

[144] D. L. Donoho and M. Elad, Optimally sparse representation in general (nonorthogonal) dictionaries via L1 minimization, *Proceedings of the National Academy of Sciences* (2003).

[145] E. J. Candes and T. Tao, Near-optimal signal recovery from random projections: Universal encoding strategies?, *IEEE Transactions on Information Theory,.* **52**(12), 5406–5425 (2006).

[146] W. Maly. Realistic fault modeling for vlsi testing. In *IEEE Design Automation Conf. (DAC)* (1987).

[147] M. Tahoori. Defects, yield, and design in sublithographic nano-electronics. In *IEEE Defect and Fault Tolerance in VLSI Systems* (2005).

[148] Y. Zhao, S. Khursheed, and B. Al-Hashimi. Cost-effective TSV grouping for yield improvement of 3D-ICs. In *IEEE Asian Test Symposium (ATS)* (2011).

[149] G. S. May and C. J. Spanos, *Fundamentals of semiconductor manufacturing and process control.* John Wiley & Sons (2006).

[150] J. A. Tropp and A. C. Gilbert, Signal recovery from random measurements via orthogonal matching pursuit, *IEEE Trans. on Information Theory,.* **53**(12), 4655–4666 (2007).

[151] B. Noia, S. Panth, K. Chakrabarty, and S. K. Lim. Scan test of die logic in 3D ICs using TSV probing. In *IEEE Int. Test Conf. (ITC)* (2012).

[152] M. C. Hansen, H. Yalcin, and J. P. Hayes, Unveiling the ISCAS-85 benchmarks: A case study in reverse engineering, *IEEE Design and Test of Computers.* **16**(3), 72–80 (1999).

[153] I. Hamzaoglu and J. H. Patel. Testset compaction algorithms for combinational circuits. In *IEEE Int. Conf. on Computer-Aided Design (ICCAD)* (1998).

[154] H. Ichihara *et al.* Test compression based on lossy image encoding. In *IEEE Asian Test Symp. (ATS)* (2011).

[155] A. Chandra and K. Chakrabarty. Test data compression for system-on-a-chip using golomb codes. In *IEEE VLSI Test Symp.* (2000).

[156] B. Black, M. Annavaram, N. Brekelbaum, J. DeVale, L. Jiang, G. Loh, D. McCauley, P. Morrow, D. Nelson, D. Pantuso, P. Reed, J. Rupley, S. Shankar, J. Shen, and C. Webb. Die stacking (3D) microarchitecture. In *IEEE/ACM Int. Symp. on Microarchitecture* (2006).

[157] F. Li, C. Nicopoulos, T. Richardson, Y. Xie, V. Narayanan, and M. Kandemir. Design and management of 3D chip multiprocessors using network-in-memory. In *ACM/IEEE Int. Symp. on Computer Architecture (ISCA)* (2006).

[158] H. Sun, J. Liu, R. Anigundi, N. Zheng, J.-Q. Lu, K. Rose, and T. Zhang, 3D DRAM design and application to 3D multicore systems, *IEEE Design Test of Computers (DTC).* **26**(5), 36–47 (Sept, 2009).

[159] P. Jacob, A. Zia, O. Erdogan, P. Belemjian, J.-W. Kim, M. Chu, R. Kraft, J. McDonald, and K. Bernstein, Mitigating memory wall effects in high-clock-rate and multicore CMOS 3-D processor memory stacks, *Proceedings of the IEEE.* **97**(1), 108–122 (Jan, 2009).

[160] G. H. Loh, Y. Xie, and B. Black, Processor design in 3D die-stacking technologies, *IEEE Micro.* **27**(3), 31–48 (May, 2007).

[161] Y. F. Tsai, Y. Xie, N. Vijaykrishnan, and M. J. Irwin. Three-dimensional cache design exploration using 3DCacti. In *IEEE Int. Conf. on Computer Design (ICCD)* (2005).

[162] P. Ghosal, H. Rahaman, and P. Dasgupta. Thermal aware placement in 3D ICs. In *Int. Conf. on Advances in Recent Technologies in Communication and Computing* (2010).

[163] H. Wang, Y. Fu, T. Liu, and J. Wang. Thermal management via task scheduling for 3D NoC based multi-processor. In *Int. SoC Design Conf. (ISOCC)* (2010).

[164] O. Karim, J.-C. Crebier, C. Gillot, C. Schaeffer, B. Mallet, and E. Gimet. Heat transfer coefficient for water cooled heat sink: application for standard power modules cooling at high temperature. In *IEEE Power Electronics Specialists Conf.* (2001).

[165] D. Brooks and M. Martonosi. Dynamic thermal management for high-performance microprocessors. In *Int. Symp. on High-Performance Computer Architecture* (2001).

[166] J. Donald and M. Martonosi. Techniques for multicore thermal management: Classification and new exploration. In *Int. Symp. on Computer Architecture (ISCA)* (2006).

[167] A. Kumar, L. Shang, L.-S. Peh, and N. Jha, System-level dynamic thermal management for high-performance microprocessors, *IEEE Trans. on Computer-Aided Design of Integrated Circuits and Systems (TCAD).* **27**(1), 96–108 (Jan, 2008).

[168] T. Ebi, M. Faruque, and J. Henkel. TAPE: Thermal-aware agent-based power econom multi/many-core architectures. In *IEEE/ACM Int. Conf. on Computer-Aided Design (ICCAD)* (2009).

[169] D. Li, S. X.-D. Tan, E. H. Pacheco, and M. Tirumala, Parameterized architecture-level dynamic thermal models for multicore microprocessors, *ACM Trans. Des. Autom. Electron. Syst. (TODAES).* **15** (2), 16:1–16:22 (Mar, 2010).

[170] D. H. Kim, K. Athikulwongse, and S. K. Lim. A study of through-silicon-via impact on the 3D stacked IC layout. In *IEEE/ACM Int. Conf. on Computer Aided Design (ICCAD)* (2009).

[171] D. B. Tuckerman and R. F. W. Pease, High-performance heat sinking for VLSI, *IEEE Electronic Device Letters.* **EDL-2**(5), 126–129 (May, 1981).

[172] J.-M. Koo, S. J. Im, L. Jiang, and K. E. Goodson, Integrated microchannel cooling for three-dimensional electronic circuit architectures, *J. of Heat Transfer.* **127**(1), 49–58 (Feb, 2005).

[173] B. Dang, P. Joseph, M. Bakir, T. Spencer, P. Kohl, and J. Meindl. Wafer-level microfluidic cooling interconnects for GSI. In *IEEE Int. Interconnect Technology Conf. (IITC)* (2005).

[174] A. Coskun, D. Atienza, T. Rosing, T. Brunschwiler, and B. Michel. Energy-efficient variable-flow liquid cooling in 3D stacked architectures. In *IEEE/ACM Design, Automation Test in Europe Conf. (DATE)* (2010).

[175] M. Bakir, B. Dang, and J. Meindl. Revolutionary nanosilicon ancillary technologies for ultimate-performance gigascale systems. In *IEEE Custom Integrated Circuits Conf. (CICC)* (2007).

[176] Y.-J. Lee, R. Goel, and S. K. Lim. Multi-functional interconnect co-optimization for fast and reliable 3D stacked ICs. In *IEEE/ACM Int. Conf. on Computer-Aided Design (ICCAD)* (2009).

[177] S. Sastry. Networked embedded systems: From sensor webs to cyber-physical systems. In *Hybrid Systems: Computation and Control*, vol. 4416, pp. 1–1. Springer Berlin Heidelberg (2007).

[178] E. Lee. Cyber physical systems: Design challenges. In *IEEE Int. Symp. on Object Oriented Real-Time Distributed Computing* (2008).

[179] L. Sha, S. Gopalakrishnan, X. Liu, and Q. Wang. Cyber-physical systems: A new frontier. In *IEEE Int. Conf. on Sensor Networks, Ubiquitous and Trustworthy Computing* (2008).

[180] ANSYS Multiphysics http://www.ansys.com.

[181] COMSOL Multiphysics http://www.comsol.com.

[182] W. Huang, M. Stan, K. Skadron, K. Sankaranarayanan, S. Ghosh, and S. Velusamy. Compact thermal modeling for temperature-aware design. In *IEEE/ACM Design Automation Conf. (DAC)* (2004).

[183] Y. J. Kim, Y. K. Joshi, A. G. Fedorov, Y.-J. Lee, and S.-K. Lim, Thermal characterization of interlayer microfluidic cooling of three-dimensional

integrated circuits with nonuniform heat flux, *J. of Heat Transfer.* **132**(4), 214–219 (Feb, 2010).

[184]   H. Mizunuma, C.-L. Yang, and Y.-C. Lu. Thermal modeling for 3D-ICs with integrated microchannel cooling. In *IEEE/ACM Int. Conf. on Computer-Aided Design (ICCAD)* (2009).

[185]   A. Sridhar, A. Vincenzi, M. Ruggiero, T. Brunschwiler, and D. Atienza. 3D-ICE: Fast compact transient thermal modeling for 3D ICs with inter-tier liquid cooling. In *IEEE/ACM Int. Conf. on Computer-Aided Design (ICCAD)* (2010).

[186]   A. Sridhar, A. Vincenzi, M. Ruggiero, T. Brunschwiler, and D. Atienza. Compact transient thermal model for 3D ICs with liquid cooling via enhanced heat transfer cavity geometries. In *Int. Workshop on Thermal Investigations of ICs and Systems (THERMINIC)* (2010).

[187]   E. N. Seider and G. E. Tate, Heat transfer and pressure drop of liquids in tubes, *Industrial & Engineering Chemistry.* **28**(12), 1429–1435 (Dec, 1936).

[188]   R. Shah and A. London, *Laminar Flow Forced Convection in Ducts.* Newyork Academic Press (1978).

[189]   N. Binkert, B. Beckmann, G. Black, S. Reinhardt, A. Saidi, A. Basu, J. Hestness, D. Hower, T. Krishna, S. Sardashti, *et al.*, The gem5 simulator, *ACM SIGARCH Computer Architecture News.* **39** (2), 1–7 (Aug, 2011).

[190]   S. Li. McPAT: an integrated power, area, and timing modeling framework for multicore and manycore architectures. In *IEEE MICRO* (2009).

[191]   CACTI model http://www.hpl.hp.com/research/cacti/.

[192]   T. A. Davis and E. Palamadai. KLU: Sparse LU factorization algorithm for circuit simulation. Online http://www.cise.ufl.edu/research/sparse/klu.

[193]   X. F. Peng and G. P. Peterson, Convective heat transfer and flow friction for water flow in microchannel structures, *Int. J. Heat Mass Transfer.* **39**(12), 2599–2608 (Aug, 1996).

[194]   J. D. Hamilton, *Time Series Analysis.* Princeton University Press (1994).

[195]   R. E. Kalman, A new approach to linear filtering and prediction problems, *J. of Fluids Engineering.* **82**(1), 35–45 (Mar, 1960).

[196]   Wattch version 1.02 http://www.eecs.harvard.edu/~dbrooks/wattch-form. html.

[197]   SuiteSparse http://www.cise.ufl.edu/research/sparse/SuiteSparse/.

[198]   H. Oertel, ed., *Prandtl-Essentials of Fluid Mechanics.* Springer (2009).

[199]   W. Davis, J. Wilson, S. Mick, J. Xu, H. Hua, C. Mineo, A. Sule, M. Steer, and P. Franzon, Demystifying 3D ICs: the pros and cons of going vertical, *IEEE Design and Test of Computers (DTC).* **22**(6), 498–510 (Nov, 2005).

[200]   R. Ramanujam and B. Lin, A layer-multiplexed 3D on-chip network architecture, *IEEE Embedded Systems Letters.* **1**(2), 50–55 (Aug, 2009).

[201]   G. Van der Plas, P. Limaye, A. Mercha, H. Oprins, C. Torregiani, S. Thijs, D. Linten, M. Stucchi, K. Guruprasad, D. Velenis, D. Shinichi, V. Cherman, B. Vandevelde, V. Simons, I. De Wolf, R. Labie, D. Perry, S. Bronckers, N. Minas, M. Cupac, W. Ruythooren, J. Van Olmen, A. Phommahaxay, M. de Potter de ten Broeck, A. Opdebeeck, M. Rakowski, B. De Wachter, M. Dehan, M. Nelis, R. Agarwal, W. Dehaene, Y. Travaly, P. Marchal,

and E. Beyne. Design issues and considerations for low-cost 3D TSV IC technology. In *IEEE Int. Solid-State Circuits Conf. (ISSCC)* (2010).

[202] D. H. Woo, N. H. Seong, D. Lewis, and H.-H. Lee. An optimized 3D-stacked memory architecture by exploiting excessive, high-density TSV bandwidth. In *IEEE Int. Symp. on High Performance Computer Architecture (HPCA)* (2010).

[203] W. Kim, M. Gupta, G.-Y. Wei, and D. Brooks. System level analysis of fast, per-core DVFS using on-chip switching regulators. In *IEEE Int. Symp. on High Performance Computer Architecture (HPCA)* (2008).

[204] R. Bondade and D. Ma, Hardware-software codesign of an embedded multiple-supply power management unit for multicore SoCs using an adaptive global/local power allocation and processing scheme, *ACM Trans. Des. Autom. Electron. Syst. (TODAES)*. **16**(3), 31:1–31:27 (Jun, 2011).

[205] Y. Panov and M. Jovanovic, Design considerations for 12-V/1.5-V, 50-A voltage regulator modules, *IEEE Trans. on Power Electronics*. **16**(6), 776–783 (Nov, 2001).

[206] Y. Cho and N. Chang, Energy-aware clock frequency assignment in microprocessors and memory devices for dynamic voltage scaling, *IEEE Trans. on CAD of Integrated Circuits and Systems (TCAD)*. **26**(6), 1030–1040 (Jun, 2007).

[207] W. Kim, D. Brooks, and G.-Y. Wei. A fully-integrated 3-level DC/DC converter for nanosecond-scale DVS with fast shunt regulation. In *IEEE Int. Solid-State Circuits Conf. (ISSCC)* (2011).

[208] S. Kose and E. Friedman, Distributed on-chip power delivery, *IEEE J. on Emerging and Selected Topics in Circuits and Systems*. **2**(4), 704–713 (Dec, 2012).

[209] D. Ma, W.-H. Ki, C. ying Tsui, and P. Mok, Single-inductor multiple-output switching converters with time-multiplexing control in discontinuous conduction mode, *IEEE J. of Solid-State Circuits*. **38**(1), 89–100 (Jan, 2003).

[210] M.-H. Huang and K.-H. Chen, Single-inductor multi-output (SIMO) DC-DC converters with high light-load efficiency and minimized cross-regulation for portable devices, *IEEE J. of Solid-State Circuits*. **44**(4), 1099–1111 (Apr, 2009).

[211] M. Sabry, A. Coskun, D. Atienza, T. Rosing, and T. Brunschwiler, Energy-efficient multi-objective thermal control for liquid-cooled 3D stacked architectures, *IEEE Trans. on CAD of Integrated Circuits and Systems (TCAD)*. **30**(11), 2015–2029 (Nov, 2011).

[212] X. Huang, C. Zhang, H. Yu, and W. Zhang, A nano-electro-mechanical-switch based thermal management for 3D integrated many-core memory-procesor system, *IEEE Trans. on Nanotechnology (TNANO)*. **11**(3), 588–600 (May, 2012).

[213] M. Sai and H. Yu. Cyber-physical management for heterogeneously integrated 3D thousand-core on-chip microprocessor. In *IEEE Int. Symp. on Circuits And Systems (ISCAS)* (2013).

[214]  L. Benini, A. Bogliolo, and G. De Micheli, A survey of design techniques for system-level dynamic power management, *IEEE Trans. on VLSI Systems.* **8**(3), 299–316 (Jun, 2000).

[215]  Y. Tan, W. Liu, and Q. Qiu. Adaptive power management using reinforcement learning. In *ACM/IEEE Int. Conf. on Computer-Aided Design (ICCAD)* (2009).

[216]  M. Shafique, B. Vogel, and J. Henkel. Self-adaptive hybrid dynamic power management for many-core systems. In *ACM/IEEE Design Automation and Test in Europe Conf. (DATE)* (2013).

[217]  W. Lee, Y. Wang, and M. Pedram. VRCon: Dynamic reconfiguration of voltage regulators in a multicore platform. In *ACM/IEEE Design Automation and Test in Europe Conf. (DATE)* (2014).

[218]  S. Samii, M. Selkala, E. Larsson, K. Chakrabarty, and Z. Peng, Cycle-accurate test power modeling and its application to soc test architecture design and scheduling, *IEEE Trans. on CAD of Integrated Circuits and Systems (TCAD).* **27**(5), 973–977 (May, 2008).

[219]  Y. H. Katz, D. A. Culler, S. S. S. Alspaugh, Y. Chen, S. Dawson-haggerty, P. Dutta, M. He, X. Jiang, L. Keys, A. Krioukov, K. Lutz, J. Ortiz, P. Mohan, E. Reutzel, J. Taneja, J. Hsu, and S. Shankar, An information-centric energy infrastructure: The berkley view, *Sustainable Computing: Informatics and Systems.* **1**(1), 7–22 (Mar, 2011).

[220]  S. Garg, D. Marculescu, R. Marculescu, and U. Ogras. Technology-driven limits on DVFS controllability of multiple voltage-frequency island designs: a system-level perspective. In *ACM/IEEE Design Automation Conf. (DAC)* (2009).

[221]  ILP solver 5.5 http://lpsolve.sourceforge.net/5.5/.

[222]  Autoregression analysis http://paulbourke.net/miscellaneous/ar/.

[223]  MIPS processor cores http://www.mips.com/products/processor-cores/.

[224]  A. Vahidsafa and S. Bhutani. SPARC M6: Oracle's Next Generation Processor for Enterprise Systems. In *HOT CHIPS* (2013).

[225]  R. Kumar, V. Zyuban, and D. M. Tullsen. Interconnections in multi-core architectures: Understanding mechanisms, overheads and scaling. In *IEEE Int. Symp. on Computer Arch.* (2005).

[226]  S. Rusu, H. Muljono, D. Ayers, S. Tam, W. Chen, A. Martin, S. Li, S. Vora, R. Varada, and E. Wang. Ivytown: A 22$nm$ 15-core enterprise xeon® processor family. In *IEEE Solid-State Circuits Conf. (ISSCC)* (2014).

[227]  I. Loi, P. Marchal, A. Pullini, and L. Benini. 3D NoCs — unifying inter & intra chip communication. In *IEEE Int. Symp. on Circuits and Systems (ISCAS)* (2010).

[228]  D. Kim, K. Athikulwongse, M. Healy, M. Hossain, M. Jung, I. Khorosh, G. Kumar, Y.-J. Lee, D. Lewis, T.-W. Lin, C. Liu, S. Panth, M. Pathak, M. Ren, G. Shen, T. Song, D. Woo, X. Zhao, J. Kim, H. Choi, G. Loh, H.-H. Lee, and S. Lim, Design and analysis of 3D-MAPS (3D massively parallel processor with stacked memory), *IEEE Trans. on Computers.* **64**(1), 112–125 (Jan, 2015).

[229] Y.-J. Lee and S. K. Lim, Ultra high density logic designs using monolithic 3D integration, *IEEE Trans. on Computer-Aided Design of Integrated Circuits.* **32**(12), 1892–1905 (Dec, 2013).

[230] Hybrid Memory Cube Consortium http://hybridmemorycube.org/tool-resources.html.

[231] J. Zhang, M. O. Bloomfield, jian Qiang Lu, R. D. Gutmann, and T. S. Cale, Thermal stresses in 3D IC inter-wafer interconnects, *Elsevier J. of Microelectronic engineering.* **82**(3), 534–547 (Dec, 2005).

[232] C. Erdmann, D. Lowney, A. Lynam, A. Keady, J. McGrath, E. Cullen, D. Breathnach, D. Keane, P. Lynch, M. De La Torre, R. De La Torre, P. Lim, A. Collins, B. Farley, and L. Madden. A heterogeneous 3D-IC consisting of two 28nm FPGA die and 32 reconfigurable high-performance data converters. In *IEEE Int. Solid-State Circuits Conf. (ISSCC)* (2014).

[233] T. Zhang, K. Wang, Y. Feng, X. Song, L. Duan, Y. Xie, X. Cheng, and Y.-L. Lin. A customized design of DRAM controller for on-chip 3D DRAM stacking. In *IEEE Custom Integrated Circuits Conf. (CICC)* (2010).

[234] B. Akesson, P.-C. Huang, F. Clermidy, D. Dutoit, K. Goossens, Y.-H. Chang, T.-W. Kuo, P. Vivet, and D. Wingard. Memory controllers for high-performance and real-time mpsocs requirements, architectures, and future trends. In *ACM/IEEE Int. Conf. on Hardware/Software Codesign and System Synthesis (CODES+ISSS)* (2011).

[235] C. Weis, N. Wehn, L. Igor, and L. Benini. Design space exploration for 3D-stacked DRAMs. In *ACM/IEEE Design, Automation Test in Europe Conf. (DATE)* (2011).

[236] C. Weis, I. Loi, L. Benini, and N. Wehn. An energy efficient DRAM subsystem for 3D integrated SoCs. In *ACM/IEEE Design, Automation Test in Europe Conf. (DATE)* (2012).

[237] K. Chandrasekar, C. Weis, B. Akesson, N. Wehn, and K. Goossens. System and circuit level power modeling of energy-efficient 3D-stacked wide I/O DRAMs. In *ACM/IEEE Design, Automation Test in Europe Conf. (DATE)* (2013).

[238] Denali Software Inc. Databahn DRAM memory controller IP (2009). https://www.denali.com/en/products/databahndram.jsp.

[239] S. Bayliss and G. A. Constantinides. Methodology for designing statically scheduled application-specific SDRAM controllers using constrained local search. In *IEEE Int. Conf. on Field-Programmable Technology (FPT)* (2009).

[240] M. Paolieri, E. Quiones, F. Cazorla, and M. Valero, An analyzable memory controller for hard real-time CMPs, *IEEE Embedded Systems Letters.* **1**(4), 86–90 (Dec, 2009).

[241] L. Steffens, M. Agarwal, and P. Wolf. Real-time analysis for memory access in media processing SoCs: A practical approach. In *IEEE Euromicro Conf. on Real-Time Systems* (2008).

[242] J. Tschanz and N. Shanbhag. A low-power, reconfigurable adaptive equalizer architecture. In *Asilomar Conf. on Signals, Systems, and Computers* (1999).

[243] N. Tzartzanis and W. Walker, Differential current-mode sensing for efficient on-chip global signaling, *IEEE J. of Solid-State Circuits.* **40**(11), 2141–2147 (Nov, 2005).

[244] J. sun Seo, R. Ho, J. Lexau, M. Dayringer, D. Sylvester, and D. Blaauw. High-bandwidth and low-energy on-chip signaling with adaptive pre-emphasis in 90nm CMOS. In *IEEE Int. Solid-State Circuits Conf. (ISSCC)* (2010).

[245] S. Park, M. Qazi, L.-S. Peh, and A. P. Chandrakasan. 40.4fJ/bit/mm low-swing on-chip signaling with self-resetting logic repeaters embedded within a mesh NoC in 45nm SOI CMOS. In *ACM/IEEE Design, Automation Test in Europe Conf. (DATE)* (2013).

[246] I. Foster, A. Roy, and V. Sander. A quality of service architecture that combines resource reservation and application adaptation. In *IEEE Int. Workshop on Quality of Service* (2000).

[247] V. Raghunathan, M. B. Srivastava, and R. K. Gupta. A survey of techniques for energy efficient on-chip communication. In *ACM/IEEE Design Automation Conf. (DAC)* (2003).

[248] K. Sewell, R. Dreslinski, T. Manville, S. Satpathy, N. Pinckney, G. Blake, M. Cieslak, R. Das, T. Wenisch, D. Sylvester, D. Blaauw, and T. Mudge, Swizzle-switch networks for many-core systems, *IEEE J. on Emerging and Selected Topics in Circuits and Systems.* **2**(2), 278–294 (2012).

[249] A. Hilton, S. Nagarakatte, and A. Roth. iCFP: Tolerating all-level cache misses in in-order processors. In *IEEE Int. Symp. on High Performance Computer Architecture (HPCA)* (2009).

[250] M. K. Jeong, M. Erez, C. Sudanthi, and N. Paver. A QoS-aware memory controller for dynamically balancing GPU and CPU bandwidth use in an MPSoC. In *ACM/IEEE Design Automation Conf. (DAC)* (2012).

[251] M. K. Jeong, D. H. Yoon, and M. Erez. DrSim: A platform for flexible DRAM system research. http://lph.ece.utexas.edu/public/DrSim.

[252] SPEC CPU2006 Benchmark http://www.spec.org/cpu2006/.

[253] PARSEC Benchmark http://parsec.cs.princeton.edu/.

[254] C. Ranger, R. Raghuraman, A. Penmetsa, G. Bradski, and C. Kozyrakis. Evaluating mapreduce for multi-core and multiprocessor systems. In *IEEE Int. Symp. on High Performance Computer Architecture (HPCA)* (2007).

[255] A. Gosavi, A reinforcement learning algorithm based on policy iteration for average reward: Empirical results with yield management and convergence analysis, *Machine Learning Journal.* **55**(1), 5–29 (Apr, 2004).

[256] N. Mastronarde and M. van der Schaar, Online reinforcement learning for dynamic multimedia systems, *IEEE Trans. on Image Processing.* **19**(2), 290–305 (Feb, 2010).

[257] M. Triki, A. Ammari, Y. Wang, and M. Pedram. Reinforcement learning-based dynamic power management of a battery-powered system supplying multiple active modes. In *IEEE European Modelling Symp. (EMS)* (2013).

[258] H. Shen, Y. Tan, J. Lu, Q. Wu, and Q. Qiu, Achieving autonomous power management using reinforcement learning, *ACM Trans. on Design Automation of Electronic Systems (TODAES).* **18**(2), 24:1–24:32 (Apr, 2013).

[259] N. Mastronarde, K. Kanoun, D. Atienza, P. Frossard, and M. van der Schaar, Markov decision process based energy-efficient on-line scheduling for slice-parallel video decoders on multicore systems, *IEEE Trans. on Multimedia.* **15**(2), 268–278 (Feb, 2013).

[260] L. P. Kaelbling, M. L. Littman, and A. W. Moore, Reinforcement learning: A survey, *ACM J. of Artificial Intellignece Research.* **4**(1), 237–285 (May, 1996).

[261] C. J. Watkins and P. Dayan, Q-learning, *Machine learning Journal.* **8**(3–4), 279–292 (May, 1992).

[262] M. L. Littman, Value-function reinforcement learning in markov games, *ACM Cognitive Systems Research.* **2**(1), 55–66 (Apr, 2001).

[263] M. Ghavamzadeh, H. J. Kappen, M. G. Azar, and R. Munos. Speedy Q-learning. In *Advances in Neural Information Processing Systems.* Curran Associates, Inc. (2011).

[264] E. E-Dar and Y. Mansour, Learning rates for Q-learning, *J. of Machine Learning.* **5**, 1–25 (2003).

[265] S. K. Das, S. K. Sen, and R. Jayaram. Call admission and control for quality-of-service provisioning in cellular networks. In *IEEE Int. Conf. on Universal Personal Communications Record* (1997).

[266] R. A. Shafik *et al.* On the extended relationships among EVM, BER and SNR as performance metrics. In *IEEE Int. Conf. on Electrical and Computer Engineering* (2006).

[267] C. T. Ko and K. N. Chen, Wafer-level bonding/stacking technology for 3D integration, *Microelectronics Reliability.* **50**(4), 481–488 (Apr, 2010).

[268] C. S. Tan, J. Fan, D. F. Lim, G. Y. Chong, and K. H. Li, Low temperature wafer-level bonding for hermetic packaging of 3D microsystems, *J. of Micromechanics and Microengineering.* **21**(7) (Jun, 2011).

[269] E. Beyne. 3D system integration technologies. In *Int. Symp. on VLSI Technology, Systems, and Applications* (2006).

[270] C. S. Tan, Thermal characteristic of Cu-Cu bonding in layer in 3-D integrated circuits stack, *Microelectronics Engineering.* **87**(4), 682–685 (Apr, 2010).

[271] G. Fedder, R. Howe, T.-J. K. Liu, and E. Quevy, Technologies for cofabricating MEMS and electronics, *Proceedings of the IEEE.* **96**(2), 306–322 (Feb, 2008).

[272] J. Smith, S. Montague, J. Sniegowski, J. Murray, and P. McWhorter. Embedded micromechanical devices for the monolithic integration of MEMS with CMOS. In *IEEE Int. Electron Devices Meeting* (1995).

[273] J. Fan, D. F. Lim, L. Peng, K. H. Li, and C. S. Tan, Low temperature Cu-to-Cu bonding for wafer-level hermetic encapsulation of 3D microsystems, *Electrochemical and Solid-State Letters.* **14**(11), H470–H474 (2011).

[274] R. Nadipalli, J. Fan, K. Li, K. Wee, H. Yu, and C. Tan. 3D integration of MEMS and CMOS via Cu-Cu bonding with simultaneous formation of electrical, mechanical and hermetic bonds. In *IEEE Int. 3D Systems Integration Conf. (3DIC)* (2012).

[275] S. Sherman, W. Tsang, T. Core, and D. Quinn. A low cost monolithic accelerometer. In *IEEE Symp. on VLSI Circuits* (1992).

[276] S. H. Lee, J. Mitchell, W. Welch, S. Lee, and K. Najafi. Wafer-level vacuum/hermetic packaging technologies for MEMS. In *Reliability, Packaging, Testing, and Characterization of Mems/Moems and Nanodevices* (2010).

[277] L. Peng, H. Li, D. F. Lim, S. Gao, and C. S. Tan, High-density 3-D interconnect of Cu-Cu contacts with enhanced contact resistance by self-assembled monolayer (SAM) passivation, *IEEE Trans. on Electron Devices.* **58**(8), 2500–2506 (Aug, 2011).

[278] J. Lau, C. Lee, and C. Premachandran, *Advanced MEMS Packaging.* NewYork McGraw-Hill (2009).

[279] F. O'Mahony *et al.* The future of electrical I/O for microprocessors. In *Int. Symp. on VLSI Design, Automation and Test (VLSI-DAT)* (2009).

[280] J. Zerbe, B. Daly, W. Dettloff, T. Stone, W. Stonecypher, P. Venkatesan, K. Prabhu, B. Su, J. Ren, B. Tsang, B. Leibowitz, D. Dunwell, A. Carusone, and J. Eble. A 5.6Gb/s 2.4mW/Gb/s bidirectional link with 8ns power-on. In *Symp. on VLSI Circuits (VLSIC)* (2011).

[281] S.-H. Chung and L.-S. Kim. 1.22mW/Gb/s 9.6Gb/s data jitter mixing forwarded-clock receiver robust against power noise with 1.92ns latency mismatch between data and clock in 65nm CMOS. In *Symp. on VLSI Circuits (VLSIC)* (2012).

[282] M. Hossain and A. Carusone. A 6.8mW 7.4Gb/s clock-forwarded receiver with up to 300MHz jitter tracking in 65nm CMOS. In *IEEE Int. Solid-State Circuits Conf. (ISSCC)* (2010).

[283] Y.-J. Kim and L.-S. Kim. A 12Gb/s 0.92mW/Gb/s forwarded clock receiver based on ILO with 60MHz jitter tracking bandwidth variation using duty cycle adjuster in 65nm CMOS. In *VLSI Circuits (VLSIC), 2013 Symposium on* (2013).

[284] K. Hu, T. Jiang, J. Wang, F. O'Mahony, and P. Chiang, A 0.6 mW/Gb/s, 6.4-7.2 Gb/s serial link receiver using local injection-locked ring oscillators in 90 nm CMOS, *IEEE J. of Solid-State Circuits (JSSC).* **45**(4), 899–908 (April, 2010).

[285] K. Hu, R. Bai, T. Jiang, C. Ma, A. Ragab, S. Palermo, and P. Chiang, 0.16-0.25 pJ/bit, 8 Gb/s near-threshold serial link receiver with super-harmonic injection-locking, *IEEE J. of Solid-State Circuits (JSSC).* **47**(8), 1842–1853 (Aug, 2012).

[286] B. Casper and F. O'Mahony, Clocking analysis, implementation and measurement techniques for high-speed data links-a tutorial, *IEEE Trans. on Circuits and Systems I: Regular Papers (TCAS-I).* **56**(1), 17–39 (Jan, 2009).

[287] M. Meghelli, S. Rylov, J. Bulzacchelli, W. Rhee, A. Rylyakov, H. Ainspan, B. Parker, M. Beakes, A. Chung, T. Beukema, P. Pepeljugoski, L. Shan, Y. Kwark, S. Gowda, and D. Friedman. A 10Gb/s 5-Tap-DFE/4-Tap-FFE transceiver in 90nm CMOS. In *IEEE Int. Solid-State Circuits Conference (ISSCC)* (2006).

[288] S. Gondi, J. Lee, D. Takeuchi, and B. Razavi. A 10Gb/s CMOS adaptive equalizer for backplane applications. In *IEEE Int. Solid-State Circuits Conference (ISSCC)* (2005).

[289] J. Lee, J. Weiner, and Y.-K. Chen, A 20-GS/s 5-b SiGe ADC for 40-Gb/s coherent optical links, *IEEE Trans. on Circuits and Systems I: Regular Papers (TCAS-I).* **57**(10), 2665–2674 (Oct, 2010).

[290] S. Yamanaka, K. Sano, and K. Murata, A 20-Gs/s track-and-hold amplifier in InP HBT technology, *IEEE Trans. on Microwave Theory and Techniques.* **58**(9), 2334–2339 (Sept, 2010).

[291] J. A. Plaza, A. Collado, E. Cabruja, and J. Esteve, Piezoresistive Accelerometers for MCM Package, *IEEE J. of Microelectromechanical Systems.* **11**(6), 794–801 (2002).

[292] A. Deutsch, Electrical Characteristics of Interconnections for High-Performance Systems, *Proceedings of the IEEE.* **86**(2), 315–357 (1998).

[293] M. Chen, J. Silva-Martinez, M. Nix, and M. Robinson, Low-voltage low-power LVDS drivers, *IEEE J. of Solid-State Circuits (JSSC).* **40**(2), 472–479 (2005).

[294] D. K. Bhattacharryya and S. Nandi. An efficient class of SEC-DED-AUED codes. In *Int. Symp. on Parallel Architectures, Algorithms, and Networks* (1997).

[295] S. Ma, J. Wang, H. Yu, and J. Ren. A 32.5-GS/s two-channel time-interleaved CMOS sampler with switched-source follower based track-and-hold amplifier. In *IEEE Int. Microwave Symp. (IMS)* (2014).

[296] Z. Yu, K. You, R. Xiao, H. Quan, P. Ou, Y. Ying, H. Yang, M. Jing, and X. Zeng. An 800MHz 320mW 16-core processor with message-passing and shared-memory inter-core communication mechanisms. In *IEEE Int. Solid-State Circuits Conf. (ISSCC)* (2012).

[297] P. Ou, J. Zhang, H. Quan, Y. Li, M. He, Z. Yu, X. Yu, S. Cui, J. Feng, S. Zhu, J. Lin, M. Jing, X. Zeng, and Z. Yu. A 65nm 39GOPS/W 24-core processor with 11Tb/s/W packet-controlled circuit-switched double-layer network-on-chip and heterogeneous execution array. In *IEEE Int. Solid-State Circuits Conf. (ISSCC)* (2013).

[298] T. Mattson, R. van der Wijngaart, M. Riepen, T. Lehnig, P. Brett, W. Haas, P. Kennedy, J. Howard, S. Vangal, N. Borkar, G. Ruhl, and S. Dighe. The 48-core SCC processor: the programmer's view. In *Int. Conf. for High Performance Computing, Networking, Storage and Analysis (SC)* (2010).

[299] S. Bell, B. Edwards, J. Amann, R. Conlin, K. Joyce, V. Leung, J. MacKay, M. Reif, L. Bao, J. Brown, M. Mattina, C.-C. Miao, C. Ramey, D. Wentzlaff, W. Anderson, E. Berger, N. Fairbanks, D. Khan, F. Montenegro, J. Stickney, and J. Zook. TILE64$^{TM}$ processor: a 64-core SoC with mesh interconnect. In *Int. Solid-State Circuits Conf. (ISSCC)*, IEEE (2008).

[300] A. DeOrio, D. Fick, V. Bertacco, D. Sylvester, D. Blaauw, J. Hu, and G. Chen, A reliable routing architecture and algorithm for NoCs, *IEEE Trans. on Computer-Aided Design of Integrated Circuits and Systems (TCAD).* **31**(5), 726–739 (May, 2012).

[301] M. Kinsy, M. Pellauer, and S. Devadas. Heracles: Fully synthesizable parameterized MIPS-based multicore system. In *Int. Conf. on Field Programmable Logic and Applications (FPL)* (2011).

[302] K. Tsuchida *et al.* A 64Mb MRAM with clamped-reference and adequate-reference schemes. In *Proc. Int. Solid State Circuit Conf. (ISSCC)*, pp. 258–259 (2010).

[303] D. Schinke *et al.*, Computing with novel floating-gate devices, *Computer.* **44**(2), 29–36 (2011).

[304] S. Dietrich *et al.*, A nonvolatile 2-mbit cbram memory core featuring advanced read and program control, *IEEE Journal of Solid-State Circuits (JSSC).* **42**(4), 839–845 (2007).

[305] Y. Wang and H. Yu. Ultralow-power memory-based big-data computing platform by nonvolatile domain-wall nanowire devices. In *Proceedings of the 2013 ACM/IEEE international symposium on Low power electronics and design* (2013).

[306] B. Lee *et al.*, Phase-change technology and the future of main memory, *Micro.* **30**(1), 143–143 (2010).

[307] Y. Wang and H. Yu. Design exploration of ultra-low power non-volatile memory based on topological insulator. In *Nanoscale Architectures (NANOARCH), 2012 IEEE/ACM International Symposium on*, pp. 30–35 (2012).

[308] D. Halupka *et al.* Negative-resistance read and write schemes for STT-MRAM in 0.13 um CMOS. In *Proc. Int. Solid State Circuit Conf. (ISSCC)*, pp. 256–257 (2010).

[309] X. Wu *et al.* Hybrid cache architecture with disparate memory technologies. In *Intl. Symp. Computer Architecture (ISCA)*, vol. 37, pp. 34–45 (2009).

[310] S. Yu and H. Wong, Compact modeling of conducting-bridge random-access memory (cbram), *IEEE Trans. on Electron Devices.* **58**(5), 1352–1360 (2011).

[311] S. Sheu *et al.* A 4Mb embedded SLC resistive-RAM macro with 7.2ns read-write random-access time and 160ns MLC-access capability. In *Proc. Int. Solid State Circuit Conf. (ISSCC)*, pp. 200–202 (2011).

[312] C. Gopalan *et al.* Demonstration of Conductive Bridging Random Access Memory (CBRAM) in Logic CMOS Process. In *Proc. Memory Workshop (IMW)*, pp. 1–4 (2010).

[313] J. Jameson *et al.*, Quantized Conductance in $Ag/GeS_2/W$ Conductive-Bridge Memory Cells, *IEEE Electron Device Lett.* **33**(2), 257–259 (2012).

[314] Y. Wang *et al.* Design of low power 3d hybrid memory by non-volatile cbram-crossbar with block-level data-retention. In *Proceedings of the 2012 ACM/IEEE international symposium on Low power electronics and design*, pp. 197–202 (2012).

[315] M. Kund *et al.* Conductive bridging ram (cbram): An emerging non-volatile memory technology scalable to sub 20nm. In *Electron Devices Meeting, 2005. IEDM Technical Digest. IEEE International*, pp. 754–757 (2005).

[316] G. Loh. 3D-stacked memory architectures for multi-core processors. In *Int. Symp. on Computer Architecture*, ACM/IEEE (2008).

[317] N. Minas *et al.* 3d integration: Circuit design, test, and reliability challenges. In *Intl. On-Line Testing Symp. (IOLTS)*, p. 217–217 (2010).

[318] T. Thorolfsson, N. Moezzi-Madani, and P. Franzon, Reconfigurable five-layer three-dimensional integrated memory-on-logic synthetic aperture radar processor, *Computers & Digital Techniques, IET.* **5**(3), 198–204 (2011).

[319] Y. Wang *et al.* Design exploration of 3d stacked non-volatile memory by conductive bridge based crossbar. In *3D Systems Integration Conference (3DIC), 2011 IEEE International*, pp. 1–6 (2012).

[320] X. Huang *et al.*, A nanoelectromechanical-switch-based thermal management for 3-d integrated many-core memory-processor system, *Nanotechnology, IEEE Transactions on.* **11**(3), 588–600 (2012).

[321] C. Hua, T. Cheng, and W. Hwang. Distributed data-retention power gating techniques for column and row co-controlled embedded sram. In *Proc. Intl. Workshop on Memory Technology, Design, and Testing*, pp. 129–134 (2005).

[322] W. Lu *et al.*, A functional hybrid memristor crossbar-array/cmos system for data storage and neuromorphic applications, *Nano Letters.* **12**(1), 389 (2012).

[323] X. Dong, N. Jouppi, and Y. Xie. Pcramsim: System-level performance, energy, and area modeling for phase-change ram. In *Proc. Int. Conf. Computer-Aided Design (ICCAD)*, pp. 269–275 (2009).

[324] S. Kaeriyama *et al.*, A nonvolatile programmable solid-electrolyte nanometer switch, *Solid-State Circuits, IEEE Journal of.* **40**(1), 168–176 (2005).

[325] M. Tada *et al.*, Nonvolatile crossbar switch using tiox/tasioy solid electrolyte, *IEEE transactions on electron devices.* **57**(8), 1987–1995 (2010).

[326] M. Ben-Jamaa *et al.*, Silicon nanowire arrays and crossbars: Top-down fabrication techniques and circuit applications, *Science of Advanced Materials.* **3**(3), 466–476 (2011).

[327] W. Park *et al.*, A pt/tio2/ti schottky-type selection diode for alleviating the sneak current in resistance switching memory arrays, *Nanotechnology.* **21**(19), 195201 (2010).

[328] K. Gopalakrishnan *et al.* Highly-scalable novel access device based on mixed ionic electronic conduction (miec) materials for high density phase change memory (pcm) arrays. In *VLSI Technology (VLSIT), 2010 Symposium on*, pp. 205–206 (2010).

[329] A. Chen and M. Lin. Variability of resistive switching memories and its impact on crossbar array performance. In *Reliability Physics Symposium (IRPS), 2011 IEEE International*, pp. MY–7 (2011).

[330] Y. Wang, H. Yu, and W. Zhang, Nonvolatile CBRAM-crossbar-based 3-D-integrated hybrid memory for data retention, *Very Large Scale Integration (VLSI) Systems, IEEE Transactions on.* **22**(5), 957–970 (May, 2014).

[331] G. Jeong *et al.*, A 0.24-$\mu$m 2.0-v 1t1mtj 16-kb nonvolatile magnetoresistance ram with self-reference sensing scheme, *Solid-State Circuits, IEEE Journal of.* **38**(11), 1906–1910 (2003).

[332] S. Schechter *et al.* Use ecp, not ecc, for hard failures in resistive memories. In *ACM SIGARCH Computer Architecture News*, vol. 38, pp. 141–152 (2010).

[333] E. Seevinck, P. van Beers, and H. Ontrop, Current-mode techniques for high-speed vlsi circuits with application to current sense amplifier for cmos sram's, *IEEE Journal of Solid-State Circuits (JSSC)*. **26**(4), 525–536 (1991).

[334] C. Xu, X. Dong, N. Jouppi, and Y. Xie. Design implications of memristor-based rram cross-point structures. In *Proc. Design Automation and Test in Europe (DATE)*, pp. 1–6 (2011).

[335] M. Koga *et al.* First Prototype of a Genuine Power-Gatable Reconfigurable Logic Chip with FeRAM Cells. In *Proc. Intl. Conf. on Field Programmable Logic and Applications (FPL)*, pp. 298–303 (2010).

[336] J. Xie, X. Dong, and Y. Xie. 3d memory stacking for fast checkpointing/restore applications. In *Intl. 3D Systems Integration Conf. (3DIC)*, pp. 1–6 (2010).

[337] J. Hennessy and D. Patterson, *Computer architecture: a quantitative approach*. Morgan Kaufmann Pub (2011).

[338] S. Bird *et al.* Performance characterization of SPEC CPU benchmarks on Intel's core microarchitecture based processor. In *SPEC Benchmark Workshop* (2007).

[339] H. Qin *et al.* SRAM leakage suppression by minimizing standby supply voltage. In *Proc. Int. Symp. on Quality Electronic Design (ISQED)*, pp. 55–60 (2004).

[340] T. Nagai *et al.* A 65nm low-power embedded DRAM with extended data-retention sleep mode. In *Proc. Int. Solid State Circuit Conf. (ISSCC)*, pp. 567–576 (2006).

# Index

Printed in the United States
By Bookmasters